The Dynamic Genome

The Dynamic Genome

A Darwinian Approach

Antonio Fontdevila

OXFORD
UNIVERSITY PRESS

OXFORD
UNIVERSITY PRESS

Great Clarendon Street, Oxford OX2 6DP

Oxford University Press is a department of the University of Oxford.
It furthers the University's objective of excellence in research, scholarship,
and education by publishing worldwide in

Oxford New York

Auckland Cape Town Dar es Salaam Hong Kong Karachi
Kuala Lumpur Madrid Melbourne Mexico City Nairobi
New Delhi Shanghai Taipei Toronto

With offices in

Argentina Austria Brazil Chile Czech Republic France Greece
Guatemala Hungary Italy Japan Poland Portugal Singapore
South Korea Switzerland Thailand Turkey Ukraine Vietnam

Published in the United States
by Oxford University Press Inc., New York

First published 2011

British Library Cataloguing in Publication Data
Data available

Library of Congress Cataloging in Publication Data
Library of Congress Control Number: 2011920652

Typeset by SPI Publisher Services, Pondicherry, India
Printed in Great Britain
on acid-free paper by
CPI Antony Rowe, Chippenham, Wiltshire

ISBN 978–0–19–954137–9

10 9 8 7 6 5 4 3 2 1

Table of contents

Preface

The theory of evolution is generally accepted as *the* explanation of the Earth's present biodiversity. However, since Darwin first proposed natural selection as the leading mechanism of evolution, a series of other mechanisms have been periodically advanced to explain the observed variability. Some of these are proposed as complementary to natural selection but others constitute alternative explanations. The purpose of this book is to analyse some of the most widely publicized of these mechanisms in the context of our present genomic understanding.

The ever-increasing knowledge of whole genome sequences is unveiling a variety of structures and mechanisms that impinge on current evolutionary theory. The origin of species, the evolution of form, and the evolutionary impact of transposable elements are just a few of the many processes that have been revolutionized by ongoing genome studies. These novelties, among others, are examined in this book in relation to their general significance for evolution, emphasizing their human relevance. For example, the assessment that most of the human genome (and of other higher organisms as well) does not encode proteins demands an evolutionary explanation. Moreover, a high fraction of this non-coding DNA (almost 50% in humans) belongs to a class of mobile, repetitive sequences that has been qualified as 'junk DNA' ever since its inception. However, a more thorough analysis of parts of this mobile fraction is revealing their long-term adaptive role in the evolution of new functions. On the other hand, the small genomic differences between humans and chimps (about 1.2%) challenges our understanding of what makes us humans in terms of genetic differences. Certainly, neither the increase in number of genes nor, probably, the changes in coding sequences are the key evolutionary differences that define our humanity. Most probably the relevant steps towards the evolution of higher forms, humans among them, have arisen in regulatory and assembling processes that decide when, where, and in which combination the already existing genetic blocks operate. These are just a few glimpses of these controversial issues that this book examines in the context of Darwinian evolution.

Genome studies generate information, often dispersed in different disciplines, that deserves a continuous effort of synthesis if we are to understand its basic significance. Evolution provides the perfect framework to focus this synthesis. This book aims to provide a synthetic, readable account of some widely debated evolutionary issues in the context of our growing understanding of functional genomics. Recently, this debate has centred on arguments extracted from genomic and molecular information that have been intended to 'deconstruct' the Darwinian Theory. The purpose of this book is to show that whilst genome dynamism is providing new, and previously unanticipated, sources of variability, there is no reason to dismiss the role of natural selection as the mechanism that sorts out these potentialities. In other words, this genome potential provides new possibilities (and also constraints) for evolution, but the realization of this potential is driven by natural selection.

There are many examples to justify the argument to 'reconstruct', rather than to 'deconstruct', the Darwinian Theory. Exploration of whole genomes has allowed us to depict the genome as an ecosystem-like landscape where gene factors are species-like elements to be adapted (if not domesticated) to the genome environment. This analogy has revived old evolutionary controversies, such as the role of internal constraints as leading factors in evolution versus the pre-eminence of the external environment in Darwinian organismal selection. Perhaps the case for genome level selection is nowhere more relevant than in the role of mobile elements as selfish units operating in genomes. Evolution of form is another arena where the old Darwinian gradualist processes are being challenged. Goldschmidt's 'hopeful monsters' have been brought back to life by those that see developmental changes elicited by big-leap mutations of developmental genes. These are only a few of many instances in which the new understanding of genomes has been interpreted by some as supportive of non-Darwinian processes. I argue, however, that most (if not all) of these contentions can be shown to be inconsistent because Darwinian reasoning continues to provide a sufficient explanation for our new genomic knowledge.

While I was writing this text I realized how much I and most of my generation fellows were influenced by the ideas from those extremely bright evolutionists that produced the Modern Synthesis of Evolution. Without their ability to synthesize the evolutionary knowledge from their time the theory of evolution would never have experienced the current advance. Yet, as time went by, I developed a sense of critical admiration that led me to appraise the tenets of the synthesizers and also, in view of the new scientific advances, their shortcomings. In this book I have often impinged on some principles of the Modern Synthesis in an attempt to incorporate the current wisdom that stems from the genome studies. This endeavour may seem at times rather misleading, if not presumptuous, something that I never mean to be. On the contrary, I intend to give all the credit to the Modern Synthesis without disregarding how the new genome evidence obliges to change, enlarge, or modify some of its tenets. The discussion of the concept of species is an example of how, in my opinion, the new evidence of gene transfer impinges on the concept of biological species, an iconic concept in the Modern Synthesis.

Currently, the science of evolution emerges as a new synthesis waiting for the synthesizers. From the ashes of many old tenets, like some that underpinned the Modern Synthesis, new pillars will certainly emerge that are largely bolstered by the genome understanding. There is no doubt that some principles will have to be reshaped or even removed, introducing new concepts, but the fundamental question remains whether Darwin's principle of 'descent with modification' by natural selection as the main driving force withstands the thrust of genome studies. This book is an abridged answer to this. Many fundamental lessons learnt from the genome are presented and their evolutionary significance is discussed. When all these contentious issues are taken together a picture emerges which shows, although many mechanisms need to be updated and new concepts need to be incorporated to the theory of evolution, that there is no reason to debunk Darwin's fundamental principles. In fact, the genome shows the same fingerprints of the action of natural selection as other organism structures do: that is, a force working like a tinkerer for immediate survival.

The Dynamic Genome addresses a wide range of readers. Firstly, it is intended to fill the gap that many graduate students in evolutionary biology experience between their general interests in evolution and their specific research work, helping them to focus the present controversies within both a historical and a conceptual framework. I have often been aware in discussions with students that they lack the historical background of many controversial issues, which gives them an incomplete understanding of their conceptual roots. Secondly, any scientist with a background and interest in evolution will find much insight through learning about the new contributions of genome studies to the more controversial issues of the present evolutionary debate and how they apply to anti-Darwinian arguments. I use a precise and not overly academic language, and provide a comprehensive glossary to clarify every scientific term. An understanding of the more complex concepts is enhanced by a number of illustrations. In summary, this book is intended to be of value to graduate (as well as senior undergraduate) students not only by providing them with putative solutions on controversial issues in Darwinian Theory but also, and more importantly, by explaining why and how these controversies exist. But it will also find its way to a postgraduate and scholar readership (evolutionists and non-evolutionists alike), especially those interested in the present wave of attempts at discrediting Darwinian arguments under allegations of contradictory genomic data.

Acknowledgements

This book would not have been possible without the cooperation and support of many people and organizations. The initial step of this project was fuelled by Michael L. Arnold (University of Georgia, Athens, USA), who introduced me to Ian Sherman, Commissioning Editor at OUP. Then I started to develop a book outline to be submitted to the publishers. The outline was projected and written during a sabbatical leave sponsored by my university that I spent under the generous auspices of my colleagues Esteban Hasson (University of Buenos Aires, Argentina), Menno Schilthuizen (University Malaysia Sabah, Institute for Tropical Biology, Kota Kinabalu, Sabah, Malaysia), Ary Hoffmann (The University of Melbourne, Australia), and Marie-Louise Cariou and Pierre Capy (University of Paris-Sud XI, Orsay, and the CNRS campus, Gif-sur-Yvette, France). To all of them I owe a large gratitude for providing the perfect intellectual and social settings to my endeavour.

The following people read parts of the book manuscript and contributed with many priceless suggestions to the text that certainly enriched its content and comprehension. Chapter 1 was revised by Douglas Futuyma, Arcadi Navarro, Mauro Santos, Lluís Serra, and William Stone; Chapter 2 by Jaume Baguñà and James Valentine; Chapter 3 by Christian Biémont, María Pilar Garcia-Guerreiro, John McDonald, and Rachel O'Neill; Chapter 4 by John Avise, Menno Schilthuizen, and Alan Templeton; Chapter 5 by Enrico Rezende, Mauro Santos, Lluís Serra, and Eleutherios Zouros; and the Glossary by Mauro Santos and Emilio Valadé. To all of them I express my deep gratitude for having sequestered part of their precious time to peruse my text. Special thanks are addressed to my colleagues in the Barcelona scientific area (Baguñà, García-Guerreiro, Navarro, Rezende, Rodriguez-Trelles, Santos, and Serra), who not only read and revised parts of the manuscript but also comforted and assisted me with frequent calls and meetings to personally discuss many controversial issues that worried me. Of course, they provided me with wise recommendations, some of them accepted, and others not, by me, but the responsibility of the book's content is mine alone.

I am especially indebted to the people who helped me to illustrate the book. Among them and first of all is my technician Montserrat Peiró, who took care of downloading and, where necessary, redrawing most of the figures, and obtained permissions to reproduce them. The final clarity of the book's presentation owes much to her dedication and skill. I would also like to express my gratitude to Karen Allendoerfer, Patricia Beldade, Paul Brakefield, Oskar Brattström, Sean B. Carroll, Andre Coetzer, Jordi Fernández, Cedric Feschotte, Randy Jirtle, Mauricio Linares, Susan Linquist, Juan G. Montañes, Ginés Morata, Jeff Palmer, Barry Rice, and Susan Wessler for providing and helping me to access high-resolution colour figures.

My editors at OUP deserve a special recognition. First, Ian Sherman, who led me from the initial steps of my outline project towards the final acceptance by OUP, and later Helen Eaton, the Assistant Commissioning Editor who monitored my progress in writing the book in a highly efficient, cooperative, and friendly way. The two of them have been essential in the successful completion of this volume. I am very grateful to both for their professional and helpful task that surely eased the book's production. My gratitude also goes to Muhammad Ridwaan, Assistant Production Editor; to J. Albert André, Assistant Account Manager; to Caroline Broughton, the copyeditor of my book, and to Fiona Barry, who proofread the first proofs, all of whom did an excellent job of making my book an easier, readable product for the public.

Work on this book was also morally supported by my colleagues at my university, especially in my department of Genetics and Microbiology, and was financed in part by a grant from the local government of the Generalitat de Catalunya during my sabbatical stay at foreign institutions.

Finally, my immense gratitude goes to my family and friends. They had to experience the ups and downs of my humour alongside my writing. My wife, Maribel, was the one that more closely felt the periods of physical, albeit not mental, isolation and seclusion from my presence. Her loving, understanding, and encouraging attitude helped my endurance to persist, for which I owe her a priceless recognition and dedicate this book to her with my love.

CHAPTER 1

The dynamic genome: a general introduction

On 14 March 2000 the Office of Science and Technology Policy of the White House issued a joint statement by President Clinton of the USA and Prime Minister Tony Blair of the UK for immediate release. The statement was a strong recommendation in favour of making public all research data on the human **genome**, the set of DNA molecules in our **chromosomes**, which they qualified as 'one of the most significant scientific projects of all time'. But, why does this scientific endeavour deserve such highly appreciative attention? And, more precisely, why must it be available for public access? Immediately, this apparently innocuous statement unleashed a landslide in the technological sector of the Wall Street stock market. Again, why?

The fight between public- and private-funded research into gene patenting has a long history. This issue started in 1991 when the US National Institutes of Health, a highly prestigious public institution, applied for patents for DNA clones. Although this application, vigorously opposed by the scientific community, was rejected by the US Patents Office, it queried whether our (and other organisms') genetic endowment should be someone's property. This was opposed by many people, among them President Clinton and Prime Minister Blair who, in the March 2000 statement, explicitly stated that 'raw fundamental data on the human genome, including the human DNA sequence and its variations, should be made freely available to scientists everywhere'. However, when on 26 June both leaders simultaneously announced that a first draft of the human genome had been completed, there was no guarantee that Celera, a privately funded company, would make public its DNA sequence data.

No doubt the reluctance to make public the human genome sequence is based on the future applications (and financial benefits) that its knowledge can provide to the pharmaceutical industry. Many diseases can be associated with certain gene sequences and disease-propensity can be predicted individually. We are still far away from much of this predictive medicine for most diseases, especially the complex ones, but the journey towards personalized health care is on its way. Hopefully, personal response to medicines can be tailored in the future, thus opening new avenues for drug treatment. These and many more future opportunities for improving human health depend on free access to genome information. The White House 2000 joint document recognizes this when it explicitly states that 'unencumbered access to this information will promote discoveries that will reduce the burden of disease, improve health around the world, and enhance the quality of life for all humankind'.

While the official tenet is in favour of free access, the conflict is far from being over. Private companies, like Celera, have free access to the data produced by the publicly funded agencies, but the reverse is not true. This unleashed a competitive, albeit unfair, race. In short, the Human Genome Project (HGP), a publicly funded project initiated in 1990, presently comprising many laboratories mainly from the USA, UK, France, Japan, China, along with many other countries, was projected to sequence the human genome in 15 years. Yet, in 1999, Craig Venter, the leader of Celera, audaciously claimed that he could finish the sequencing in 2 years. A frenetic race ensued and finally both competitors published simultaneously two rough working drafts (comprising about 90% of the genome) in 2001 (Lander *et al.*, International Human Genome Consortium 2001; Venter *et al.* 2001). Some authors (Waterston *et al.* 2002) have shown that the Celera sequence was not independent from the HGP sequence since it resulted from merging its sequence with large quantities of the HGP sequencing, which was accessed freely. However, both sequences were incomplete, and it was not until 2003 that the completion of the sequence was achieved, at least officially. Then our genome was deemed to be born. Yet, since then, several releases have been reported, and even now some researchers are trying to finish the genome by filling gaps, correcting mistaken sequences, and producing alternative sequences for high-variability regions (see below).

The genesis of the human genome sequencing and mapping is a hugely exciting scientific endeavour, but its complete account falls outside the realms of this present book and the reader is referred to Strachan and Read (2004) for details and to Dolgin (2009) for a review of the present work by the 'genome finishers'. Here, however, I must emphasize that it is the first time that we have been directed to the deepest dimension of our nature. Although exploration of the external universe has a long history and has produced counterintuitive and astounding perceptions, probing our internal universe, namely the instructions that make us what we are as organisms, has been rather modest, albeit significant in areas like cell and molecular biology. Now, the genome exploration is opening a new era of understanding, not only of humans but of all living beings. Moreover, there is a general consensus, shared by our leaders, that in spite of all vagaries derived from the human ego, this endeavour is comparable to the exploration of space, to say the least. This chapter (and book) aims at presenting, albeit succinctly, why, besides the practical importance of genome understanding for our welfare, there is grandeur and beauty in the knowledge that is concealed in the genome. This knowledge serves to reconstruct our heritage; to answer questions such as where we come from or why we are equal or different from our relatives, humans or not. In sum, how we did evolve?

The universe within us: a genome journey

The general characteristics of our genome structure, despite its peculiarities, may be shared by many multicellular organisms. However, some differences are worth discussing in order to understand the overall genome dynamics. Figure 1.1 shows the DNA composition of the human nuclear **genome**, which is packed in 23 pairs of chromosomes (22 **autosome** pairs plus a pair of **sex chromosomes**: two Xs in the females and one X and one Y in the males). Human cells also host **mitochondrial** organelles that contain a genome whose structure is quite different. After the first human genome annotation it became clear that only about 1.5% of total nuclear DNA (about 2900 Mb: 1 Mb equals one million DNA **bases**) codes for protein sequences. The rest is distributed among different classes. Interestingly, some sequences (like **introns** and other **RNA-coding sequences**) are not translated into proteins but are involved in coding. Among them a significant fraction, perhaps close to 5% or more, specifies RNA genes involved in regulatory processes and protein synthesis. Recently, a new kind of regulatory RNA gene, globally designated sRNAs (**small RNAs**), is being characterized that represents an important fraction of these untranslated genes (see Chapter 3 for their description).

Our proteins are, on average, as long as those found in the majority of **eukaryotes**, but this does not hold true with gene length. In most eukaryote genes protein-coding DNA is made of pieces (the **exons**) interspersed by non-coding DNA stretches (the introns). In humans (and in most mammals) the average intron is quite large (>4 Kb, 1 Kb equals one thousand DNA bases), larger than in most invertebrates, while human exons almost never exceed 300 base pairs (bp) in length. This architecture yields a human gene landscape in which tiny exons look like islands in an immense DNA-intronic ocean. Does this peculiar organization have any functional meaning? Remember that after **DNA transcription** the ensuing primary transcript of the **messenger RNA** (mRNA) must be processed (spliced), which consists of a cut-and-paste operation that joins the transcribed exons after eliminating the transcribed introns. Splicing depends on precise mechanisms of exon-boundary recognition and can be tissue-specific. Some evolutionists argue

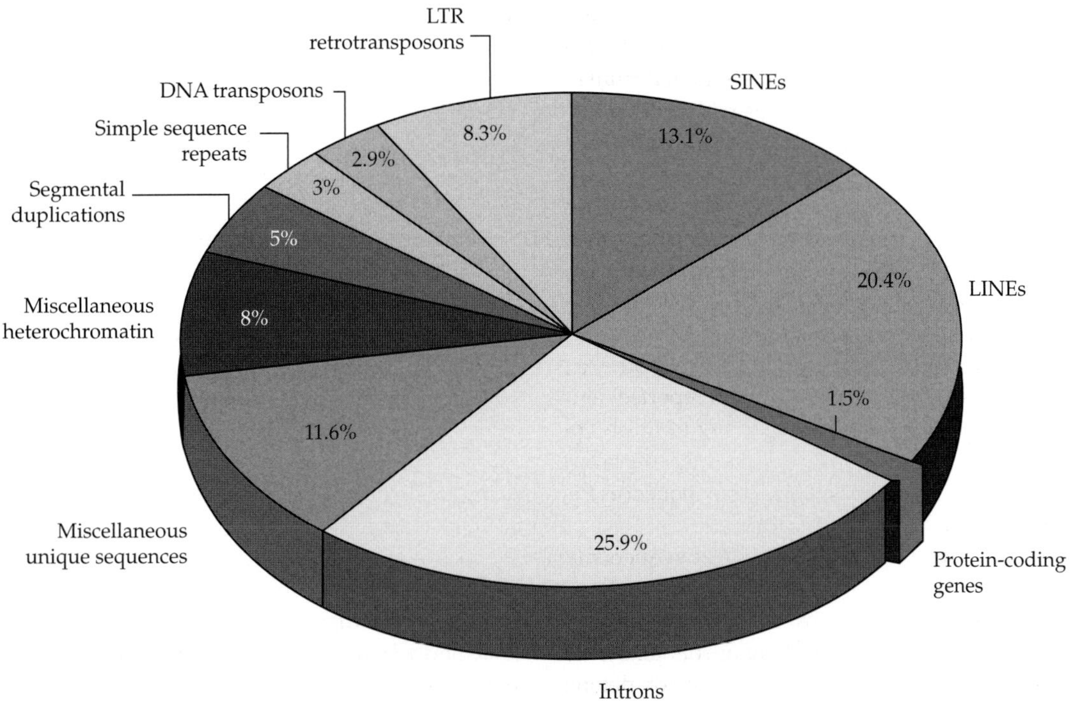

Fig. 1.1. A graph summary of the composition of the human genome. Note that the largest component of the genome is protein non-coding (over 98%) in which transposable or mobile elements (DNA transposons, LTR retrotransposons, and SINEs and LINEs) amounts to the most abundant proportion (almost 45%). In the human genome and in many mammals, albeit not in other organisms such as Drosophila, introns represent a large fraction (nearly 26%). From Gregory (2005a) with permission from Nature Publishing Group.

that since exon recognition is favoured by short exons, and because more than half of human genes experience alternative splicing (Fig. 1.2), which is often tissue-specific, this intron-long gene architecture favours the high output transcript, and then protein, diversity produced per human gene (see Chapter 3).

Alternative splicing has been advocated as one mechanism for increasing complexity, but it is not the only one. Gene regulation is also a likely candidate. The huge amount of non-coding DNA in humans (about 98.5%), and also in other complex organisms, provides the potential for regulatory DNA sequences (see below). Admittedly, non-coding human DNA consists of a large fraction (about 50%) of mobile genetic elements (stretches of DNA that can copy themselves and insert their copies in different genome sites), whose role in evolution is discussed in Chapter 3 and which we are not going to deal with here, other than mentioning their putative regulatory deployment. Contrary to coding regions, identified by their amino acid-translation capacity, regulatory sequences do not show a general key for identification. However, evolutionists have reasoned that comparing genomes from different species, where some non-coding regions show strong conservation, would be an indication that **purifying selection** is weeding out all deleterious mutations, and these conserved regions may yield minimal estimates for the regulatory fraction in the genome. Thus, when human and mouse genomes are compared, the conservative intergenic regulatory DNA amounts to 1.5–2%. This refers to large sequence stretches located within intergenic regions that are often under more selective pressures than most coding DNA. In the past decade a new class of regulatory small RNA transcripts (20–30 bp long), named **microRNAs** (miRNAs), have been characterized, and it has been found that these could be the regulatory agents of at least 20% of human genes (Xie *et al.* 2005). **Small interfering**

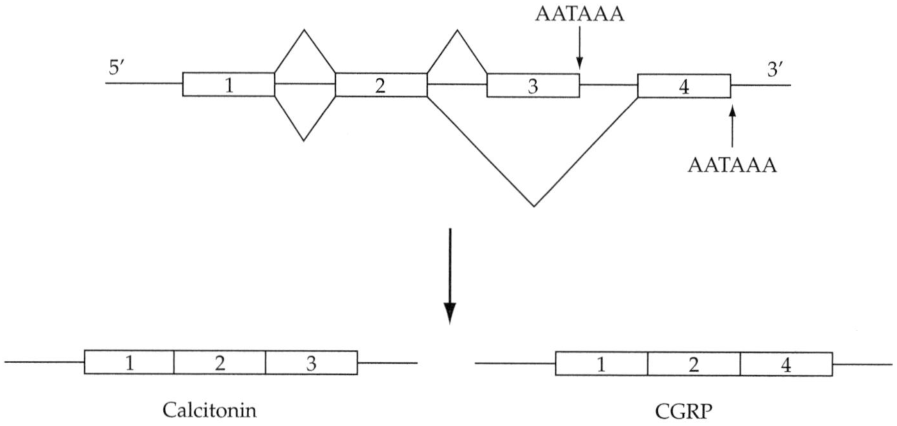

Fig. 1.2. Alternative splicing in the 'calcitonin' gene. The scheme of the gene (top) depicts 4 exons (1–4) and two excision polyadenilated signals (AATAAA). In the thyroid the gene is spliced using the exon3 signal as the end of transcription, generating the calcitonin isoform involved in calcium metabolism. Alternatively, in the hypothalamus the mRNA is processed by exon3 excision using the exon4 signal as the transcription final. This splicing produces a functionally different isoform named CGRP peptide.

RNAs (siRNAs), another class of small RNAs discussed in Chapter 3 as silencing agents of mobile elements, are likely to be an important component of the regulatory machinery, not only in humans but also in many other eukaryotes, and may be a universal mechanism in the genome. More recently, large amounts of new RNA classes have been discovered, indicating that the total transcription products (the transcriptome) represent a large genome fraction (about 93% of DNA in the human genome). The transcriptome comprises mostly non-coding RNA transcripts, the putative regulatory function of which is discussed below.

In sum, all classic functional DNA (protein-coding and classic RNA-coding) explains some 5% of genomic DNA in humans; the remaining DNA still accounts for a huge 95% fraction. What is the evolutionary explanation for this 'excess' DNA? One of the earliest surprises produced by the early genome studies 50 years ago was that genome sizes vary by 200,000-fold. Intriguingly, this size variation is often not correlated to organismic complexity. Moreover, when the post-genomic era started after whole genomes were routinely sequenced, the realization that the gene number was lower than previously estimated by more indirect methods, also came as a huge surprise. So, what is the solution to these conundrums? Where does this extra DNA come from and what is its function, if any? Why do two organisms that differ strikingly in complexity, such as a nematode and a human, have about the same number of genes, but vastly different amounts of DNA?

Does genome size matter?

In the middle of the twentieth century, prior to the whole-genome sequence era when genome sizes were indirectly estimated, it became immediately clear that the educated guess of DNA-content increasing with organismal complexity did not always hold. Some organisms far less complex than humans (genome size: about 2900 Mb), ranging from some microbial eukaryotes such as *Amoeba dubia* (675,500 Mb), to many fishes and amphibians (exceeding 100,000 Mb), and even to many flowering plants such as some lilies (120,000 Mb), exhibit a DNA content that is outstandingly larger. Not only that, the genome-size variation within eukaryotic groups was enormous (see Fig. 1.3) and not necessarily related to organismal complexity. For example, insects show a genome-size variation of over 100-fold, from 90 Mb in some flies to 15,000 Mb in the genomes of many grasshoppers (about five times the human genome size). The same scaling applies to flowering plants, whose genome size varies about a 1,000-fold, and to amphibians and bony fishes. The genome size variation in fishes is most spectacular. It varies from the 370 Mb size in the

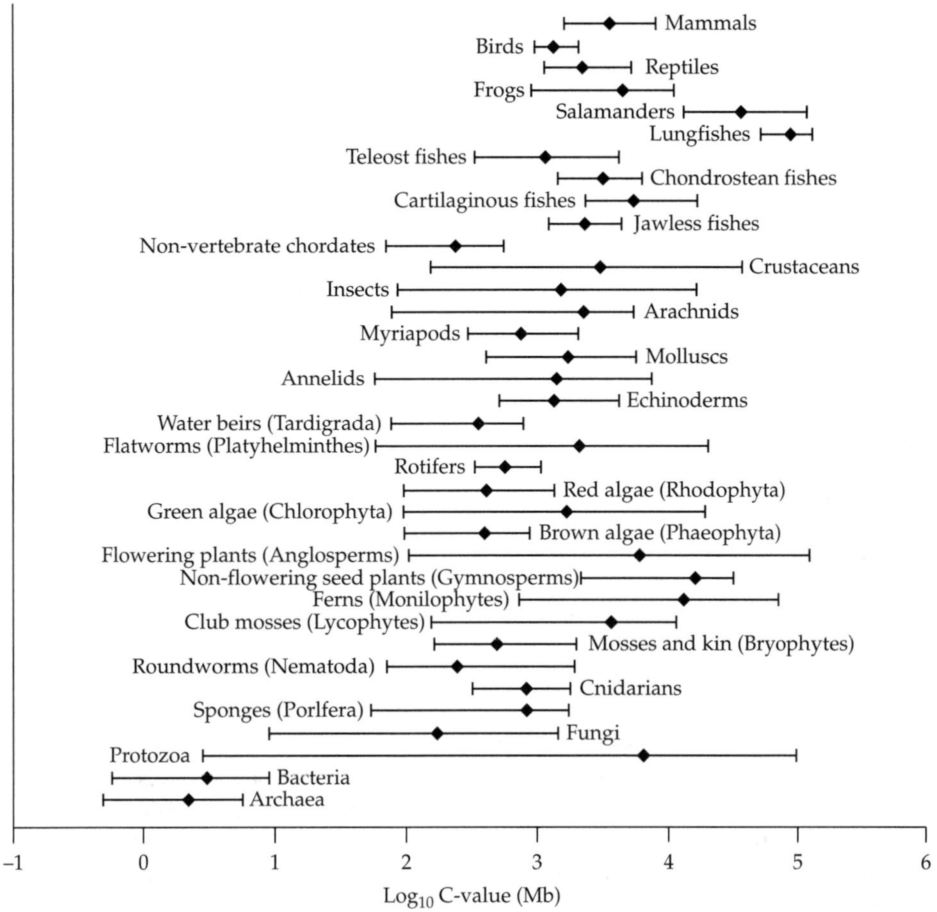

Fig. 1.3. The C-value paradox. Depiction of means and overall ranges of genome size in different groups of organisms, arranged by complexity, that illustrates the lack of correlation between complexity and genome size. (Mb means megabase = 1 million of DNA base pairs). From Gregory (2005a) with permission from Nature Publishing Group.

smallest described vertebrate genome (*Fugu rubripes)* up to 10,000 Mb (over three times the human genome) in many bony fishes.

Since DNA is the blueprint of the organism—its genetic instructions—this lack of parallelism between organic complexity and DNA content seemed like a paradox, often referred as the '**C-value paradox**'. C-value stands for the total (constant) amount of DNA in a haploid cell. Why then should an increase in DNA content not imply the building of a more complex organism? And what are the causes, if any, governing the relationship between genome size and Darwinian adaptation? Several advances in the understanding of the genome architecture that were achieved during the 1960s and 1970s helped to partially answer these questions. First, molecular biology showed that the eukaryotic genome hosts a large amount of **repetitive DNA**, and second, that coding DNA in genes consists of pieces (exons) interspersed with DNA stretches (introns) without any coding function. In addition, by the 1970s, genome researchers were acquainted with the presence of many genes without a function (**pseudogenes**) that evolved from functional genes. This was a time of excitement: basic discoveries like the formulation by Kimura (1968) of the neutral theory of molecular evolution (see Box 1.1) were under way and mobile genetic

Box 1.1 Some hints on the neutral theory of evolution

In 1968, Kimura, a theoretical population geneticist, proposed that the majority of DNA base substitutions in evolution are driven by a random process, known as **genetic drift**, instead of by natural selection. This formally developed principle led to the neutral theory of the molecular evolution, as it became widely known. To understand this theory of highly sophisticated mathematical formalism, it is necessary to define several underlying concepts. First, when mutations occur, we talk about their rate of appearance (their mutation rate, u) to be distinguished from their rate of fixation, which equates to the probability that a mutation will be fixed in the population (θ_f) (see below). If the new mutation does not change the fitness of the organism that carries it, which means that it is not selected in favour or against, we call it 'neutral'. **Neutral mutations** can become fixed in finite populations, however, due to the random effect of the finite sample of gametes that pass to the next generation. This random process is named 'random drift', because gene frequencies drift across generations until genes are fixed or lost. Provided that all variants of a gene (alleles) are equally fit (i.e. neutral) their ultimate probability of fixation is equal to their initial frequency. Thus, when a new neutral mutant **allele** appears in a diploid population (with two gene alleles per individual) of N individuals, its frequency in the population is $1/(2N)$, which is exactly its ultimate probability of replacing the other allele (i.e. its fixation probability θ_f).

Population genetics theory comprises a highly formalized body of predictive equations for genetic drift. For instance, if the genetic variability is neutral, genetic drift promotes loss of variability (i.e. fixation) at the rate of $1/(2N)$ per generation in an idealized population of N diploid, randomly mating individuals with equal fertility. However, in reality populations behave differently. Among other particulars their individuals may not mate randomly, may reproduce with unequal success, the sex ratio may often deviate from 1:1, and population size may vary seasonally. For these reasons, not all individuals in a population contribute genes equally to the next generation and the **effective population size** (N_e) may be lower than the census size (the number we count). Thus, the loss of variability due to drift in these real populations must be equated to $1/(2N_e)$, where N_e stands for the size of an idealized population that experiences the same drift effect as the real population. Population geneticists use N_e instead of N in order to make different populations comparable.

Second, note that we are interested in the rate of substitution (k) that is the rate at which a new neutral mutant replaces the ancestral gene. Note also that the frequency of new neutral mutants per generation is $2N_e u$, but only a fraction of them (their fixation probability $1/(2N_e)$) will be fixed, thus the rate of substitution equals the product of both, that is u. Note that this parameter is independent of N_e because the two quantities, the frequency of new mutants and the probability of fixation of each mutant, depend on N_e in exactly opposite ways. Thus, the rate of neutral substitution only depends on the rate of neutral mutation (u). Interestingly, provided that the mutation rate is constant through time, molecular biologists have a kind of biological clock that allows them to estimate the relative time of divergence by computing the number of nucleotide differences between two species. In sum, whatever the population size might be, should the rate of neutral mutation be constant, the rate of substitution would not vary across genes.

Third, the standard neutral theory posits that a large proportion of new mutations is neutral and that the majority of the remaining mutations is deleterious and negatively selected. Only a minority of mutations, albeit adaptively important, increase the organism's fitness and are positively selected. We still do not know exactly the relative proportions of these mutations, but here we are more interested in how natural selection can detect favourable or deleterious mutations in small populations where drift may be an important

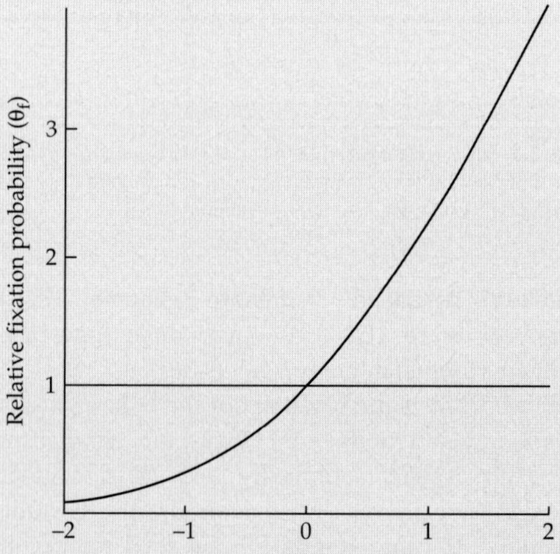

Fig. A. Probability of fixation (θ_f) of a mutant allele as a function of the effective population size (N_e) and the selection coefficient (s) relative to the neutral case ($1/2N_e$). The horizontal line depicts the neutral fixation probability ($s = 0$). After Ohta (1992).

continues

Box 1.1 *(continued)*

force for fixation. Population geneticists have estimated that the probability of fixation (θ_f) of an allele with selection coefficient *s* relative to the neutral probability of fixation is approximated by $4N_e s/(1-e^{-N_e s})$. This function, depicted in Fig. A, shows that for new slightly deleterious or favourable mutations, with values of $|N_e s| \ll 1$, where the vertical lines stand for absolute values, their fixation probability is close to the neutral case ($N_e s = 0$; $\theta_f = 1$). Note that even slightly favourable (deleterious) mutations do not become easily fixed (lost) by selection, because if everything depends on the effective population size then genetic drift is the leading force in the face of natural selection. We might state it in a more formal way by saying that for each population size there exists a set of selection coefficients that make new mutants behave as effectively neutral.

The strength of this theory, named 'nearly neutral' as an extension of the neutral theory (Ohta 1992), has influenced many explanations that range from understanding the uniformity of evolutionary rates in diverse lineages, to explaining the accumulation of much non-functional DNA in the genome. But above all, this theory has provided a 'null hypothesis' against which the observed putative adaptive patterns in nature can be tested, showing, once more, that, despite its complexity—abhorred by many students—understanding evolution is not possible outside a population genetics framework.

elements, discovered by McClintock (and almost ignored by geneticists) almost 30 years earlier (see Chapter 3 for an evolutionary account), were being found in all surveyed organisms, from single-celled to multicellular species. All these discoveries led to the general idea that genomes host a lot of non-coding DNA (introns, pseudogenes, repetitive, mobile elements, etc) that explains the 'excess' DNA in the C-value paradox.

The recent explosion of genome sequencing has confirmed the C-value paradox and has refined it. In particular, whole-sequence knowledge has produced better gene number estimates and confirmed that genome size differences are not due to variation in gene number (Table 1.1). Overall, eukaryotes vary by about 10-fold in the number of genes, much less than the 200,000-fold variation in genome size. Puzzlingly, organisms that contain the same number of genes often show very different genome sizes, such as the mouse (2,500 Mb) and the flowering plant *Arabidopsis thaliana* (125 Mb). In sum, since an increase in genome size does

Table 1.1. Genome size and number of protein-coding genes in eukaryotes (after Lynch 2007)

	Genome size (Mb)	Gene number
UNICELLULAR SPECIES		
Saccharomices cerevisiae	12.05	6,213
Plasmodium falciparum	22.85	5,268
Entamoeba histolytica	23.75	9,938
Trypanosoma spp.	39.20	10,000
OLIGOCELLULAR SPECIES		
Aspergillus nidulans	30.07	9,541
Neurospora crassa	38.64	10,082
ANGIOSPERM PLANTS		
Arabidopsis thaliana	125.00	25,498
Oryza sativa	466.00	60,256
ANIMALS		
Caenorhabditis elegans	100.26	21,200
Drosophila melanogaster	137.00	16,000
Fugu rubripes	365.00	38,000
Gallus gallus	1,050.00	21,500
Mus musculus	2,500.00	24,000
Homo sapiens	2,900.00	24,000

not imply a parallel increase in gene number, the 'excess' DNA should be accounted for by the non-coding, intergenic DNA. Apparently, this solves the C-value paradox, but it explains neither the origin nor the mechanisms that promote the huge genome differences encountered. To answer these questions the non-genic DNA was considered useless by earlier researchers, like 'junk' carried passively by the genome, which it was unable to get rid of. Once the evidence was clear that genomes were composed in a large part of mobile elements (almost 50% in the human genome; see Table 1.2), some researchers largely identified '**junk DNA**' as these elements. Because of their invasive ability through transposition, they were qualified as 'selfish' elements, and their persistence in the genome was viewed as the result of a dynamic balance between their accumulative invasive impulse and the cell mechanisms that get rid of them (Doolittle and Sapienza 1980; Orgel and Crick 1980). While the evolutionary significance of mobile elements as selfish genome components is still a contentious issue that is amply discussed in Chapter 3, the 'selfish' DNA hypothesis does not wholly explain the huge differences in genome size, nor the mechanisms, adaptive or stochastic, that cause such differences.

Table 1.2. Transposable or mobile element content in species genomes

	%
UNICELLULAR SPECIES	
Saccharomices cerevisiae	3–5
ANGIOSPERM PLANTS	
Arabidopsis thaliana	14
Oryza sativa	16
Zea mays	60–80
Lilium	95–99
ANIMALS	
Caenorhabditis elegans	6
Drosophila melanogaster	15–22
Drosophila simulans	5
Anopheles gambiae	16
Tetraodon nigroviridis	0.14
Fugu rubripes	2
Xenopus laevis	37
Rana esculenta	77
Gallus gallus	27
Mus musculus	40
Homo sapiens	45

Source: Biémont and Vieira (2005); Kidwell (2005), where references can be found.

Why are some organisms more tolerant of '**selfish**' (or junk) DNA than others? This question has been answered in two ways. The adaptive hypothesis postulates that extra DNA, regardless of its non-coding character, has a functional value itself because of its direct effect on phenotype. The long-observed correlation between genome size and cell size is corroborated by direct, causal observations. It is known that organisms that double their chromosomes (polyploids) also increase their cell size. Additionally, when there is within-species variation in DNA content, or when subspecies and closely-related species with differences in genome size are compared, a significant correlation between cell size and genome size occurs. But how may an increase in cell size influence adaptive life history traits?

Some evolutionists (see, for instance, Cavalier Smith 1985) have proposed adaptive functional explanations derived from cell size increase, such as buffering concentration fluctuations of the regulatory proteins or the protection of coding DNA from mutation. Others have concentrated in establishing close relationships between cell size (as a proxy for genome size) and some complex traits that impinge on phenotype fitness. Since amphibians show a large range of genome size among vertebrates, some researchers (Roth *et al.* 1994, 1997) have chosen these organisms to test the hypothesis that brains of large-cell amphibians would contain fewer cells than brains of small-cell relatives, and thus the brains of the former would deploy less complexity than the latter. The brain area under study was the optic tectum, the region where visual information is mainly processed. Among the salamanders, the members of the family Plethodontidae show the largest genomes (and cells), especially within the tribe Bolitoglossini, whose members exhibit genome sizes approaching 100,000 Mb. These salamanders develop the least complexity in their optic tecta, leading some authors to affirm that cell size (and genome size) is a predictor of morphological complexity in the brain. In sum, as the investigators explain (Roth *et al* 1994):

small salamanders with large cells have the simplest tecta, whereas large salamanders with small cells exhibit the most complex tectal morphologies. Increases in genome, and consequently cell size, are associated with a decrease in the differentiation rate of nervous tissue, which leads to the observed differences in brain morphology.

Increased genome size in salamanders also has profound influences on developmental rates. Since large genomes seem to retard, disturb, or inhibit genetic expression, perhaps by slowing down rates of cell proliferation and differentiation, development is truncated and terminal stages fail to occur. This, in turn, produces sexually mature adults that retain juvenile characteristics, a condition termed '**paedomorphosis**'. Plethodontids show all kinds of developmental processes, from full metamorphosis, through facultative metamorphosis—depending on the environmental circumstances—and paedomorphosis, to direct development with no metamorphosis at all. Genome size is usually correlated to the ability to metamorphose: large genome sizes tend to be associated with direct development whereas small genome sizes are found in those organisms that experience full metamorphosis. That Plethodontids, salamanders with large genomes, usually have direct development comes as no surprise. Moreover, this relationship between direct development (no full metamorphosis) and increased genome size is also found in other amphibians, like frogs, and in insects as well (Gregory 2005c).

Of course, the crucial question on this organism- and genome-level relationship is what comes first? Does genome size increase in response to changes in organism-level selection (the top-down hypothesis)? Or does genome size increase by intragenomic processes (selective or not) that lead to effects on organismal fitness (the bottom-up hypothesis)? In the bolitoglossines, at least, both processes may seem to be at work, albeit at quite different intensities. Many genera in these organisms show miniaturized body sizes (reaching 20 mm at maximum), the evolutionary reason for which has been explained as an organismal adaptation to exploit specific niches in highly competitive microhabitat communities. Since bolitoglossines have the largest genome sizes of all terrestrial vertebrates, body miniaturization and large cell sizes conflict to decrease their complexity in adaptive traits, such as visual acuity for predation. Genome size reduction would have been a plausible response in order to ease this conflict, but did not occur. Regardless of whether it is very difficult to reduce genome size, or whether large genomes have some advantage, natural selection has compensated for the decrease in complexity by fostering substitutive adaptations that include a highly specialized tongue, a new behaviour of waiting predation, and, above all, a series of special features in the visual system that enhance the predatory adaptation of the bolitoglossines. In fact, these salamanders are characterized by having the fastest and most precise feeding mechanisms found among amphibians. This feeding mechanism is paralleled by highly developed stereoscopic vision. In addition, many bolitoglossines, particularly the arboreal species, are true acrobats, making use of limb and tail specializations. These compensatory adaptations to predation may not be unique. Another negative effect of large cells may have been the problem of how to efficiently circulate huge erythrocytes in the tiny blood vessels of these miniaturized organisms. Again, the solution might be not to reduce the genome size but to ease the blood circulation by evolving enucleated erythrocytes.

The salamander study exemplifies how genomic research has to be coupled with hierarchical research at different levels, ranging from genomes to organisms, if we are to thoroughly understand evolutionary processes. While still a contentious issue, the evolutionary meaning of the C-value paradox depends on genome dynamics as much as on population dynamics. In the bolitoglossine case, a top-down process does not seem very relevant to genome size changes, whereas the increase in genome size has elicited, through cell size enlargement, profound changes in organismal complexity that have had to be compensated for by evolving new adaptive traits. Moreover, this worked-out case reveals that organic complexity results from a multilevel interaction of hierarchical relationships in which genome size is just one factor, albeit an important one, to be considered (Fig. 1.4). If we accept this principle, the observation that an increase in organismal complexity does not necessarily follow from an increase in genome size should not come as a big surprise.

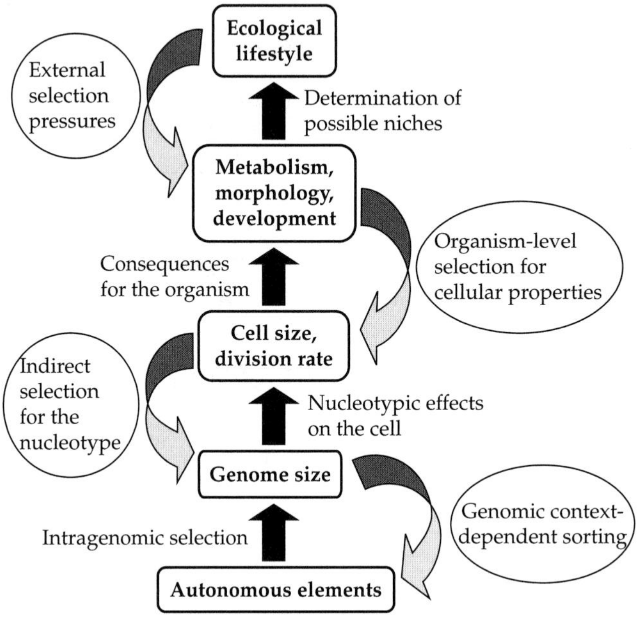

Fig. 1.4. A view of hierarchical relationships between different levels ranging from subgenomic dynamics to organismal ecology. In the bottom-up flow (black arrows) intragenomic selection can determine the expansion or elimination of DNA elements affecting the genome size, which can influence cell size and cell division rate. At the upper levels all these genomic and cell variables may determine organism traits, such as metabolism, morphology (i.e. body size), and developmental rate, that may act on the organism ecological lifestyle. But the flow of level interactions can also proceed top-down (curved grey arrows). The organism populations are subjected to the laws of population biology, namely external and organism selection, which can determine life history traits such as developmental rate, body size, or indirectly even genome size. However, since the effectiveness of natural selection depends on the population size, genetic drift emerges as an important factor to be considered in the persistence of genomic fractions that may not be adaptive at the moment of their appearance, moulding the genome size and its architecture (see text for details). Adapted from Gregory (2005b) with permission from Nature Publishing Group.

Does population size matter?

It can be argued that the C-value paradox cannot only be explained by a single life history component or cell trait. The 'excess' DNA may be the combined result of several intragenomic and organismic processes, and its study might seem to be best approached on a case-by-case basis. Yet, it cannot be denied that some general explanations, like universal correlations between multilevel parameters, have long been reported. Among them, the relationship between genome size and developmental rate, discussed above, is well supported. But genome size may also be related to higher level parameters, such as population size. Species with large genome sizes, such as humans, tend to show small population sizes, but unicellular eukaryotes and other organisms with lower genome sizes usually form large populations (Fig. 1.5). This wide-ranging relationship, albeit imperfect, has been hypothesized by Lynch, a population geneticist, as the main causative reason to explain the 'excess' DNA presence in large, complex organisms (usually showing lower population sizes). Namely, he posits that most DNA in large genomes, comprising non-coding DNA such as intergenic regions, introns, untranslated regions (UTRs), and mobile elements, was not originally acquired by selective mechanisms but is the outcome of non-selective, stochastic forces, which operate in populations due to their finite size. This mechanism, known as '**genetic drift**' (see Box 1.1), postulates that the incorporation of a new mutation in a genome, its fixation probability, depends not only on its selective value (s) but also on the population size (N). When N is large, selection is prime among the forces that decide the mutation fate, but in small populations, even rare less fit mutations may be fixed. This is because the gene

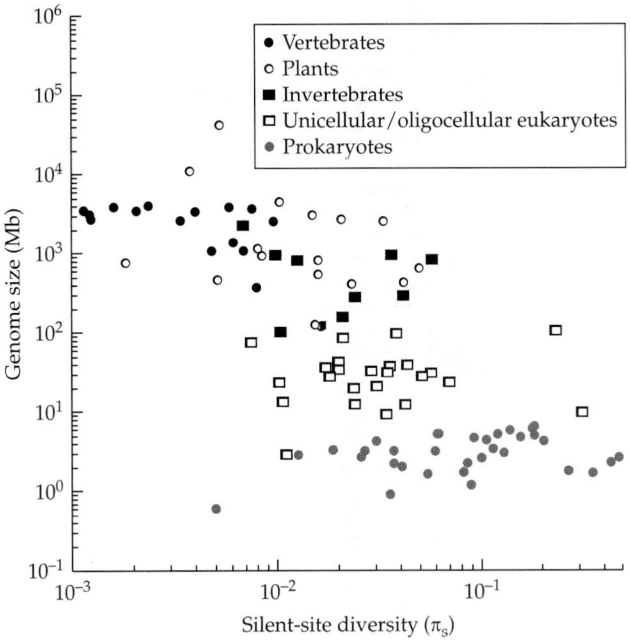

Fig. 1.5. Genome size is negatively correlated with effective population size (estimated by $\pi_s = 2N_e u$, see p. 12 in the text) for many taxonomic groups. From Lynch (2007) with permission from the author.

transmission probability from one generation to the next may by chance allow one rare variant (i.e. a rare less fit gene) to pass if the number of individuals (and the ensuing transmitted gametes) per generation is small; as for instance when from a box containing 10 balls, 9 white and one black, a few extractions with replacement of balls—which may represent the small number of transmitted gametes—results in all balls that are picked-up being black (the rare variant analogue) purely by chance.

Population genetics theory comprises a highly formalized body of predictive equations for genetic drift. For instance, if the genetic variability is neutral (all variants are equally fit) genetic drift promotes loss of variability (i.e. fixation) at the rate of $1/(2N)$ per generation in an idealized population of N diploid, randomly mating individuals with equal fertility. However, in reality populations behave differently and not all individuals contribute genes equally to the next generation. The loss of variability in these real populations is equated to $1/(2N_e)$, where N_e stands for effective population size, that is, the size of an idealized population that experiences the same drift effect as the real population (see Box 1.1). Interestingly, although the fixation probability for a neutral mutation is also $1/(2N_e)$ (the initial frequency of a new mutant), the probability of fixation of a selected mutant must be different, greater than $1/(2N_e)$ if the mutant is favourable and less than $1/(2N_e)$ if it is detrimental. The question is, how much different?

For non-neutral mutations the probability of fixation is also quite dependent on the effective population size, especially in small populations. Population genetics theory tells us that the fate (fixation or elimination) of a selected **allele** (favourable or deleterious) relative to the neutral fixation depends on $N_e s$ by a non-linear function (see Figure A in Box 1.1), where s is the increase (or decrease) in fitness due to the allele presence (selection coefficient). Now, suppose that a new slightly deleterious allele ($s = -10^{-5}$: the minus sign in front stands for deleterious) occurs in a large population of $N_e = 5\times10^5$ individuals. Perusal of the function in the figure tells us that for $N_e s = -5$ the relative fixation probability is negligible. However, if the same mutation occurs in a population of 500 individuals then $N_e s = -5\times10^{-3} = -0.005$, a value close to the neutral fixation probability, which is 0.001 (1/2N = 1/1000) (See Box 1.1). In general, if the product $N_e s$ is in absolute value much less than 1 then a new mutation behaves practically as neutral (quasi neutral) and is fixed, despite opposition from selection in the case of being deleterious.

These basic findings of population genetics inspired Lynch's ideas. Provided with the negative correlation between effective population size and genome complexity, the latter of which is correlated to increasing genome size, Lynch argues that 'excess' DNA did not enter the genome as a selfish nor as a favourable DNA, but as slightly deleterious DNA sequences that could not be eliminated by selection in small populations subjected to drift. Before getting involved in the following discussion about Lynch's tenet, I must clarify that adaptive positive natural selection is by no means excluded from genome evolution in this challenging hypothesis. On the contrary, Lynch is not prejudiced against adaptive evolution—he recognizes that organic evolution promoted by natural selection is basic to understanding evolution—but he states that 'by enhancing the permissiveness of the population genetic environment for the passive emergence of gene architectural complexity, the non-adaptive force of random genetic drift sets the stage for future paths of adaptive evolution in novel ways that would not otherwise be possible'.

Because effective population size matters for the fixation probability of a new deleterious mutant, the estimation of N_e is crucial for the validation of any non-adaptive drift-fuelled theory in genome evolution. Though not an easy task, population genetics also provides a formal way to estimate N_e by teaching us that, at equilibrium, neutral variability (π_s) equals 4 $N_e u$, where π_s (also known as silent-side nucleotide diversity) stands for the proportion of nucleotides at neutral sites that differ between two randomly chosen sequences in the population. The overall comparative value of this relationship depends on several assumptions, among which equal mutation rates across taxa comes first. While mutation rates per generation seem to be positively related to genome size—smaller in **bacteria** than in Drosophila, and lower in Drosophila than in humans—genome size does not seem to be directly involved. Rather, the positive scaling of per-generation mutation rate with large genome size, generally associated with long generation time organisms, could be accounted for simply by the increase in opportunities for germ-line replication errors in organisms with long life cycles (and large genome sizes). When these mutation rate differences are considered, the true range of N_e expands relative to that found with estimates from π values, but the correlation between N_e and estimates of π persists according to Lynch (2007).

A much more problematic assumption in long-term N_e estimation from extant molecular variability is the disregarding of population size fluctuations through evolutionary time. Differences among taxa in their nucleotide variability at silent DNA sites, provided that they are neutral and differences in mutation rate are negligible, are taken as indicative of differences in N_e. Yet, the observed nucleotide differences are greatly influenced by recent demographic histories. For instance, they may have accumulated only since the last **bottleneck** in the population history of each species, and then do not estimate the long-term N_e. However, we are not interested in the present-day N_e, but in the N_e at the time when the genome increased in size, that likely preceded the last bottleneck (**coalescent**) demographic event. Moreover, estimates of N_e derived from neutral molecular variability apply only to the average time to fixation of a neutral mutation, namely the past $2N_e$ generations. This should not be fatal for species with large N_e, but, since the evolutionary time for acquiring many genome features often requires hundreds of millions of years, estimates in species of small N_e, on the order of less than 10^5, may easily be flawed.

These and other objections (see below) to the non-adaptive hypothesis of the origin of genomes have been considered irrelevant by its defenders. Lynch (2007, p.100) argues that 'because most aspects of genomic evolution develop on time scales of millions of years, clear associations (between genome size and N_e) are unlikely to be observed at the level of species within genera'. This tenet has been criticized because of its ambiguity in defining a level at which a real test of hypothesis can be performed. While a wide-range negative relation between N_e

and genome size might be real, despite uncertainties in the scaling, correlation does not imply direct causation. A final test for this hypothesis must come from independent case studies in which related taxa, differing in few features to avoid cause-confounding parameters, can be compared. At present, we still do not have enough genome sequences of related species pairs to provide a large set of independent contrasts with enough statistical power to falsify this hypothesis. But some efforts are slowly emerging.

Case studies on genome size evolution

The non-adaptive hypothesis of genome size evolution rests on the causal significance of a long-term pattern of correlation between genome size and population size. Brian Charlesworth (2008), a leading population geneticist, on reviewing Lynch's 2007 book, commented sharply that 'there are reasons to expect that rigorous comparative tests of hypothesis about genome evolution will come to be based on careful contrasts of related taxa, differing in far fewer features that those used by Lynch'. One of these tests was attempted by Yi and Streelman (2005) using ray-finned fish species (Actinopterygii). This monophyletic group diverged from other vertebrates more than 400 million years ago (mya) and since then has diversified into freshwater and marine species. The authors confirm that freshwater species have larger genomes than marine species and, using **microsatellite** markers to estimate variability, conclude that genome size and genetic variability (and thus N_e) are negatively correlated. Interestingly, when they controlled for other traits that might co-vary with variability, such as body size and generation time, and also for phylogenetic non-independence among lineages, genome size and population size are still highly negatively correlated.

This experiment has been cited as an example of the careful contrast of related taxa that bolsters the non-adaptive origin of genome size (Yi 2006; Lynch 2007: p. 100). Yet, some critical voices have challenged the validity of several assumptions and conclusions in this fish test. Gregory and Witt (2008), in a close re-analysis of this work, found several misinterpretations in it, mainly because a) ancient **polyploidy** is likely the major determinant of ray-finned fish genome size, and not the marine or freshwater habitats that determine large or small population sizes, respectively; and b) microsatellite variability, due to high mutation rates of microsatellites, only estimates short-term population effective sizes and, moreover, is highly influenced by recent demographic events that are likely to have occurred during the recent **Pleistocene** glaciations. Gregory and Witt's main concern applies to the way the data have been retrieved and combined by Yi and Streelman. For example, polyploids, found only in freshwater fishes, constitute a large proportion (5 out of 17) of the analysed sample and influence the relationship significantly. The authors (Yi and Streelman) try to explain the polyploidy increase in freshwater by resorting to the unfounded argument that reduced N_e in freshwater fishes 'might have permitted fixation of otherwise deleterious mutations that led to genome **duplications**', in their own words. In fact, when the freshwater polyploids are separated from non-polyploids the regression analysis shows that the relationship between genome size and microsatellite variability (heterozygosity) in non-polyploids turns out to be positive (Fig. 1.6), contrary to the expectation of the non-adaptive hypothesis. Moreover, many analysed freshwater fishes inhabit northern latitudes, where they experienced severe bottlenecks in Pleistocene glacial **refuges**, a caveat in long-term population estimates as stated above. Since a large proportion of freshwater fishes in the analysed sample show ancient polyploidy and recent demographic fluctuation, one wonders how much of a fish-wide correlation between genome size and N_e supports the idea that population size alone can explain changes in genome size.

Because the evolution of genome size is so complex, it should not come as a big surprise that real genome contrasts of related taxa do not produce results that fit expectations. Among mammals, carnivores, with population sizes smaller than those of herbivores, should present larger genome sizes, but the opposite is actually the case. Also, the streamlined genomes of many **endosymbiont** and obligate parasites with extremely reduced population sizes oppose the non-adaptive model of genome evolution, prompting many evolutionists to propose genome-directed natural selection as

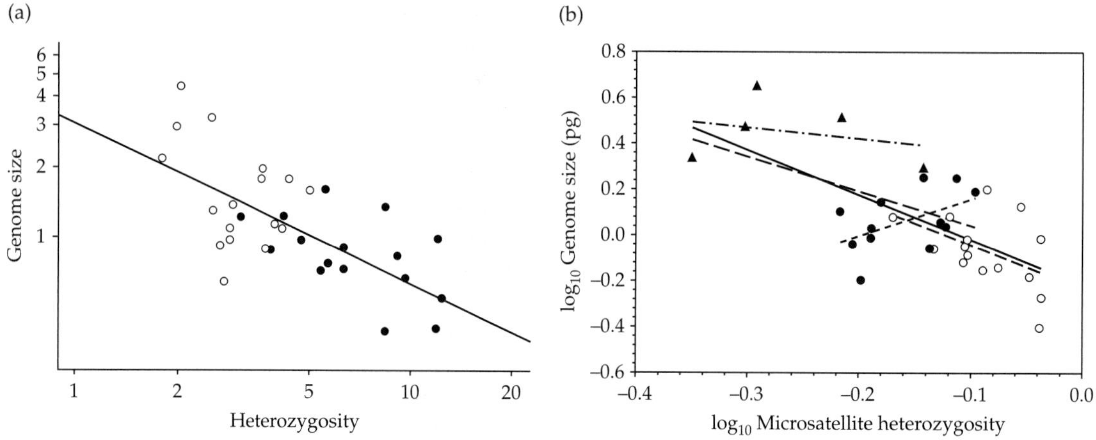

Fig. 1.6. (a) Overall negative relationship (log-transformed) between genome size and microsatellite heterozygosity, as a proxy for effective population size, in ray-finned fish. Black circles indicate marine species, and white circles refer to freshwater species. (From Yi and Streelman 2005 with permission from Elsevier). (b) Relationships between genome size and microsatellite heterozygosity in the same fish data set of Yi and Streelman, separated into polyploids (black triangles), freshwater non-polyploids (black circles), and marine species (white circles). Regressions are shown for all fishes (solid line), all freshwater species (large dashed line), all marine species, except one questioned datum, (medium-dashed line), freshwater non-polyploids (short-dashed line), and polyploids (dashed-dotted line). Note that regression in freshwater non-polyploids (black circles) becomes positive. From Gregory and Witt (2008) with permission from NRC Research.

the explanation (see Gregory and Witt 2008 for references). These and other case studies have been deemed irrelevant by defenders of non-adaptive genome evolution because, as stated above, clear associations between N_e and genome size are unlikely to be observed at the level of species within genera due to the long-term effect of genome size evolution. Since the number of well-worked-out case studies is still small, it may be too early to appreciate the falsification strength of these contradictory examples. But the scale at which the model can be falsified must be clearly defined if we are to consider the model a real scientific hypothesis, and not a 'bullet-proof caveat', as it has already been called (Gregory and Witt 2008).

The multi-parameter approach

While the above population approach to explaining the origin of genome diversity (and complexity) has great merit, this single-parameter theory, based on the inefficiency of selection in the face of genetic drift in small populations, raises some concerns. At least two types of objection have been reported. First, the long-term population-size estimates from molecular variability are not always correct because, as explained above, population demography and short-term validity of observed diversity are often not considered. Moreover, we do not have enough real sequence comparison tests to validate the non-adaptive hypothesis, and some of those that have been recently reported do not support the predictions (see below). Second, since the evolution of genome size is considered a highly complex process, the parameters of which may be multiple and varied even between related species, a single-parameter explanatory hypothesis seems rather simplistic to some evolutionists. As stated earlier, many different components of the species biology, like genome size, cell size, developmental rate, body size, and population size, among others, are confounded by their close relationships and what is cause and what effect is rather difficult to discern. Even the genome can be the target of selection; in particular its size may not only expand but also shrink (as in parasitic eukaryotes, to adapt to their opportunistic lifestyle), and more obviously, genomes are subjected to many dynamic processes that, as in mobile-element induced transposition, could impose their own intragenomic rules.

While it could still be too early to understand in full the dynamic genome, it appears, after the present accelerated improvement in genome sequencing efficiency, that this may soon be remedied. We are approaching a time in which I believe that there is a consensus that comparative genomics is vindicating the central role of population genetics to the understanding of genome evolution, as Lynch (2007) explains. Yet, genome dynamics comprises a set of mechanisms that go beyond mere nucleotide substitutions and must be understood if we are to relate them to the basic evolutionary forces of selection, mutation, migration, **recombination**, and drift. These mechanisms, including large and small DNA **duplications**, mobile element transposition, genome acquisition by lateral transfer, intron insertions, and alternative splicing, are foremost in the dynamics of genomes and introduce a putative way to fit the Darwinian view to present knowledge. In particular, the role of natural selection, so cherished by Darwin, must be contrasted with that of the other evolutionary forces to decide whether or not, and how, selection is still the channelling agent in organic evolution at the genomic level. Most of this chapter's argument deals with the importance of genomic mechanisms in shaping the structure of evolutionary theory 150 years after Darwin.

Genome evolution by duplication

Having discussed the increase in genome size mainly through the incorporation of repetitive non-coding DNA, it is time to analyse the dynamics that underlies the increase of gene content along the evolutionary scale. The number of genes has steadily accrued on the genome throughout evolutionary time, though at a rate that does not mirror the rate of organismic complexity. As stated above, more complex organisms, like humans, do not show a proportionate increase of number of genes compared to less complex ones, such as nematodes or flies. Why is this?

There is no doubt that Ohno's seminal work, published in his 1970 book *Evolution by **Gene Duplication***, set the stage for the still highly favoured idea that DNA duplication is the major mechanism for incorporating new functional genes in the genome. Although some earlier evolutionists (see Taylor and Raes 2004 for a review) had suggested that duplication of whole (or parts of) chromosomes could contribute to evolution, Ohno was first to assign a central role to duplication in adaptive evolution. Originally, the idea was received with reluctance, I believe, because it was considered a challenge to the gradualist neo-Darwinian synthesis. Even though polyploidization, the doubling of the whole chromosomal set, was a long-documented process of great evolutionary value in plants (see Chapter 4), chromosome doubling in animals was far from being accepted in the 1970s. Yet Ohno's great intuition, based on limited molecular data, posited that two rounds of polyploidization underlie the evolutionary origin of jawed vertebrates (see Chapter 2). This proposition, received with great scepticism by most neo-Darwinians, was very difficult to prove in the pre-genomic era, but once comparative genomics became established as everyday, this '2R (2 round) hypothesis', as it is now named, gained much credibility (Fig. 1.7). In the past decade, analysis of the ever increasing numbers of sequenced genomes has not only confirmed many of the earlier indirect inferences of gene and genome duplications, but has also unveiled many unanticipated episodes of duplication at points during the evolution of most lineages. Now it is firmly accepted that the majority of genes in eukaryotic organisms evolved from a duplicate copy of an ancestral gene. Although the two copies (known as paralogs, a kind of homologs) diverge after the duplication event, they may still show much sequence similarity (homology; see Chapter 2) in spite of their often different functions. Moreover, many genes form large sets of paralogs (**gene families**), the members of which have diversified toward different, complex functions. Gene families are common in multicellular organisms and some examples are well characterized, ranging from basic physiological functions, such as the immunological and the oxygen transporter (the globin family) systems in mammals, to complex developmental functions—as in morphogenetic networks in most animals (see Chapter 2). This repetitious gene landscape vindicates Ohno's pioneering view of evolution by gene (and genome) duplication.

Once it is accepted that the gene evolutionary pattern in the genome has largely originated through duplication, the next step is to understand how a new duplicate copy evolves to the fixation of a new functional gene. Origination of a new copy in

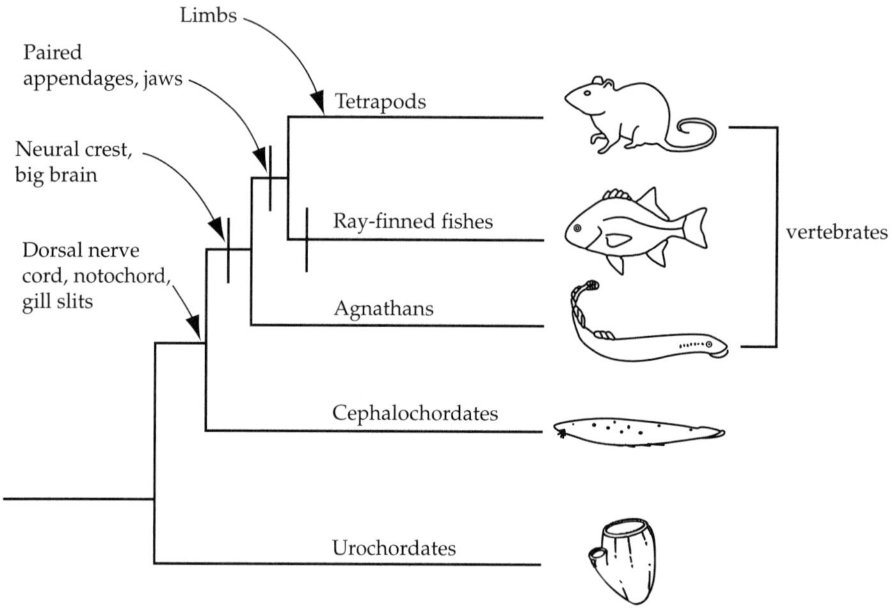

Fig. 1.7. Simplified tree of chordates showing the two whole genome duplications (depicted by vertical bars) that occurred before the evolution of jawed vertebrates. The bar located deep in the ray-finned fish lineage denotes a polyploidization event that was followed by secondary polyploidization in various fish lineages. Some evolutionary novelties that likely accompanied duplications and nodes are also depicted. Adapted from Lynch (2007), with permission from Sinauer Associates.

a single individual must be followed by an increase in population frequency driven by population genetic forces: directional (selection) and/or random (drift). Duplicate genes may deploy new functions, but the moment of function acquisition is not easy to deduce from the available genome information. Thus, if the initial gene expansion were due to rare original beneficial mutations, selection will be more effective in species with larger population sizes, and prokaryotes and many unicellular eukaryotes with enormous population sizes should contain the highest numbers of genes, which is not the case. In fact, organisms that live in comparatively smaller populations, like many mammals, host the highest number of genes, many of them originated by duplication (see Table 1.1). The tentative guess is that other population forces, like drift, could be at work, and that beneficial gene functions may then be acquired in later stages. But this tenet must be tested before being accepted.

Recent whole-genome studies picture the genome as a dynamic arena where new duplicate genes are continually being produced. However, the majority of them are eliminated right away by drift and/or natural selection, and only a few of them experience new functional adaptive changes that allow them to persist in coexistence with their parental copies. One way to probe duplicate-induced genome dynamics is to study the evolutionary demography of duplicate genes. By computing the silent-site divergence between duplicate gene pairs in a genome, which estimates the time since duplication (see Box 1.1), the age distribution of all duplicate genes can be depicted and used to estimate their average origin and elimination rates in different taxa. While no significant differences in birth rate estimates exist between unicellular organisms and animals (gene duplication average probability of 0.004 per 1% divergence for silent sites), loss rate (D) estimates of new duplicates are significantly lower in multicellular organisms (average D values of 0.34, 0.18, and 0.13 for unicellular species, invertebrates, and vertebrates, respectively) (Lynch 2007, p. 199). Since multicellular organisms tend to have larger genomes, a positive correlation between half-life (persistence) of duplicate genes and genome size is established. Remember that genome size and effective population size (N_e) are negatively correlated, so there seems to be a greater ability for a new duplicate gene to survive the deleterious effect of mutations (quantified by *s*) in small populations (large genomes). Thus, following Lynch and Conery's (2003) reasoning, lower $N_e s$ should be more permissive to the evolution of new genes by duplication (see Figure A in Box 1.1).

This population analysis is important because it sheds light on the three-phase evolutionary dynamics of a new duplicate gene: its origin, its expansion, and its optimization. The most popular, and simple, mechanism, known as neofunctionalization, proposes that once a duplicate gene is produced there is a low, albeit significant, probability that the duplicated pair be retained for a long time in the population, mainly due to random drift. A mutation conferring a new beneficial function appears in one copy of the pair, normally after duplication and during the expansion of the duplicated pair. Fixation of this new gene is accomplished by rapid **directional selection** favouring the new function, followed by purifying selection. While this is an attractive model, note that the above conclusion, that lower N_e facilitates evolution by duplication, looks paradoxical if neofunctionalization is the right mechanism. In a small population there is a high probability that deleterious mutations that inactivate the gene are retained in the duplicate copy before the favourable mutation appears, because purifying selection is less effective than in large populations. Contrarily, in large populations the probability of acquiring a new favourable mutation before gene inactivation is much higher, because purifying selection will efficiently eliminate deleterious mutations. In sum, 'neofunctionalisation of a duplicate gene is a large-population phenomenon', as Lynch said (2007: p. 212).

So, how can the greater number of gene families in multicellular eukaryotic organisms be explained? Isn't the low effective population size of multicellular, complex eukaryotes an impediment to gene evolution by neofunctionalization? Although some case studies may support the neofunctionalization model (see below), many gene duplication instances rather favour another model, in which preservation of duplicate gene pairs is accomplished by degenerative mutations that inactivate different functional aspects in each gene pair (the subfunctionalization model). Thus, each of the two genes is partially degraded in different complementary functions. They are then not wholly inactivated, but their joint expression is needed to reconstitute the whole function provided by the ancestral gene. Lynch and Conery (2003) think that subfunctionalization may be the predominant mechanism in small populations, because the longer retention time of duplicate genes in small populations allows the progress of subfunctionalization and their ulterior persistence.

Besides fitting into the non-adaptive hypothesis of gene-number increase in multicellular organisms, subfunctionalization seems to be supported by worked-out studies. In multicellular eukaryotes the rule is that each gene has several modular regulatory regions, which control gene expression in different tissues or cell environments, and/or possess the ability to perform several functions through the alternative means of gene splicing, or because it codes for different functional **domains** (see Chapter 2). This functional versatility means that a mutation could affect only one specific function, for instance in a tissue, a substrate, or a **receptor**, without altering another function. Once alternate degenerative mutations occur, likely in sequential order, in duplicate genes, they become subfunctionalized, complement each other, and are thereafter maintained by purifying selection. However, subfunctionalization may be just the beginning of new evolutionary novelties driven by **positive selection** on new favourable mutations that optimize the function in each gene (see Box 1.2).

Many cases of subfunctionalization have been revealed in regulatory and/or substrate specificity gene regions in ray-finned fishes. Studies with zebrafish, an ancient ray-finned polyploid, the genome of which duplicated several times before the ray-finned radiation (see above), show several examples of subfunctionalization. Among them, two genes for microphthalmia-associated **transcription factors** (*mitfa* and *mitfb*) evolved by duplication and posterior subfunctionalization through deletions in both coding and regulatory regions that conferred on them tissue-specificity expression: *mitfa* to the neural crest, and *mitfb* to the epiphysis and olfactory bulb. Interestingly, the products of each duplicate are **homologous** to each of the two

Box 1.2 Modelling the tortuous route to new genes via duplication

The classic model proposed by Ohno (1970) of new gene evolution by duplication relies on the release of one duplicate copy from the previous selective constraints. Duplication initiates a phase of redundancy in which new deleterious mutations can be accumulated in one copy, avoiding the eliminating effects of **purifying selection**. Although the commonly-favoured thought is that the new copy is the one that acquires the new function, this may not be true because both copies are subject to the same mutational fate. After gene duplication in an individual the gene pair must be maintained in the population until a beneficial mutation appears. This route to fixation can be accomplished either because the duplication is beneficial (for instance if it provides a dosage advantage) or, as seems to be more likely, due to random drift. Incidentally, the most probable fate of duplication is loss. Another possible fate is inactivation, which drives the duplicated copy towards a pseudogene. In a few cases, however, a mutation that opens a new function appears and positive selection drives the mutation-containing copy to fixation. Figure B (a) depicts the whole process of this model called 'neofunctionalization'.

This model depends on the maintenance of the active allele that is eventually modified by a favourable mutation. In large populations, purifying selection can likely maintain functionality, but in small populations selection is ineffective in detecting slightly deleterious mutations, which finally inactivate the duplicate gene (see text for details). Theoretical population genetics considerations (see Lynch 2007) support this reasoning in full, assigning neofunctionalization mostly to large populations. Since the majority of new functions that evolved via duplication are found in high eukaryotes with small population sizes, some authors thought that preservation of duplicated genes could be the outcome of subfunctionalization of complementary ancestral functions in each duplicated copy. While not the only one, the DDC (Duplication-Degeneration-Complementation) model proposed by Force *et al.* (1999) is the most popular. In DDC, degenerative mutations inactivate different subsets of the original function from each gene-pair duplicate. These mutations behave as neutral because their deleterious effect in one copy is cancelled by the complementary function still performed by the other copy (see Fig. B (b)).

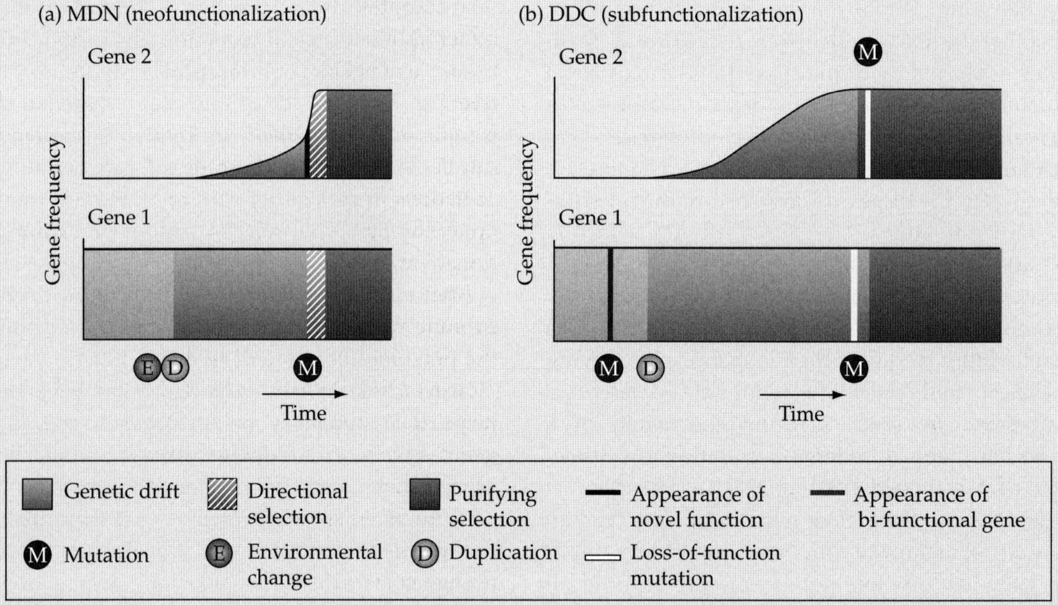

Fig. B. Schematic time-flow of a duplication in terms of the population frequencies of both the original gene 1 and the duplicated gene 2. Before the duplication, the original gene is fixed in the population, then two models of how a duplicate gene could first become fixed and later become preserved are shown. (a) Neofunctionalization model. At point E an environmental change makes the new function functional. The duplication appears later and it drifts until a neofunctionalizing mutation appears, which, if favourable, is driven to fixation by directional selection. (b) Subfunctionalization model (DDC). Here duplication is driven to fixation by drift. Then degenerative mutations at different subsets of the original function occur that escape purifying selection because they complement each other. However, after this subfunctionalization a process of neofunctionalization can occur. Adapted from Conant and Wolfe (2008) with permission from Nature Publishing Group.

continues

Box 1.2 *(continued)*

Though these models may seem simplistic, the route to new genes is complex. Timing of mutational and selective events varies and defines the route. For instance, in neofunctionalization functional divergence occurs after duplication, whereas in classic subfunctionalization the gene function is already dual before duplication. Each duplicate gene just separates and, perhaps, optimizes functions (Fig. B), but no brand new function evolves. The subfunctionalization concept, however, may comprise more than one single route that often opens a new evolutionary pathway to new functions. So, neofunctionalization and subfunctionalization are not necessarily mutually exclusive processes (see text and Fig. B).

The case of divergence of the aldosterone and cortisol receptors in tetrapods (discussed in the text; see Fig. 5.3) exemplifies how a minor non-adaptive ancestral function (aldosterone sensitivity) has been co-opted by the opportunism of natural selection. Thus, a new function has been resurrected through subfunctionalization of an already established function (receptor affinity for cortisol).

alternative splicing products coded by the ancestral homologous gene, found in tetrapods, illustrating that subfunctionalization in the zebrafish genes allowed the adoption of one splicing variant by each duplicate gene. Another zebrafish example is provided by two cytochrome aromatase genes: one expressed in the ovaries, the other in the brain. As predicted, the homologous tetrapod gene expresses in both tissues. There are many similar examples in fishes which follow the simple rule that an ancient dual role in expression has been split in two separated roles by the subfunctionalization of a duplicated gene pair (see Lynch 2007 for references).

The plausibility of subfunctionalization notwithstanding, because this model implies that no extremely new functions evolve there is still a great controversy as to how and when novelties emerge. There is a tendency to present neofunctionalization and subfunctionalization as mutually exclusive, but, as Conant and Wolfe (2008) state, 'this dichotomy is only valid when considering the mechanism of (duplication) preservation'. They go on to explain that 'After preservation, duplication genes continue to evolve, meaning that subfunctionalisation can contribute to novelty simply by enabling duplicate genes to survive for long periods, increasing the chances of a neofunctionalizing mutation.' One interesting example of the **co-option** of a minor function that appeared without an initial selective value is provided by the steroid hormone **receptor** family. Two genes encoding two of these receptors (MR and GR) evolved by duplication of an ancestral gene before the split between cartilaginous fishes and the lineage that led to tetrapods. MR receptors bind the hormone aldosterone, and GR receptors bind cortisol. Both genes are present in teleosts and tetrapods, but tetrapod MR binds aldosterone whereas teleost MR does not, because aldosterone is not present outside tetrapods. Thus, the guess is that the binding of aldosterone appeared as a non-adaptive mutation before the duplication and was later co-opted (see p. 34 in Chapter 2 and glossary) by tetrapods when cortisol affinity was reduced in tissues where aldosterone signalling was favoured (see Fig. 5.3). The lesson to take away from this complex example is that persistence of duplications can be favoured by subfunctionalization, but new mutations of present or future adaptive value can appear along the way and be co-opted as evolutionary novelties.

After all the above examples it might seem that subfunctionalization is a mechanism exclusive of the polyploid lineages. This is far from the reality. However, what is an accepted fact is that polyploidy helps the persistence of duplications when the genes involved are affected by dosage balance. Many genes, like those whose coded products interact with other genes' products, are very sensitive to changes in dosage proportions. Except perhaps for dosage compensation in organisms with chromosomal sex-determination, whole-genome duplication (WGD) does not change the relative dosage production among genes, so duplication is not detected as detrimental. This could explain why genes evolved by polyploidization generally show greater longevity than those arising by small-scale duplications (SSD) (in yeast: 27×10^6 years in genes by WGD versus 17×10^6 years by SSD; 2×10^6 years in SSD-originated vertebrate genes versus 45×10^6 years in WGD-originated ray-finned genes, as cited by Lynch 2007). The dosage-balance hypothesis may also justify the high functional similarities of WGD-originated genes across taxa: including ribosomal proteins, protein kinases, transcription factors, and other highly dosage-dependent genes.

Many unicellular eukaryotic species also show the signatures of subfunctionalization in WGD-originated duplicate genes. In this case functional partitioning cannot affect tissue-specific expression, but occurs at the metabolic or subcellular level. WGDs have been documented in the genomes of the yeast *Saccharomyces cerevisae* and the unicellular ciliate *Paramecium tetraurelia*, the latter of which experienced three rounds of WGD yielding about 40,000 genes. An interesting study case in *S. cerevisae* involves subfunctionalization of a galactose gene regulatory circuit **promoted** by WGD (Hittinger and Carroll 2007). The ancestral protein is represented by Gal1, a single protein present in the yeast *Kluyveromyces lactis* that has a dual role as a regulatory protein and as an enzyme that metabolizes galactose. This dual role has a negative interacting effect that lowers the dynamic range of transcription due to the use of the same promoter for both functions. But, the duplicate gene pair products Gal1 and Gal3 in *S. cerevisae* specialized into an enzyme and a regulatory protein, respectively. The specialization involved changes in promoter structure that optimized the transcription according to the galactose concentration in the cell. In this case it is not clear when the change from an enzymatic to a regulatory function occurred, but the change is possibly adaptive by co-opting an ancestral dual function.

Genome evolution by transposition

Mobile elements, also known as **transposable elements** (TEs) due to their transposition ability, are responsible for the composition and structure of a large genome fraction. Chapter 3 gives a wide perspective of TE-induced genome evolution, particularly focused on regulatory and genetic innovation. Here, I want to give a brief introductory overview of aspects that relate TE dynamics to some topics discussed above, such as genome size and duplication.

TE amplification rivals genome duplication as a major force in genome expansion. As explained above, evolution by polyploidization is a well-documented process in plants and also in animals. The evolutionary history of many organisms, especially plants, shows a repeated, cyclical process of genome duplications interspersed by deletions, divergences, and rearrangements in which TE activity is often present. When genome duplication is ancient this complex process makes polyploidization recognition a difficult task. For instance, yeast experienced a whole genome duplication 100 mya that has been barely recognized because only about 8% of the duplicated genes persists today, and many chromosomal segments containing blocks of duplicated genes have been moved across chromosomes subsequent to polyploidization (Fig. 1.8).

A typical pattern of duplication followed by extensive gene loss and proliferation of TEs is found in recent maize evolution. Maize experienced a polyploidization event 20 mya that was followed by gene loss and a series of TE transposition bursts over millions of years. Dating of TE insertions in some maize genome regions indicates that a large increase in TE transposition, that occurred in the past 5 million years (my), is likely to be responsible for a two-fold genome expansion. In sum, recent TE activity in maize has generated a huge genome size due to TE intergenic insertions (Fig. 1.9). Although the maize story exemplifies how TE activity drives genome expansion, it is not the only case study in which powerful, episodic TE bursts have shaped large genomes. In the human genome, for instance, the insertion episodes have occurred over the past 70 my or so, albeit at a non-uniform rate. Thus, the majority of *Alu* elements, a class of retroelements (see Chapter 3), were inserted about 40 mya, but they continued to expand the human genome to date by inserting more than 1 million fixed copies that constitute more than 10% of our genome. Smaller genomes are also expanded by TE insertions. In *Drosophila melanogaster*, with a 137 Mb size genome, even though fixation of TEs is rare, a burst of thousands of **retrotransposons** has been documented 5–10 mya.

However, although polyploidization and TE transposition episodes are powerful mechanisms of genome size expansion, it is unlikely that they could

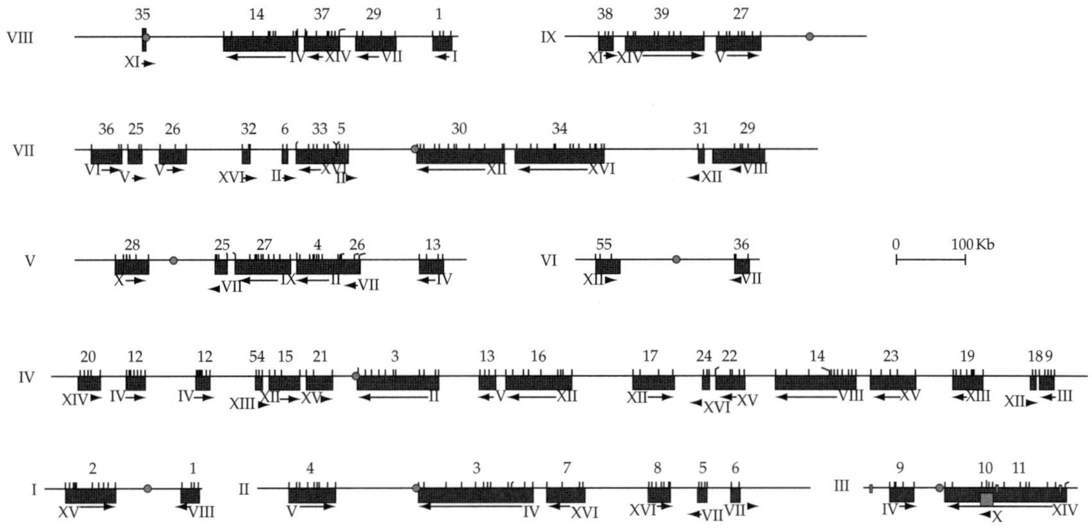

Fig. 1.8. Some chromosomal duplicated segments that persist today derived from a whole-genome duplication in the yeast *S. cerevisae*. Each duplicate block pair is designated by a number. Roman numerals below each block indicate the chromosome location of the other copy. Arrows depict the relative orientations of the two copies of a block, usually conserved with respect to the centromere. Thus, for instance, block 29 in chromosome VIII has its copy in chromosome VII in the same orientation. Adapted from Wolfe and Shields (1997) with permission from Nature Publishing Group.

explain on their own the enormous variation in genome size, mainly because small genomes, present in many disparate lineages, must then have been able to avoid TE expansions and genome duplications for millions of years, which is very improbable. As an example, *Arabidopsis*, a model flowering plant, has experienced at least two, and probably three, genome-wide duplications followed by massive loss of DNA, which has likely kept its genome size at the small value of 125 Mb. In previous sections I presented some putative mechanisms that may explain the large observed disparity in genome size and structure, and pointed out that even intragenomic rules can also be decisive. The TE intragenomic dynamic generates a series of processes that impinge on genome evolution, ranging from local scale mutations to large rearrangements, some of which may induce deletions. This TE dynamic is discussed in detail in Chapter 3, but here I want to present those processes that are more directly related to genome size disparity.

Genome size contraction is well documented and supports the idea that mechanisms that eliminate DNA must exist. We can cite, among others, the DNA loss during repair of DNA breaks and also as a product of recombination (homologous or not). Interestingly, TEs may be active in these mechanisms. For instance, recombination between dispersed homologous TE copies can duplicate and also invert chromosomal regions. This process, often referred to as **ectopic recombination**, generates deficient sequences (deletions) as well (Fig. 1.10). In humans TE recombination-mediated deletions have been documented for **retrotransposons** (*Alus* and *L1s*). Many deletions are highly deleterious, being the cause of cancer and other genetic impairments, but others have persisted in the human genome since the split from the chimpanzee lineage. These studies also show that these recombination events have likely eliminated as much as 1 Mb of the human genome over the past 5 my. There are other deletional processes and overall they may strongly oppose, albeit at different organismal rates, genome size expansion. Thus, in Drosophila, an organism in which the deletion rate is higher than in mammals, it is estimated that small deletions could eliminate half of the 'junk' DNA in about 10–15 my (Petrov 2002).

Many evolutionists agree that a model that postulates a stable equilibrium, reached at the point

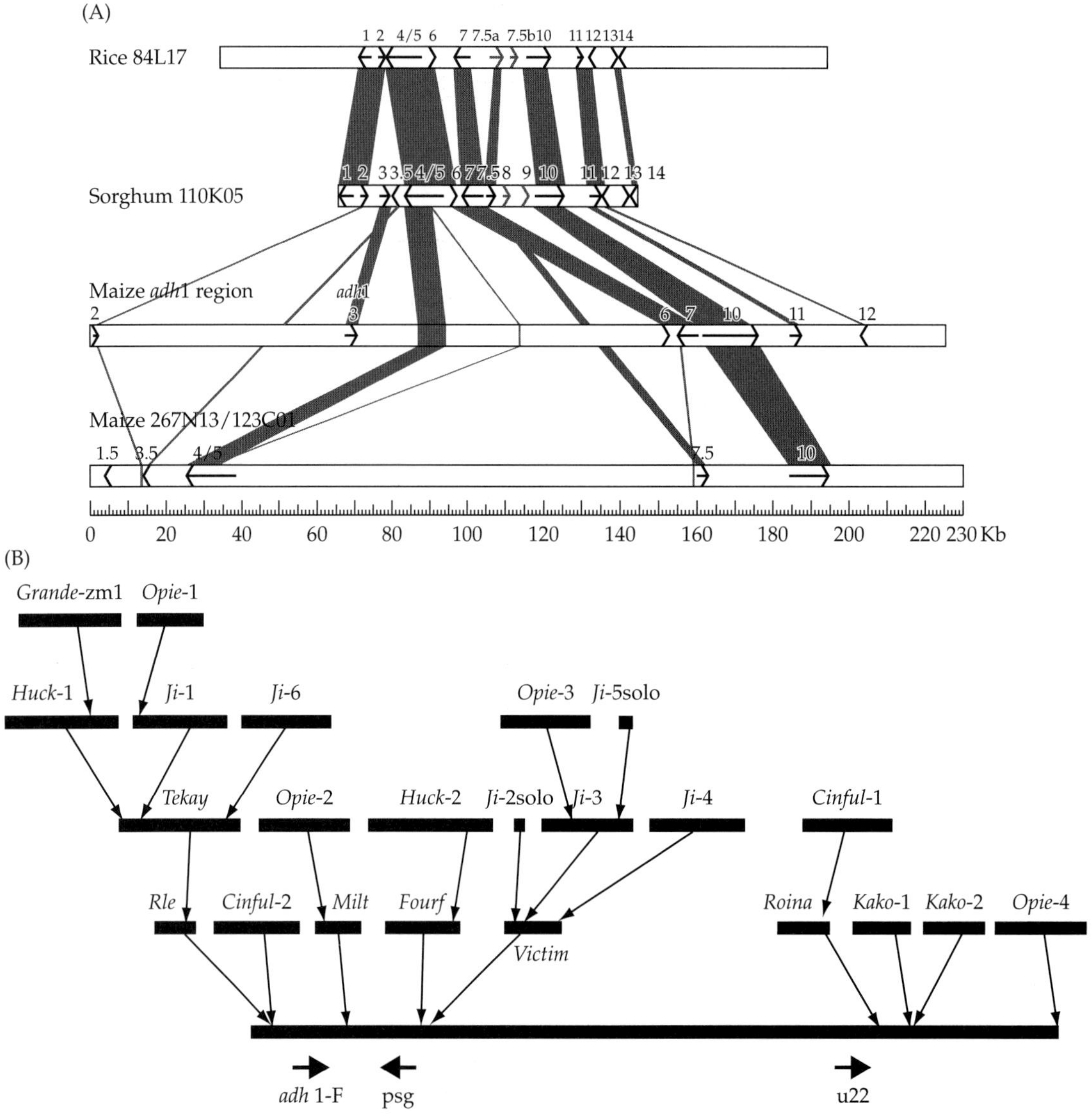

Fig. 1.9. Maize genome expansion by transposition. (A) The figure depicts the homologous genome segments, shown as shaded areas, of maize, sorghum, and rice in a region close to the adh-1 site. The maize region is represented twice because of the polyploidization event that occurred 20 mya. Genes are denoted by numbers and arrows indicate orientation. Note the expansion of the region in maize relative to sorghum, evidenced by the greater distances between genes due to insertion of TEs since the divergence of both species. Also, note that many genes are lost in the duplicate region (for example, 2, 3, 6, 7, 11, 12). From Petrov and Wendel (2006) with permission from Oxford University Press. (B) In this 240 Kb region of the maize genome, two-thirds of the region is made up by 23 TE inserts, many of them (14) nested one inside the other. It is estimated that 15 of them were inserted in the last 3 my. The figure shows the family names of the TEs and indicates the positions of three genes (bottom). From SanMiguel *et al.* (1998) with permission from Nature Publishing Group.

where the insertion rate equals the elimination rate, often lacks convincing support because insertion rates exceed loss rates by 100-fold (Maside *et al.* 2001). It has been postulated, however, that the deleterious effects of insertions might increase with the copy number in an invaded genome. The underlying basis of such a view comes from mutation accumulation experiments in which the continuous

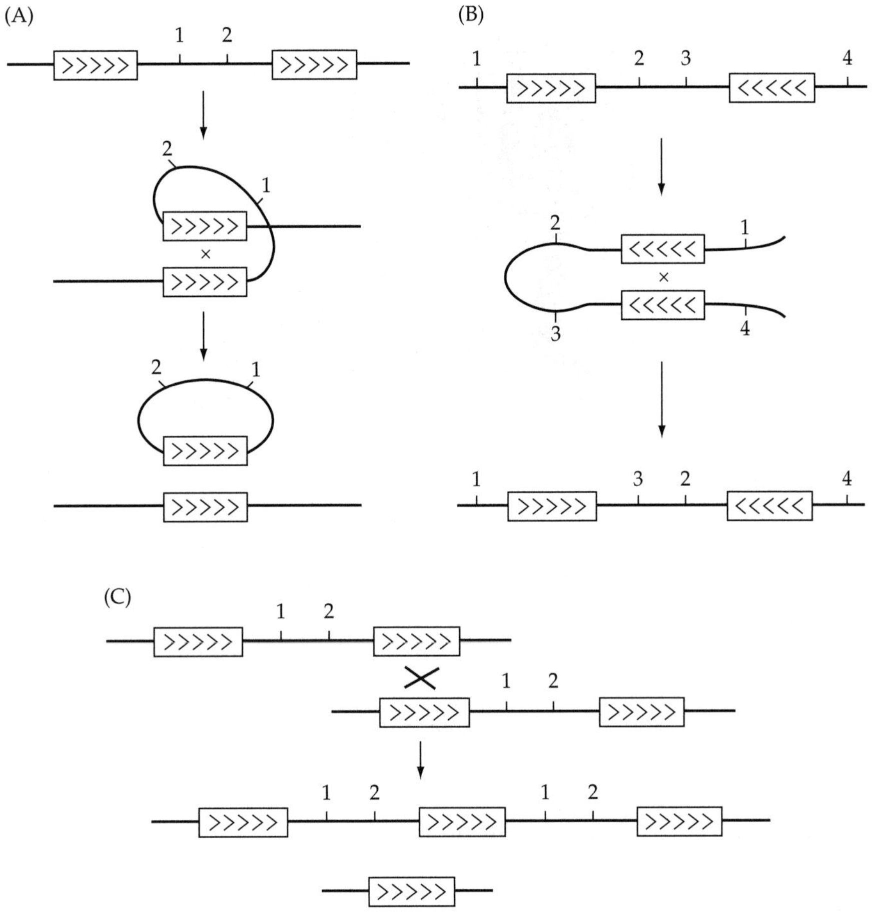

Fig. 1.10. Scheme of how ectopic recombination between TEs generates deletions (A: when both insertions show the same orientation), inversions (B: when in opposite orientations), and duplications (C: when in different chromosomes).

acquisition of deleterious mutations reaches a point where the fitness decrease is accelerated due to a kind of synergy among consecutive mutations. These results may be true for conventional mutations, but they remain unproved for TE insertions. On the other hand, ectopic recombination between dispersed TEs (see previous paragraph) may provide a time-increasing control mechanism for TE elimination, because its effectiveness scales up with the copy number increase. If ectopic recombination controls the accumulation of TEs, low recombination chromosomal regions should host more elements than normally recombining regions. This is true, albeit not always, in some *D. melanogaster* surveys, but since natural selection is also less efficient in low recombination regions we cannot be sure that ectopic recombination, and not purifying selection, is the real cause of TEs removal.

Population geneticists, theorists and experimentalists alike, found difficulties in explaining the observed population distributions of insertion frequencies by applying the conventional models. When genomes became available, however, a new picture of TE persistence emerged. The first genome-wide studies have unexpectedly shown that most full length **LTR** retrotransposons are very young. For instance, John McDonald and his collaborators (Jordan and McDonald 1998; Bowen and McDonald 1999, 2001) produced evidence that the average ages of LTR retrotransposons in fruit flies (<500,000 yr), yeast (<100,000 yr), and nematodes (<500,000 yr) are much younger than the age of the host species. Later, similar works (Bartolomé *et al.* 2009) that surveyed all kinds of TEs in the genomes of the fruit fly species *D. melanogaster*, *D. simulans*, and *D. yakuba* have confirmed that a large fraction (about one-third) of all TE families are much younger than the time of species split. The age of most young TEs, measured by interspecies TE divergence and within-species TE diversity, ranges between 30,000 and 40,000 years, a period prior to the world-wide colonization of *D. melanogaster* and *D. simulans* from their original African territories about 15,000 years ago. This unanticipated result is at odds with the stable equilibrium model for copy number. Interestingly, while copy-number equilibrium and vertical transmission cannot explain the youth of TEs, these surveys are consistent with the hypothesis of TE acquisition by **horizontal (or lateral) gene transfer (HGT)** from other species.

Once thought very improbable, except for prokaryotic organisms, HGT is today becoming a widely-accepted mechanism across eukaryotic taxa (see Chapter 4 for a detailed treatment). In Drosophila, the horizontal transmission of *P* element from *D. willistoni* to *D. melanogaster* in recent historical times is widely documented (see Box 3.3), but this is not the only case; other TEs like *hobo*, *mariner* (that belong to the class 2 like *P*), and also LTR elements as *copia* and even non-LTR elements, albeit at a lower rate, have also been horizontally transmitted to Drosophila species and to other insects as well (see Burt and Trivers 2006, pp. 267–71; and Chapter 3). With all the current information a new dynamic TE pattern is emerging that, instead of favouring an exclusive equilibrium model between insertion and excision rates, proposes periodic cycles of TE bursts followed by large TE expansions that initially overcome the strength of selective loss. As time goes by the insertion rate declines, in part because of deleterious mutation accumulation in TEs, and finally the fate of most TEs is their loss. However, occasionally new horizontal invasions introduce new waves of foreign TEs that initiate a new TE-expansion episode. Incidentally, TE bursts can be induced not only by horizontally transmitted TEs but also by other environmental and/or genomic changes that generate stress in the organism (see Chapter 3 for a longer discussion).

This burst–invasion–loss cyclic model, although well supported in Drosophila, is not universally accepted for other organisms. Thorough genome studies in organisms like maize and humans unveiled the fact that full-length TEs are much older. For instance, in humans, HERVs (Human Endogenous Retrovirus) are >25 my old, exceeding the age of the human split from the chimpanzee lineage (about 6 my). Table 1.3 summarizes the TE insertions that occurred in the past 75 my compared to the total human TE composition. In addition, maize experienced a series of TE expansions since a polyploidization event about 20 mya (see above Fig. 1.9), after its divergence from the sorghum sister lineage 15–20 mya. The sequence of a 240 Kb region close to the *Adh*-1 gene has shown that it contains 23 TE inserts (21 LTR retrotransposons and 2 solo LTRs), 15 of which were inserted in the past 3 my, totalling about 50% of this DNA region. Thus the persistence of maize TEs is apparently much longer than in Drosophila and, together with genome duplication, largely accounts for the increased genome size of maize (2,400 Mb) compared to sorghum (750 Mb), its close relative. Several explanations for these contrasting TE dynamics in genome persistence have been proposed. Since **methylation** is an inactivating mechanism well documented in plants and mammals, but

Table 1.3. Transposable element composition of the human genome (from Burt and Trivers 2006, p. 278, with permission from The Belknap Press of Harvard University Press)

		% of genome	
	No. of inserts	**Total**	**Acquired in the last 75 my**
DNA transposons	400,000	3	1
LINEs	1,000,000	21	8
SINEs	2,000,000	14	11
LTR retroelements	600,000	9	4
Total	4,000,000	46	24

much less present in invertebrates, methylation-induced TE inactivation has been invoked as the main mechanism supplying the need for selective elimination of deleterious TEs. On the other hand, Drosophila, and many organisms with compact genomes but without an effective methylation mechanism, must rely heavily on deletion processes to control the deleterious TE load. This by no means implies that deletion is not a universal process; rather it means that the DNA loss attributable to deletions is much higher in invertebrates (estimated as a 75-fold higher in Drosophila) than in mammals.

The RNA machine in the genome

While whole-genome studies have produced continuous surprises ever since their inception, we are beginning to form an image of genome structure that allows us to understand how evolution has proceeded from simple to complex genomes. Some features that have been discussed above, such as changes in genome size, duplications, and TE insertions, are important mechanisms in genome evolution. Yet there could be other processes, including those possibly not yet discovered, that are less well understood. Among these, the abundant non-coding genome transcripts, whose role in gene regulation is being investigated intensively, occupy first row importance. For instance, it has recently become apparent that about 85% of the nonrepeat portion of the Drosophila genome is transcribed and **processed** into mature transcripts during embryogenesis (Manak *et al.* 2006). Moreover, recent technically advanced studies of the transcript content—the transcriptome—in the eukaryotic cells revealed that the vast majority of the genome is transcribed, mostly as non-coding RNA (ncRNA), ranging from about 70% in *C. elegans* (a nematode worm) to 93% in humans. Intriguingly, transcript functions are largely unknown.

For decades the eukaryotic transcriptome was thought to consist mainly of mRNA molecules, the transcription products from gene protein coding DNA, plus other RNA molecules, including ribosomal RNA (rRNA) and transfer RNA (tRNA) that perform specific tasks in the protein translation cell machinery. Yet, when advanced genomic techniques, like RNA-sequencing (RNA-seq), arrived, hundreds of millions of short **cDNA** sequences, obtained by retrotranscribing cell RNAs molecules into DNA, could be obtained in a single experiment. Assembling all these short cDNA sequences, often only tens of nucleotides long, has been a challenging task, mainly because a DNA region may produce a plethora of ncRNAs that overlap because they use multiple transcription starting sites (TSS). At first, these ncRNA molecules were thought to be just 'junk' DNA transcripts that originated as artefacts prone to degradation and without functional value, but throughput genomic approaches have discovered patterns and revealed functional mechanisms that are unlikely be explained by an artefact. Briefly, the complexity of the transcription in eukaryotes surpasses the mRNA production because many, if not most, non-coding DNA regions give rise to many RNA molecules in a 'pervasive' transcription process. First, several molecules known collectively as interfering RNAs, because they regulate the expression of genes by inhibiting protein translation or by degrading the mRNAs, were discovered (see Chapter 3 for a detailed account). But this was just the tip of the iceberg. In 2007, a long, genome-wide ncRNA class was first described in mice by sequencing cDNA libraries. Later, the higher sensitivity of genome techniques revealed that the non-coding sequence is at least four-fold longer than the coding sequence and that the transcriptome is much more complex than anticipated. Namely, **long ncRNAs** were often overlapping with coding and non-coding transcripts, implying that a sequence could be transcribed into both kinds of transcripts. Moreover, since the expression of many mouse long ncRNAs is restricted to embryonic cell differentiation or to specific tissues like the brain, the idea that ncRNAs could participate in regulatory processes was advanced. Now we know that the functional repertoire of long ncRNAs comprises roles of regulatory gene expression in chromatin modification, transcription, and post-transcriptional processing (Fig. 1.11).

Diverse RNA products of pervasive transcription associated with promoter regions in eukaryotes have been described recently. Figure 1.12 depicts a simplified version of the transcription complexity reported in the latest studies with mammals and yeasts. Among other characteristics, ncRNAs are

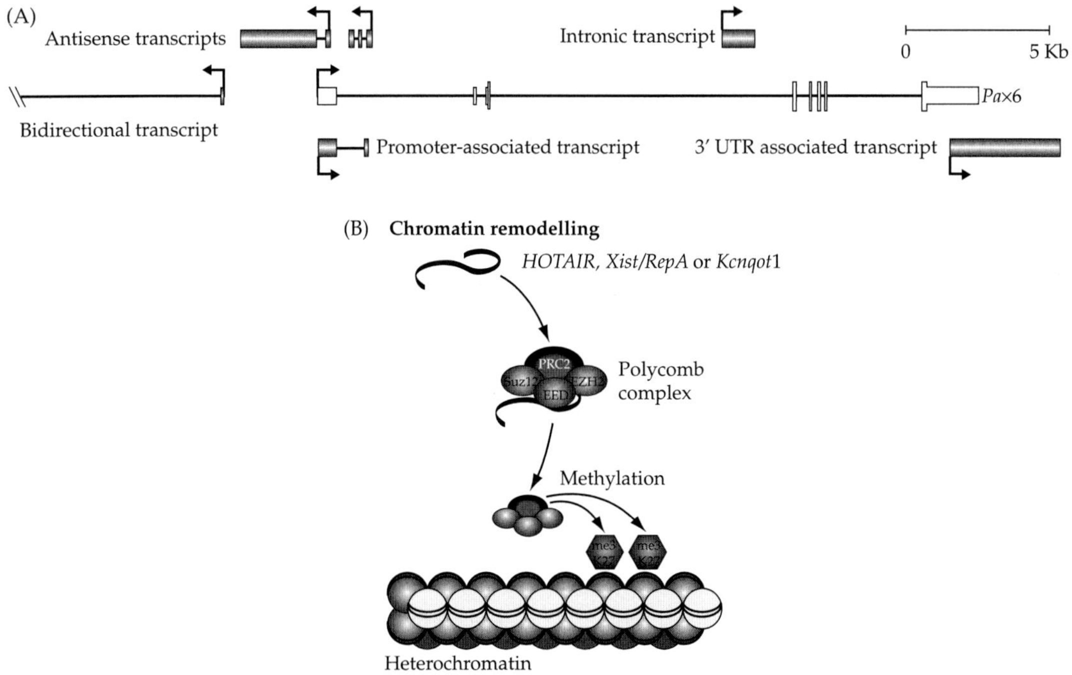

Fig. 1.11. Illustration of organization and function of long non-coding RNAs. (A) Long non-coding transcripts (grey boxes) associated with *Pax6* (boxes in white represent exons; straight line depicts intronic regions). This diagram illustrates the complexity of the overlapping of ncRNAs and coding regions. (B) A schematic example of how ncRNAs (*HOTAIR*, *Xist/RepA*, or *Kcnqot1*) recruit chromatin modifying Polycomb complex to the *HoxD* locus, the X chromosome, or the *Kcnq1* domain, respectively, where the trimethylate lysine 27 residues (me3K27) of histone H3 induce heterochromatin formation and repress gene expression. From Mercer *et al.* (2009) with permission from Nature Publishing Group.

often transcribed from gene promoter regions in the same or opposite orientation to genes. In yeasts, at least one third of gene boundaries transcribe into long RNAs (>200 bp), named CUTs (cryptic unstable transcripts) and SUTs (stable unannotated transcripts), that are likely involved in the regulation of gene transcription. Analogously, a new class of long ncRNAs named PROMPT (promoter upstream transcript), similar to yeast's CUT, has been discovered in humans. In fact the human promoter regions transcribe several ncRNAs, often bidirectionally. This characteristic applies not only to long RNAS (like PALRs and PROMPTs) but also to short RNAs. In many instances these short ncRNAs overlap with the long ncRNAs, suggesting that they may be the result of long RNAs cleavage followed by further processes that enhance their stability, although this is still unproved in many cases. Their function might also be related to regulation because when synthetic PASRs (promoter associated short RNAs), a mammalian short ncRNA class, were introduced into cells, they inhibited transcription of the gene whose boundary sequences (promoter or coding) were used as a template to synthesize PASRs.

The leading performance of ncRNAs in the genome dynamics was actually anticipated, and later confirmed, in the throughput analysis of 1% (30 Mb) of the human genome by the ENCODE (encyclopaedia of DNA elements) project (published in *Nature* **447**, 799–816, 2007). Contrary to the long-held view that protein-coding sequences (the 'genes') are the 'prima-donna' actors in the genome play, the ENCODE report unambiguously revealed that gene expression into proteins is preceded by a complex network of molecular decisions, likely played out by RNA molecules. Among other interesting findings, the ENCODE highlights that the human genome is pervasively transcribed into primary transcripts including many non-protein-coding transcripts, with many of them overlapping

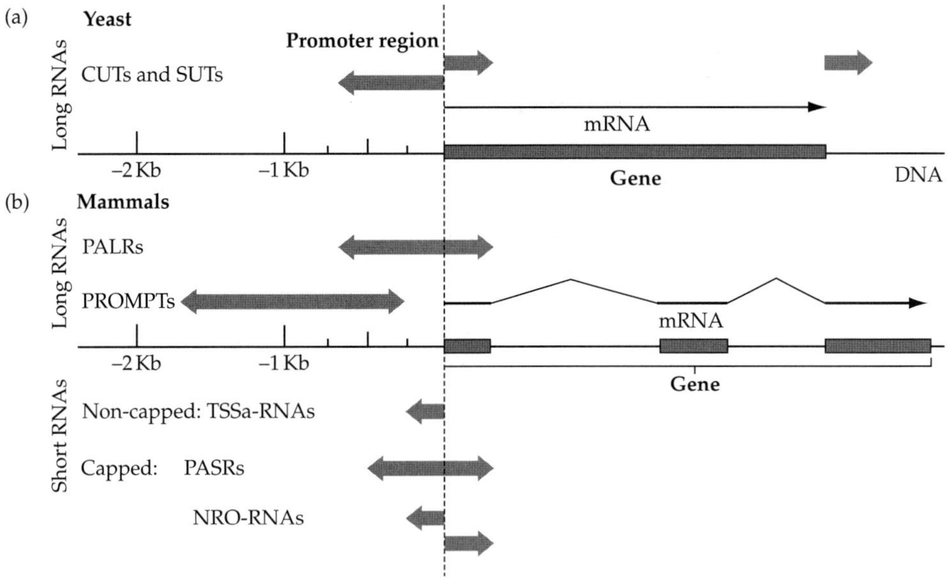

Fig. 1.12. Schematic illustration of the transcription complexity in eukaryotes. (a) In yeast, long non-coding RNAs (named CUTs and SUTs) are the transcription products of sequences in the promoter region. (b) In mammals, not only long ncRNAs (PALRs and PROMPTs) are produced in the promoter region, but some shorter (less than 200 nucleotide long) sequences are also transcribed in this region. Some ncRNAs are capped in order to increase their stability. Arrowheads indicate the direction of transcription that can be uni- or bidirectional. See text for a more detailed account of this complexity of transcription. From Carninci (2009) with permission from Nature Publishing Group.

protein-coding sequences and others mapping in regions previously thought to be silent. Other genome features revealed by this project include the findings of many new transcription start sites (TSS) totalling, with the already known TSS, almost tenfold more than the number of genes. This finding agrees with the extensive transcription observed in the genome. Moreover, although the functionality of many transcripts remains to be proved, the old view of the genome as a set of isolated loci transcribed independently seems to be at odds with the present evidence. On the other hand, it is most likely that the genome encodes a host of RNA transcripts that are linked to protein-coding transcripts and drive the expression pattern in the genome. This view, named by some as the 'genome RNA machine' (Mattick 2009), is compatible with the finding that many highly conserved sequences are not part of annotated genes and still may be fully functional. Interestingly, this genome landscape picture may also explain some of the ambiguities in biomedical studies aimed at finding the cause of inherited diseases. To such purpose medical researchers study the nucleotide variation, called **single-nucleotide polymorphisms (SNPs)**, in thousands of individual sequences to detect any association between a specific SNP and a disease. Although this method allows the detection of candidate genes involved in the disease, often, as in diabetes, many associated SNPs are located within genomic regions devoid of genes. This emphasizes the need to know the functionality of non-coding regions in the human genome if we want to understand the genetic cause of disease.

Structural variants in the genome: a population approach

In this introductory chapter I have presented some fundamental mechanisms that operate in the making of the genome. Despite all the insights acquired since the human genome presentation in 2001, we are still faced with many uncertainties. Some difficulties derive from the still unfinished condition of the human genome, in spite of the serial announcements to the contrary. Apart from difficulties in sequencing chromosome heterochromatic regions (mainly **centromeres** and **telomeres**), there is evidence of missing bits, mistaken DNA stretches, and complex rearrangements that resist sequencing. Although the April 2003 release fulfilled the technical definition of completion—less than 1 error per 10,000 nucleotides and 95% of the gene-coding stretches—there were still gaps. Moreover, structural variation was poorly detected. Initially, there was a consensus that differences between individual genomes largely consisted of single nucleotide polymorphisms (SNPs) amounting to 0.1%, but we now know that individual genomes differed by hundreds of DNA regions that can be deleted, repeated, inserted, or inverted, involving millions of nucleotides. For instance, chromosome 6 hosts the major histocompatibility complex (MHC), a >4 Mb region that contains more than 100 genes mainly involved in immune responses. When several individual sequences (**haplotypes**) from different people were compared, more than 37,000 DNA nucleotide sites were variable and about 7,000 structural variations were found, a genetic diversity of an order of magnitude larger than the genomic average. It is unquestionable that this knowledge about human variability enriches our understanding of adaptive human evolution at the population level, and also allows us to establish disease associations that were previously impossible. A tentative guess estimates that at least 400 specific locations should be analysed for alternative sequences to reach a representative spectrum of our diversity.

In general, these **structural variants** (SVs) are defined as >1 Kb genome regions that are found in variable numbers, localization, and orientation in different individuals of a species. When sufficiently large, these SVs can be detected by microscopic chromosome-staining techniques and comprise the classic karyotypic variants including translocations, inversions, deletions, duplications, and all numerical chromosomal changes. However, in the past few years genome-scanning array technologies (Fig. 1.13 and Plate 1) have revealed a DNA variation in structure that affects segments smaller (ranging from 1 Kb to 3 Mb in size) than those detected microscopically. Among these submicroscopic SVs, the **copy-number variants** (CNV) are the most studied and abundant to date. CNVs are DNA fragments that appear at a variable copy number among individuals of a population when compared with the reference human genome sequence. They include inversions, insertions, deletions, and duplications. The number of CNVs detected to date amounts to between 12,000 and 15,000 which possibly represent more than 8,000 firmly established CNVs. In a study that surveyed DNA from 40 individuals (20 Africans and 20 Europeans), each individually tested genome differed from the reference sample at more than 1,000 sites. Interestingly, the CNVs overlap about 13% of human genes in this study. This prompted the authors (Conrad *et al.* 2009) to investigate the putative role of CNVs in known human diseases, and they found three well-established cases (Crohn's disease, psoriasis, and obesity) and other strong candidates for human ailments. In sum, more than 30 disorders have been associated with CNVs. These data, although preliminary, suggest that the CNVs may affect individual fitness and have a role in evolution. This is a contentious issue, like the evolutionary role of other structural features of the genome, such as genome size, discussed in previous sections.

The fact that the CNV deletions are underrepresented in high gene-content regions suggests that CNVs are subjected to purifying selection. Moreover, there are also reports of positive selection, as in a polymorphic inversion in chromosome 17 that increases fertility. Other very convincing underpinnings of CNVs' evolution by positive selection include their gene-enriched composition, mainly in genes encoding secreted products and in environment-responsive genes. Also, the elevated rates of protein evolution for the CNV regions (CNVRs), during the long time period since human and mouse divergence, bolsters the positive selection hypothesis. Yet, as more genome-wide data sets become available, new CNV features deserved attention. Two unnoticed features in earlier surveys were that CNVRs are G+C rich, especially when they overlap with **segmental duplications (SDs)**. These duplications, formerly named low copy repeats, are DNA segments >1 Kb long that occur in two or more copies that share >90% sequence identity. They are fixed, but often they can vary in copy number, in which case they are CNVs. Because G+C-rich regions tend to show a higher gene density and are

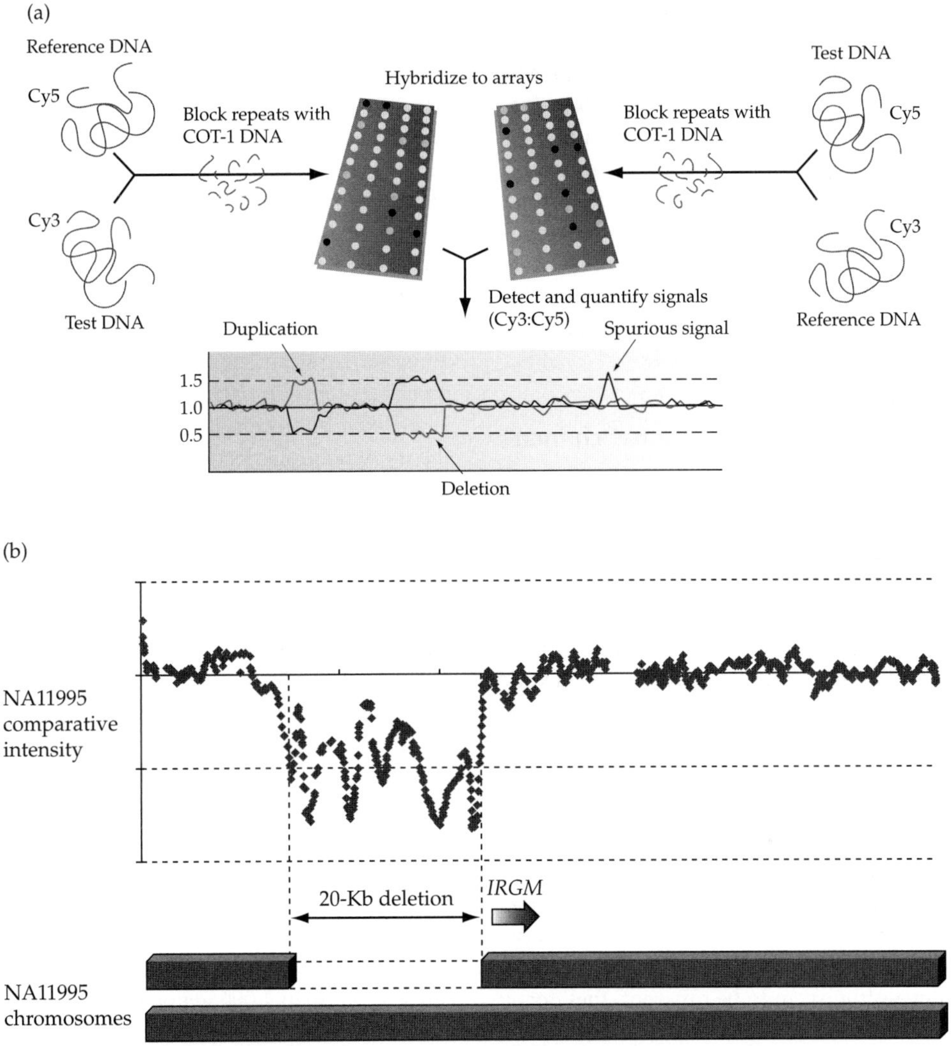

Fig. 1.13. Schematic illustration of the array-based, genome-wide method for identification of CNV. (a) Reference and test DNA samples are differentially labelled with fluorescent tags (Cy5 and Cy3, respectively), and are then hybridized to genomic arrays spotted with one of several DNA sources (BAC clones, PCR fragments, or any kind of DNA fragments) (left side). After hybridization, the fluorescence ratio (Cy3:Cy5) is determined. A lower ratio indicates an absence of DNA in the test DNA relative to reference DNA, i.e. a deletion; likewise, a higher ratio reveals a duplication. To detect spurious signals a reversed labelling test is carried out, that must show a reciprocal signal (right side). From Feuk *et al.* (2006) with permission from Nature Publishing Group. (b) A schematic detection close-up of a deletion in one of the two copies of chromosome 5 showing that when there is a different copy number the ratio of signals from the two fluorescent labelled samples, reference and test DNA, shifts to lower values. In this example the deletion locates 20 Kb upstream of the *IRGM* gene, a gene that acts by autophagy as a resistance system to intracellular parasites like mycobacteria and has been associated with Crohn's disease, a disorder that causes inflammation of the digestive tract. From Armour (2009) with permission from Nature Publishing Group (see also Plate 1).

frequently prone to copy number change, the CNV enrichment in genes and in G+C content could be elucidated by this non-adaptive explanation. Interestingly, this enrichment does not hold for CNVRs lying outside of SDs, meaning that SDs may influence the retention of genes inside CNVs. But how? Two population geneticists (Hill and Robertson) showed long ago that in regions of low recombination the strength of purifying selection is diminished, and slightly deleterious mutations can be maintained and even fixed. Since CNVRs that overlap with SDs are regions where recombination is low, the Hill–Robertson effect might explain the higher rates of evolution observed in CNVRs because the fixation rate of deleterious, rather than adaptive, mutations is fostered by low recombination. This reasoning must be further supported by more data set analyses, but again demonstrates that a population genetic reasoning is greatly needed for adaptive evolution to be ascertained.

How to build a dynamic genome: chance versus necessity

Four decades ago Jacques Monod, a Nobel Prize-winning French geneticist, wrote a landmark book: *Le hasard et la nécessité* (*Chance and Necessity*: 1970) in which he applied both concepts to the evolutionary process. Chance means to Monod the random (non-adaptively directed change) feature of genetic changes; necessity is used as a synonym for natural selection. Both concepts operating in concert lead to adaptive evolution. Darwin (1859) in *The Origin of Species* defined evolution as 'descent with modification', where descent emphasizes the genetic component of character transmission, while modification refers to evolutionary change. Darwin proposes that when both components act in concert they may also lead to adaptive evolution. Although these two views of evolution seem to be different, they focus on the same fundamental driving forces of evolution, namely mutation and selection. Mutation is a chance phenomenon that occurs regardless of the benefit or loss to the organism; it is Monod's chance component. Since mutations, or DNA changes in current parlance, are acquired by descent, they comply with the first premise of Darwin's definition. Selection, better natural selection, is Darwin's great insight, a mechanism that explains by natural means why adaptation is possible through modifications tuned to the environment. It is Monod's necessity concept: an evolutionary component that explains how 'from a noise source (chance mutations) selection is able, by itself, to extract all the music of the biosphere', in his poetic words. This apparently well-grounded reasoning still remains an unacceptable principle to a large proportion of humans, regardless of their educational background. The likely reason for this rebuttal is that natural selection excludes the intervention of forces alien to natural mechanisms in the origin of species, human species included. Thus, by accepting Darwinism, many people think that human nature becomes devoid of transcendent, supernatural meaning, a tenet too unbearable for most human beings.

Darwin's theory (dubbed Darwinism), widely accepted by scientists, was proposed at a time of great ignorance in basic biological disciplines such as genetics and ecology, two underpinnings of the theory of natural selection. An avalanche of biological knowledge has enriched our evolutionary understanding ever since, and has obliged biologists to incorporate new discoveries into Darwinism. First, in the middle of the twentieth century, advances in genetics, mainly in population genetics, and in other disciplines like palaeontology, zoology, and botany, were combined into a new evolutionary synthesis, called 'the Modern Synthesis' or 'neo-Darwinism'. Later, in the 1960s and 1970s, the birth of molecular biology incorporated a new approach that led to the neutral theory of molecular evolution (see Box 1.1). Finally, the recent arrival of the genomic and post-genomic era has allowed us to scrutinize the genome to get a deeper insight into the evolutionary mechanisms, as sketched in the above sections. At present, perusal of the more than 1,000 genomes sequenced to-date is a routine task. Still in its inception, genomics is a tantalizing endeavour that continuously reveals unanticipated insights.

While progress in genome understanding has grown amazingly over the past decade, we are still far from a scientific narrative connecting genome dynamics with the behaviour of organisms' phenotypes. However, current progress in the comprehension of the evolutionary underpinnings of developmental biology is contributing to easing this connection. Both frontier fields, genomics and developmental biology, are expected to merge eventually in one common endeavour, namely the elucidation of the mechanisms by which adaptive evolution occurs. Ever since the first whole-genome sequences were reported, new insights on genome dynamics have revealed unanticipated traits that have challenged our views. Some of them, like the C-value paradox or the pervasive RNA transcription, among others, are central topics in evolutionary arguments in favour of or against the basic pillars of Darwinism. In an attempt to summarize, I dare say that most disputes try to elucidate whether evolution is primarily channelled by external (e.g. natural selection or drift) or internal (e.g. developmental constraints) mechanisms. The theory of constraints is rooted in the eighteenth century 'Naturphilosophie', an idealistic philosophy that viewed current organisms as the transformed beings from a few archaic types following internal laws of development that constrained their evolution (see Chapter 2 for a detailed discussion). The long-held divorce between development and genetics has fostered the view of the prevalence of development constraints in channelling evolution. Yet the current advances in developmental biology are shedding light on the real nature of constraints and natural selection, vindicating the role of the latter in adaptive evolution.

Another contentious issue that has emerged within genome studies refers to the role of genetic drift and other random events, like **symbiosis**, in evolution. Above I have discussed how some evolutionists think that drift under favourable population characteristics, instead of natural selection, fuels genome evolution. In addition, horizontal gene transfer has been advocated as a driving force in the persistence of mobile elements in the genome, and even as a major force of genome building (see Chapter 4). These instances have been interpreted by some evolutionists as proof that natural selection is secondary in the evolution of novelties. I believe that in this contention there is a misunderstanding of the opportunistic nature of natural selection. Let's take, for instance, the putative role of drift in the enlargement of genome size. The current verification that this 'excess' genome can be later used, after evolutionary changes, for adaptive functions like gene regulation, gives a clue to how natural selection acts. Drift, horizontal transfer, genome reunion in symbiosis, duplications, or other non-adaptive mechanisms, are mechanisms that contribute to create variability for natural selection to act upon. They can be compared to the mutation role in the theory of natural selection. It is the 'chance' element, whereas selection is the 'necessity' element.

As we progress in our understanding of the genome architecture and function, the new knowledge surely enlarges our view of the intricacies of evolution and some think that might also oblige us to change some tenets of the Darwinian paradigm. To evolutionists it is without doubt a continuous source of challenging questions that deserve attention. This book aims to present some of these challenges in order to argue whether the new genomic knowledge conforms to the bases of Darwinism, obliges us to reconstruct it by incorporating new premises, or finally requires us to deconstruct the whole Darwinian system. Whether these challenges reconstruct or deconstruct the paradigm is the underlying topic of this work. I hope you will enjoy reading it.

CHAPTER 2

The unity of type: ancient homologies in the genome

But this unity of type through the individuals of a group, and this metamorphosis of the same organ into other organs, adapted to diverse use, necessarily follows on the theory of descent.

—(Charles Darwin, Essay 1842, in p. 40 of Francis Darwin, (ed.) 1909. *The Foundations of the Origin of Species. Two Essays Written in* 1842 *and* 1844. Cambridge: Cambridge University Press)

Scarcely six years after returning from his five-year round-world journey on the *HMS Beagle* brig, Darwin had already written an essay (the 1842 *Sketch*) in which he stated the words 'theory of descent' for the first time. In fact, less than two years after his arrival, Darwin sketched his theory of natural selection in his 'Transmutation Notebooks' D and E, written in 1838–39. These and other similar documents show convincingly that Darwin came upon his theory of descent with modification very early on. The revolutionary impact of his view on the origin of species did not escape Darwin. In an 1844 letter to his friend Hooker, a leading botanist, he states: 'I am almost convinced (quite contrary to opinion I started with) that species are not (it is like confessing a murder) immutable.' That he felt guilty publicizing his revolutionary ideas can be inferred from this sentence. Indeed, Darwin refused to publish his theory for more than 20 years until Alfred R. Wallace, a younger naturalist that arrived at a similar theory of the origin of species, sent him his famous article on natural selection. Darwin was shocked by the similarity of Wallace's ideas to his own, as made explicit in a letter to Lyell, a geologist friend, in which he complains: 'I never saw a more striking coincidence; if Wallace had my MS. sketch written out in 1842 he could not have made a better short abstract!.' This episode precipitated Darwin's public diffusion of his theory, first as a joint 1858 communication with Wallace to the Linnean Society, and one year later by the publication of *The Origin of Species*.

But why was Darwin so reluctant to publish his theory? Was it because he feared a wide refusal that could endanger his reputation as a naturalist? Or, was it because he was still not sure about his hypothesis? Or, perhaps, did he fear a great attack from the religious and social establishment? No doubt his proposal that species originated not by special creation but as the result of a process of 'transformation' (Darwin's term for evolution) involved not only a scientific revolution but also a social upheaval of the religion-based Victorian class society. Darwin felt intimately this social (and moral) challenge when his father advised him not to talk about his ideas, even with his wife Emma. Mid-nineteenth-century Western society was still founded on the biblical creationism paradigm. All living beings, created independently by an omnipotent Creator, occupy, according to this paradigm, an immutable position in a ladder, progressing from bacteria to amoeba, ascending to multicellular lower organisms, and reaching higher organisms, such as complex vertebrates, at the top. This ladder of nature (*scala naturae*), though increasing in complexity from bottom to top, does not allow any transformation among steps. Thus, man, obviously at the zenith of creation (only below the angels), was disconnected from all other living beings and given a special human status by the grace of the Creator. Moreover, the *scala naturae* not only underlies the detachment (and, of course, the superiority) of man from other living beings, but also provides a natural justification for the segregation among human beings in a society stratified from birth by class and race. The theory of descent dismantled the ladder of nature because it allowed the transformation among steps, and therefore destroyed with it all the ladder's social segregating implications, which made Darwin and his ideas (Darwinism) highly dangerous. Moreover, if living beings are related by descent, there is only one primeval type from which all organisms derive by modifications. This is the concept of the unity of type.

All these social and religious provocative issues notwithstanding, Darwin likely had other reasons to remain silent. In 1844, the same year of the letter to Hooker, Chambers, a writer and publisher, pseudonymously published *Vestiges of the Natural History of Creation* in defence of evolution. The strong criticisms against the book, from both orthodox scientists and educated readers alike, influenced Darwin's determination not to publish. *Vestiges* stirred up great interest in the 'species problem', but was plagued by scientific inaccuracies that were abused to discredit Chamber's ideas on evolution. At this point Darwin decided not to indulge himself in Chamber's amateurish approach, and therefore not to publish his theory until a solid body of evidence was available to him. To this end he devoted the next two decades to the rigorous observation of nature and performed detailed experiments with plants and animals that were later used to support his writings. If you consider that he had arrived at his theory very early, viewed under the modern urge to publish, Darwin could have paid a heavy toll for delaying publication. Fortunately, his planned delay was worthwhile. Eventually, his long argument was so firmly bolstered by evidence and reasoning that criticisms could not be directed to scientific inaccuracies, but must only appeal to lack of scientific knowledge in his time. There was total ignorance in some fields, such as in the area of genetics, and Darwin suffered from this. Since heredity is the link among evolving forms, it is hardly surprising that the unity of type could not be bolstered by any sound genetic theory. In fact, even Mendelian genetics, developed in the first half of the twentieth century, was unable to provide a unified theory of biodiversity. As we shall see below, the birth of this theory had to wait the advent of molecular genetics. Nevertheless, the work of great anatomists and embryologists of the nineteenth century was of great help in suggesting that the myriad of body forms could be reduced to a few (perhaps one) types. Their interpretation was not always in Darwinian terms, but Darwin took advantage of their work, of which he was an admirer, to support his theory.

In this chapter I will try to revive one of the most contentious tenets of evolution, namely the unity of kind in body form, to illustrate how current genomic studies support Darwin's theory of descent, with modification. Nowhere is the evidence for the unity of kind more compelling than in the recent deciphering of developmental gene networks. It is only in the past two decades that genetics has been fully incorporated into developmental studies, fostering a dramatic advance in evolution. This body of knowledge, called evolutionary developmental biology (dubbed 'Evo Devo'), is uncovering the most basic mechanisms, unknown to Darwin, that explain how natural selection acts on the origin of the 'endless forms', in Darwin's words, that have evolved on our planet. Many of these mechanisms are so unanticipated that some researchers consider them revolutionary, even justifying a new evolutionary non-Darwinian paradigm. Others, however, while agreeing with their revolutionary character, claim that the Darwinian paradigm applies perfectly to them. In the following pages I will present current evidence of the unity of kind, confirming that it follows on from the theory of descent, as Darwin anticipated in 1842. His pioneering insight, built upon numerous observations, challenged the scientific and social establishment of his epoch, and also constitutes an example of scientific rigour and self-restraint in terms of precipitate conclusions, all worthy of imitation.

Body plans: the quest for homologies

The finding of homologies, the body similarities that derive from the common descent of living beings, has always been a difficult task. The reasons are two-fold. First, if divergence time is long, changes in body structure are so large that similarities are hard to detect. Often, homologies are not revealed by observing adult structures but only by comparing early embryonic stages in development. Second, diverging changes are not only found in structure but also in function. The transformation of the reptile jaws into the middle-ear ossicles in mammals is an excellent example of this kind of homology. When an ancestral structure is modified to perform a new adaptive function it is said that this structure has been 'co-opted', a term popularized by Gould, and the co-opted new adaptive structure is named 'exaptation'.

Nowhere are the difficulties in establishing homologies more acute than when comparing the great groups of animals and plants. Currently, we recognize about 35 of these groups, named phyla (singular **phylum**), in animals, and about 12 in plants (named divisions), but these numbers are still not definitive. All organisms in each phylum (or **division**) share the same basic **body plan** (or '**bauplan**'), although they may show differences on the surface. Thus, the vertebrates, (mammals, birds, reptiles, amphibians, and fishes), belong to the phylum Chordata, despite the fact that to the naked eye they look quite different from each other. Differences in other chordates are even more striking. The lancelet (named *amphioxus* by scientists), is a two- to three-inch long, filter-feeding, burrowing, marine invertebrate chordate, devoid of head, that hardly resembles the vertebrates (Fig. 2.1).

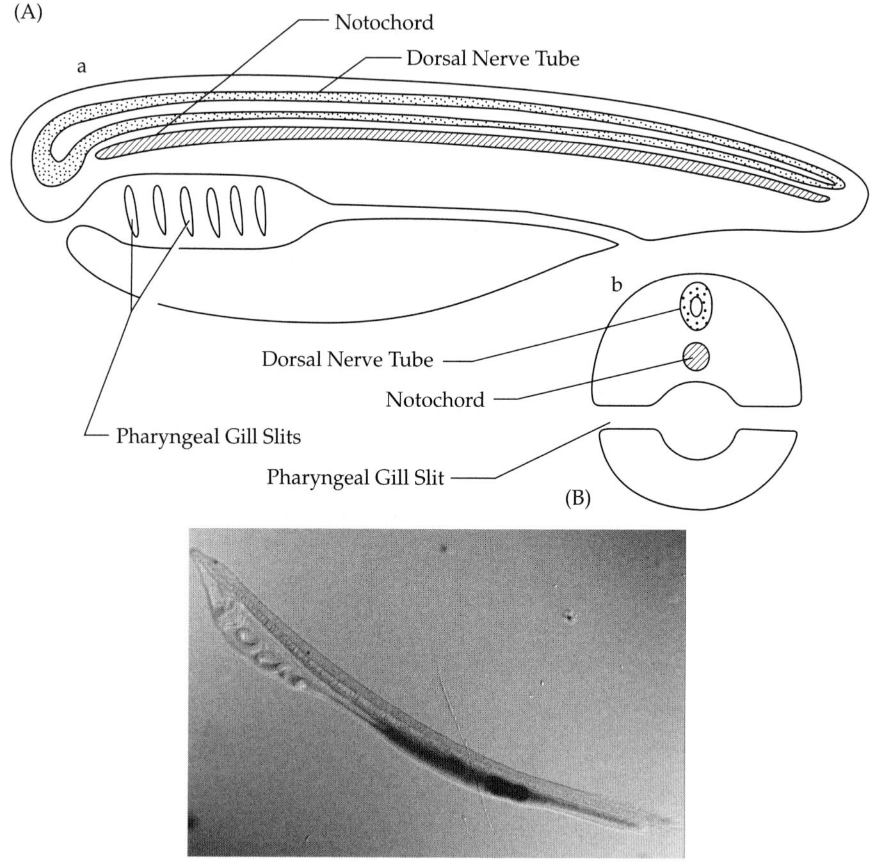

Fig. 2.1. (A) Two views (a: lateral; b: cross-section in the gill region) of a schematic chordate showing the three diagnostic body structures: a dorsal nerve tube (or nerve chord), a notochord, and gill slits. As embryos we have all three, but we loose our notochord and gill slits early in development. The nerve chord becomes our brain and spinal chord. (B) A picture of a lancelet showing the three diagnostic chordate characters. Although it has a brain-like tip at the nerve chord end, it lacks a real head. Part A from Moore (1984) with permission from Oxford University Press; part B courtesy by Jordi Garcia-Fernández.

Yet, it shares with them some fundamental body structures, namely a nerve cord, a rod-like notochord, and slits, which are also present in vertebrates, either in embryos or adults. The inherent difficulties in discovering basic homologies notwithstanding, the great anatomists of the nineteenth century did a great job in defining body plans. Among them Georges Cuvier, at the Natural History Museum in Paris, whose genius lay the foundations of comparative anatomy, was able to reduce all animal biodiversity into four body plans: Vertebrata (vertebrates), Mollusca (molluscs), Articulata (arthropods), and Radiata (animals with radiate symmetry, namely the cnidaria and ctenophore phyla). Cuvier argued that these body plans are so independent of each other, that even they cannot be placed in a linear series of complexity. 'How could we find an intermediate form between an octopus (a mollusc) with its complex structure, and a fish?' he asked, and immediately answered himself: 'Not only there was no way to transform one step into another in the chain of being [another term for *scala naturae*], but the reality of gradation at the grand scale is delusive. At most, gradation occurs only within each major body plan.'

The quest for similarities between animals was initially inspired by the early work of German biologists. They, with the poet and scientist Goethe at the forefront, constituted a school called 'Naturphilosophie' that viewed the world as made of ideal types, named archetypes, from which all observed forms were variations. Interestingly, they opposed the static Newtonian view of the universe, regulated by fixed laws. Rather, it was thought that archetypes were able to produce a variety of forms from the simple to the complex. Some have argued that this idealistic view contains the seeds of organic evolution, but this is likely untrue, at least if we consider the ideas of their defenders. Among them Richard Owen, a British anatomist, was the leader. As early as 1840 he claimed that archetypes were blueprints in God's mind and urged Darwin to look for them. While Owen saw the unfolding of life as a kind of divine 'evolution' from created archetypes, and not by way of descent, Darwin likely thought of archetypes, at least in private, as pure speculation. It is no surprise that Owen became the most aggressive detractor of Darwin's theory of descent.

Cuvier's interpretation of independent body plans was opposed to the unity of type, and, consequently, to species transformation. However, not all of his colleagues at the Paris museum agreed. Etienne Geoffroy Saint-Hilaire was able to compare the structure of corresponding parts in different body plans, and found similar structures in all of them. He even argued that a vertebrate and an arthropod share the same structures but in opposite dorso-ventral orientation. In vertebrates the nerve-cord runs down their back (dorsal), while the digestive tract is in the ventral side; in arthropods the gut is dorsal while the nerve cord is ventral. It is as if a vertebrate were just an upside down arthropod. This was too much for Cuvier, and also for Owen, who in a famous session in the French Academy demolished Geoffroy Saint-Hilaire's arguments, on the basis that structural similarity is not enough to define a body plan. Cuvier added that it is how the organs (structures) are functionally arranged with respect to one another, and not their similarity, that defines a body plan. Cuvier won the debate, but the unity of type between vertebrates and arthropods has now been established, so Geoffroy Saint-Hilaire was right, although in an unanticipated way. It is the Evo Devo discoveries over the past two decades that have shown what underlies the unity of type for all metazoans, as we will explain below.

The Cambrian explosion

Historically, a reluctance to accept the unitary view of life was not the only outcome of the difficulties in establishing homologies and interpreting them under a theory of descent. The fossil record of early forms in the Cambrian, a geological era ranging between about 543 and 488 mya, shows a rapid origin (geologically speaking), in less than 20 my, between about 540 and 520 mya, of a large variety of body plans, including almost all animal phyla. This Cambrian explosion has launched much opposition to the theory of descent with gradual modification by non-Darwinian critics, and also discrediting of evolution by creationists.

From its origin until the early Cambrian, a period nearly 3,000 my long, life was exclusively represented on our planet by prokaryotes (bacteria and the like) and unicellular amoeba-like eukaryotes. The earliest vestiges of multicellular organisms are dated from 1,800 mya, but the oldest unambiguously distinct multicellular fossils are red algae dating from 1,200 mya. Since then, two large, worldwide animal radiations have been documented: the Ediacaran fauna (named after the Ediacaran Hills in Australia where these fossils were first found) dating from about 570 mya, and the Cambrian explosion. Ediacarans consist of tube-shaped, frond-like, or radial, inch-long organisms that left marks resembling in many cases jellyfish and sea pens (phylum Cnidaria). However, their relationship to living phyla has remained controversial, and they may represent an evolutionary experiment independent from our animal ancestors. Recently, in a meeting of the American Geological Society, Loren Babcock presented evidence of footprints that might belong to an Ediacaran legged-arthropod like current centipedes, but this contention still waits further authentication. Interestingly, interspersed among the Ediacaran fossils are traces of burrows and tracks made by animals that could dig and crawl. Such animals must have been endowed with muscles and some sort of hydraulic skeleton, a grade above jellyfish. The main difference between the two kinds of animals is due to the way the embryo forms. While cnidarians develop only two body cell layers (diploblasts), the body of other animals contains three layers (triploblasts). Also, the latter show bilateral symmetry, and are known as bilaterians for that, while the former have radial symmetry. Triploblasts have better diversified organs and tissues: skin and nerves from the ectoderm (the outermost embryonic layer); muscle, bone, and other internal organs from the mesoderm (the layer in the middle); and gut from endoderm (the innermost layer). The Ediacaran fauna became extinct about 540 mya, but regardless of whether present fauna is derived or not from the Ediacarans, the evidence of bilaterians in the Precambrian seems real.

Almost immediately after the decline and disappearance of the Ediacarans the triploblasts exploded. This Cambrian explosion is perfectly well documented in the Burgess Shale. Discovered in 1909 by palaeontologist Charles Walcott in British Columbia (Canada), and first carefully described in modern terms by Harry Whittington and students at the University of Cambridge, this fauna was popularized by Stephen Jay Gould in 1989. This site, dating to about 505 mya, contains 140 animal species assigned to more than 10 living phyla. This and other sites, like the Chegjiang (China) site dating 15 my older than the Burgess Shale, show that life in the Cambrian was highly diversified worldwide. These faunas reveal an abundance of arthropods, amounting to more than one-third of the Burgess species. Arthropods are characterized by a repetitive organization with many body segments bearing specialized jointed appendages, ranging from limbs to antennas and claws. **Lobopodians**, another abundant group, also show a repetitive organization although with unjointed limbs, as their name suggests. Some paleontologists believe that lobopodians are close to the primitive form of arthropods. Some other spectacular fossils are also found: *Opabinia*, with five eyes and a proboscis topped with a claw, and *Anomalocaris*, a half-metre long animal with a round mouth skirted by tooth-like plaques, are two of the most frightening. The high diversity in body plans (disparity) of early Cambrian faunas is bolstered by the presence of *Pikaia*, a primitive chordate, in the Burgess shale. For a long while *Pikaia* was the best-characterized ancient chordate, until finds in the Chengjiang region uncovered vertebrate fossils dating to about 520 mya. While vertebrates are not abundant in the Early Cambrian assemblages, the complexity of some Chengjiang specimens, like the jawless fish *Haikouichthys*, indicates a highly advanced stage of vertebrates at that time. Surprisingly, this life explosion was chiefly produced in about 20 my. Considering that many Cambrian fossils represent extinct body plans, Gould has suggested that life in our planet shows a decline in disparity, contrary to the widely accepted idea that life has diversified since the Cambrian times.

Taken together these facts seem to challenge the tenets of Darwinism. First, the abrupt emergence of body plans would explain the apparent difficulty in finding homologies among phyla, suggesting that other non-selective, non-gradual 'forces' led to the Cambrian explosion taking place. These 'forces', historically never defined with rigour, have been referred to internal causes related to the invention of gene-batteries involved in shape construction.

This is the field of developmental genetics and embryology, two disciplines whose combined study, in just the last 20 years, has advanced our understanding of the origin of body plans. Second, the absence of evolutionary capacity (called **evolvability**) to produce new animal body plans has always been a contentious issue. Many ecology-orientated evolutionists (the externalists) have argued that evolvability is mainly conditioned by the availability of new eco-spaces. After mass extinctions new eco-spaces were generated that allowed for the increase of diversity at the level of families and orders and classes, but surprisingly no new animal phylum evolved. It may be, argue the 'externalists', that even mass extinctions emptied no eco-spaces available for new body plans. The invasion of the land constitutes another opportunity to evolve new phyla. Accordingly, plants evolved at least seven new body plans in their land invasion, but nothing similar occurred with terrestrial animals. Plants and animals have quite different adaptive strategies and developmental programmes, but still the comparison poses a challenge to the externalists. To avoid these difficulties some evolutionists posit that developmental constraints are the only explanation for the lack of evolvability in animal body plans since the Cambrian. It is not that new forms are unable to adapt to, or compete for, a niche, but it is the internal body plan integration that constrains, in Gould's words, the evolution of new developmental changes. The originality of the constraint explanation notwithstanding, discerning between this and the ecological hypothesis must be supported by strong experimental evidence. Again, it is only by studying in detail the genetic bases of development that we can understand the underlying causes of the evolution of form. In the next paragraphs I will try to explain these bases and how they highlight the Darwinian interpretation of body plan evolution.

The evolutionary genetics of body plans

The proceedings of a 1947 meeting in Princeton (USA), attended by a group of relevant evolutionists comprising specialists as diverse as palaeontologists, geneticists, and systematists, are considered the first unanimous declaration of the evolutionary synthesis or, in brief, 'the Modern Synthesis' (see Chapter 5 for a longer account). This synthesis, first summarized by Julian Huxley (1942) in: *Evolution: The Modern Synthesis*, states that evolution ensues from the operation of forces, among which natural selection is prime though not alone, on small genetic changes in populations (**microevolution**) and that the continuous action of these forces along vast time periods explains the grand changes observed in higher taxonomic levels (**macroevolution**). The spectacular advances of population genetics in the first half of the twentieth century were incorporated into early Darwinism, something that was badly needed since a sound theory of heredity was necessary to underlie the Darwinian theory of descent. Mendelian genetics, absent in Darwin's time and rediscovered at the beginning of the twentieth century, could be applied to populations due to the many theoretical and experimental works carried out by eminent evolutionists, of whom Theodosius Dobzhansky (1937) is the epitome. However, the Modern Synthesis could not incorporate a theory of developmental genetics, because it did not at this time exist. Most evolutionists think now that, in spite of its timely value, the Modern Synthesis was incomplete because it was unable to explain satisfactorily the evolution of form.

Although macroevolution as an extrapolation of microevolution was a consensus in the Modern Synthesis, nobody knew whether the genetic processes generating small genetic changes in populations were or were not the same as those responsible for large changes in body plans. Many neo-Darwinians, as fervent followers of the Modern Synthesis are called, thought that the changes in population frequency of variants of the same gene (alleles) was all that counted, and one could disregard development as a 'black box' that transforms genetic variants to traits subjected to natural selection. Remember that embryology was considered a tool to reveal homologies underlying the unity of kind, a fundamental component in the theory of descent. Yet, either because of embryologists' ignorance of developmental genes or the neo-Darwinians' failure to appreciate developmental processes in evolution, or both, genetics and development remained almost totally divorced until the molecular era.

Hopeful or hopeless monsters

The first genetically guided report of a developmentally-altered mutant was authored in 1915 by Calvin Bridges, a member of the group of geneticists led by Thomas H. Morgan at Columbia University (USA), whose research into the fruit fly (*Drosophila melanogaster*) established the basic foundation of modern genetics. Bridges obtained a spontaneous mutant fly with two pairs of wings; the second extra pair, albeit less complete than the first pair, was a modification of the tiny round-shaped halteres present in normal flies. This mutant was named 'bithorax' because part of the haltere was transformed into wing tissue as if the fly had duplicated a thorax segment. Later, by combining several mutations a complete wing pair duplicate was obtained that received the name of *Ultrabithorax* (Fig. 2.2). Ever since then many other similar mutants, in which a body structure appears in the wrong place, have been reported by Drosophila geneticists. The *Antennapedia* mutant, that replaces antenna by legs in the fly head, represents another of these frightening monsters (Plate 2). The interest of these mutants to geneticists was that, intriguingly, these complex replacements were often due to a single, or to a few, altered genes (a mutation in genetic jargon), suggesting that differentiation of body parts is controlled by a small number of 'master' genes.

These 'monsters' were not new to scientists. In 1894 William Bateson, a pioneer geneticist, published a book, *Material for the Study of Variation*, in which he described a wealth of abnormalities in organisms ranging from bumblebees and sawflies, to crayfishes, butterflies, frogs, and many others. All these monstrosities belong, according to Bateson, to two kinds: one produced by altering the number of repeated parts, the other due to structures that appear in the wrong place, like the *bithorax* and *antennapedia* mutants. Since the wrongly placed structure normally mimics another structure, the latter mutants were dubbed **homeotic** (from *homeos* that in Greek means similar). Not surprisingly these abnormalities occur also in humans. Some of them, like hands with extra digits (polydactyly), a thorax with extra ribs, or even an extra pair of small ears, do not qualify as monstrosities because of their mild effect, but others can be very frightening. Who has not watched a terror movie starring one-eyed giants (known as cyclopes) or has not read Homer's epic-book episode with the Polyphemus cyclope, that Odysseus blinds to escape from cannibalism? These cyclopes are likely the product of author's imagination, yet they exist and are well known to scientists. This condition occurs in newborn animals due to an error in development that fails to separate the primordial forebrain and the eye into two symmetrical structures. Fortunately, cyclopic newborns are non-viable. Over decades, many embryologists performed hundreds of experiments to find out what caused these errors in development, but, although they discovered that cells of some organizer

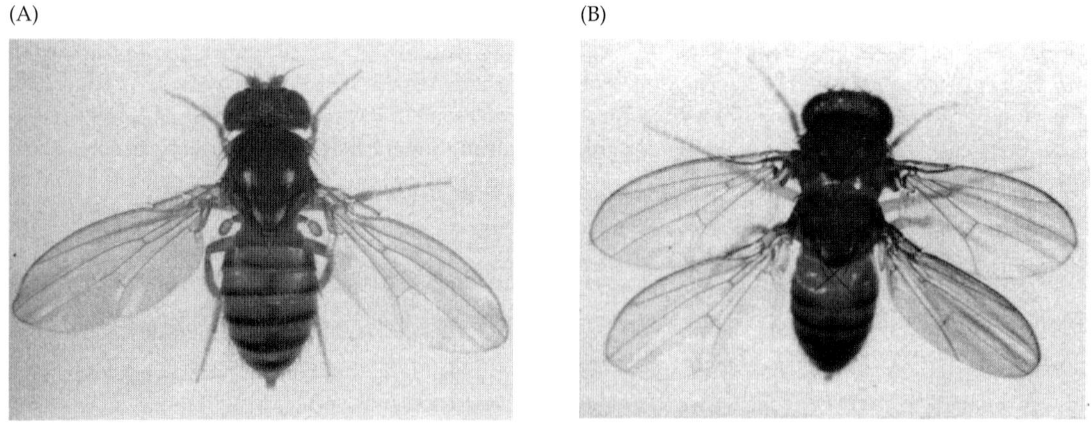

Fig. 2.2. Homeotic mutant *Ultrabithorax* (*Ubx*) of Drosophila (B) in which the third thoracic segment has been transformed into another second thoracic segment, bearing wings instead of halteres. Compare with the wild phenotype (A). After Futuyma (1998) (photos courtesy of Pamela H. Lewis, Edward B. Lewis' widow).

zones in the embryo produce substances (morphogenes) that influence the development of other cells, the underlying genetics was not revealed until recently, when molecular genetics and embryology merged.

Bateson's quest in studying these monstrosities was driven by his interest in deciphering macroevolution. He thought that morphological evolution could be produced by leaps instead of by grades, as Darwin advocated. **Homeotic mutants** were to Bateson the perfect example of macromutations. During the first half of the twentieth century, population geneticists showed that gradual changes could explain evolution, but, as stated above, extrapolation from population to above the species level was not accepted by all evolutionists. Among sceptics, Richard Goldschmidt (1940) is first. In his book: *The Material Basis of Evolution*, he asserts that evolutionary changes that determine the origin of body plans (phyla) are based on macromutations of large effect (dubbed 'systemic mutants' by him), totally uncoupled from the micromutations, which natural selection acts upon. Goldschmidt calls these highly differentiated mutants 'hopeful monsters', a term that has been much abused to explain the origin, and even the discontinuity, of body plans.

Although nobody can deny that mutations of large effect exist, the great majority of evolutionists agree that they generate unfit **genotypes** likely discarded immediately by natural selection. Their rather hopeless evolutionary nature notwithstanding, 'hopeful monsters' have provided excellent material for evolutionary studies in development. The wealth of experimental work performed mainly with homeotic mutants in fruit flies shone a new light on the evolution of form and, ironically, stirred up renewed interest, already pioneered by those geneticists at Columbia University, for a humble fly that, after all, shares many important genes with us.

Microbes, flies, and elephants: not too different

In the 1960s even the most convinced evolutionists thought it difficult to find **homologous** genes between highly diverse organisms. Ernst Mayr, one of the fathers of the Modern Synthesis, boldly stated that the 'search for homologous genes is quite futile except in very close relatives'. Rather, the dominant idea on how evolution worked to solve functional needs was that 'very different gene complexes (will) come up with the same solution', as Mayr (1963, p. 609) convincingly added. Yet, almost simultaneously with Mayr's prediction, the genius of two bacteria researchers (Jacob and Monod 1961) showed the path to reveal the rules of regulation through genetic switches that were to open a new view. They demonstrated that in the bacterium *E. coli* induction of beta-galactosidase gene, the enzyme that metabolizes lactose, is mediated by a switch that consists of two components: a repressor protein and a stretch of DNA near the enzyme gene. Normally, the repressor binds to this DNA stretch inhibiting gene expression, but when lactose is present, it attaches to the repressor causing its detachment from DNA and enabling gene transcription and enzyme expression (see Fig. G; Box 2.4). Although regulatory switch mechanisms were first described in bacteria, very soon their generality was confirmed in higher organisms, yet in a more complex way. Those who have read the illuminating book *Le hasard et la necessité* (Chance and necessity) by Jacques Monod (1970) will know what I mean and understand his famous sentence: 'What is true for *E. coli* is also true for the elephant.' While this might seem an adventurous presumption in light of the paucity of regulatory knowledge at the time, it turned out to hold true. In fact, Monod was also right in combining chance and necessity, the two fundamental components of the evolutionary process (see p. 30). I will go back to this crucial concept (see Chapter 5), but now let us continue with the quest for developmental homologies.

Twenty-five years ago, when molecular techniques became available, the eight genes known to be responsible for Drosophila homeotic mutants were cloned. They mapped on the third chromosome in two groups: the bithorax (three genes) and antennapedia (five genes) complexes. A common characteristic of these genes is a 180-nucleotide sequence (dubbed the **homeobox**), very similar to each other, that encodes a 60 amino acid-long stretch of a protein (a protein domain to biochemists), named the **homeodomain**. When this homeodomain was compared by alignment to those of other DNA-binding regulatory proteins, like repressors in bacteria and yeasts, they were surprisingly similar. Therefore, homeotic genes were suspected of encoding regulatory proteins that take part in animal development. Evolutionarily speaking, the most important discovery was, however, the amazing homology among genes belonging to organisms as distant as Drosophila and yeasts. This homology was later confirmed in other organisms, including humans. For example, all the 60 homeodomain aminoacids but one were identical in mice and Drosophila, two organisms whose lineages likely split before the Cambrian explosion, more than 500 mya. No one, including Mayr, would ever have anticipated that flies (phylum arthropod) and mice (phylum chordate) were built by the same genes. The unity of kind seemed wholly proved, at least for genes with the homeobox (dubbed Hox **genes** by geneticists).

But *Hox* genes are not alone in the building of organs and bodies. They occupy intermediate positions in the **Gene Regulatory Networks** (GRNs; Davidson 2006) of developmental genes. *Hox* genes occur in corresponding clusters and are expressed in homologous regions along the antero-posterior axis of fruit fly and vertebrate embryos. So, the similarity between animals was not just present in the sequences, but in *Hox* gene organization and spatial usage (Fig. 2.3 and Plate 3). But, although common *Hox* genes are essential to regulate the spatial expression of anatomical structures in all animals except sponges, nobody would think that fly genes to build an eye or any other fly organ could be useful to form similar organs in us or in any other vertebrate. And yet they could.

Eyes are organs that serve the same function, but their origin in different phyla was long thought to have originated independently, at least 40 times from scratch (Salvini-Plawen and Mayr, 1977). Yet this idea was discredited after Walter Gehring and his team (Quiring *et al.* 1994) isolated the *eyeless* gene, necessary for the Drosophila eye formation, and since then other similar eye-essential genes have been isolated in other organisms such as humans (*Anniridia*) and mice (*Small eye*). Surprisingly, all these genes encoded the same protein, later named Pax-6. The extremely conserved homology of the *Pax*-6 gene was demonstrated when researchers activated the mouse *Pax*-6 gene in Drosophila to induce fly eye tissue. Later, this functional exchangeability was also shown between organisms as diverse as octopuses and flatworms. One obvious interpretation of these experiments is that an early common ancestor likely used *Pax*-6 to build a primitive eye-like structure. That this evolutionary exchangeability was not unique to eye development was bolstered by the finding of similar cases in other organs. Sean Carroll and his collaborators (Panganiban *et al.* 1997) studied the *Distal-less* (*Dll*) gene, so named because it is essential to develop the outer (distal) parts of fly limbs. In fact, this gene exists not only in all the studied arthropods, confirming the unity of their kind, but in organisms ranging from vertebrates to marine worms (expressed in their appendages), to sea squirts (in their siphons and ampullae), and to sea urchin feet. Again, the most sensible interpretation is that an early ancestor of all these phyla already possessed this gene. These genes have been named 'master genes' because of their director role in the animal evolution of form. There are many of these genes, and, importantly, all of the proteins they encode contain a homeodomain, similar but not identical to *Hox* homeodomains.

Now that the deep homology between developmental genes has been proved, we can better understand the unity of kind. The descent of all extant body plans from an ancestral organism endowed with a fundamental set of master genes is not a delusion. Yet, the origin of the wealth of forms that exist using an apparently limited common set of genes deserves explanation. Moreover, when analysing the genomes, it becomes clear that organisms not only share many genes, but the number of coding genes do not increase proportionally to complexity. Thus, it has been estimated that we host about 24,000 genes, Drosophila about 16,000, and *C. elegans*, a nematode worm, about 21,200. It is clear that the much higher complexity of a chordate does not correspond with less than twice the number of genes in a fly or a worm. This paradox, discussed in Chapter 1, can be extended to the whole range of organisms (see Table 1.1) If all complex animals share a gene set so similar in number and kind, how can the evolution of current 'endless forms' be explained, whose beauty was so praised by Darwin?

Changes in latitudes, changes in longitudes

The function of the homeodomain is to bind to DNA stretches (signatures) that surround the

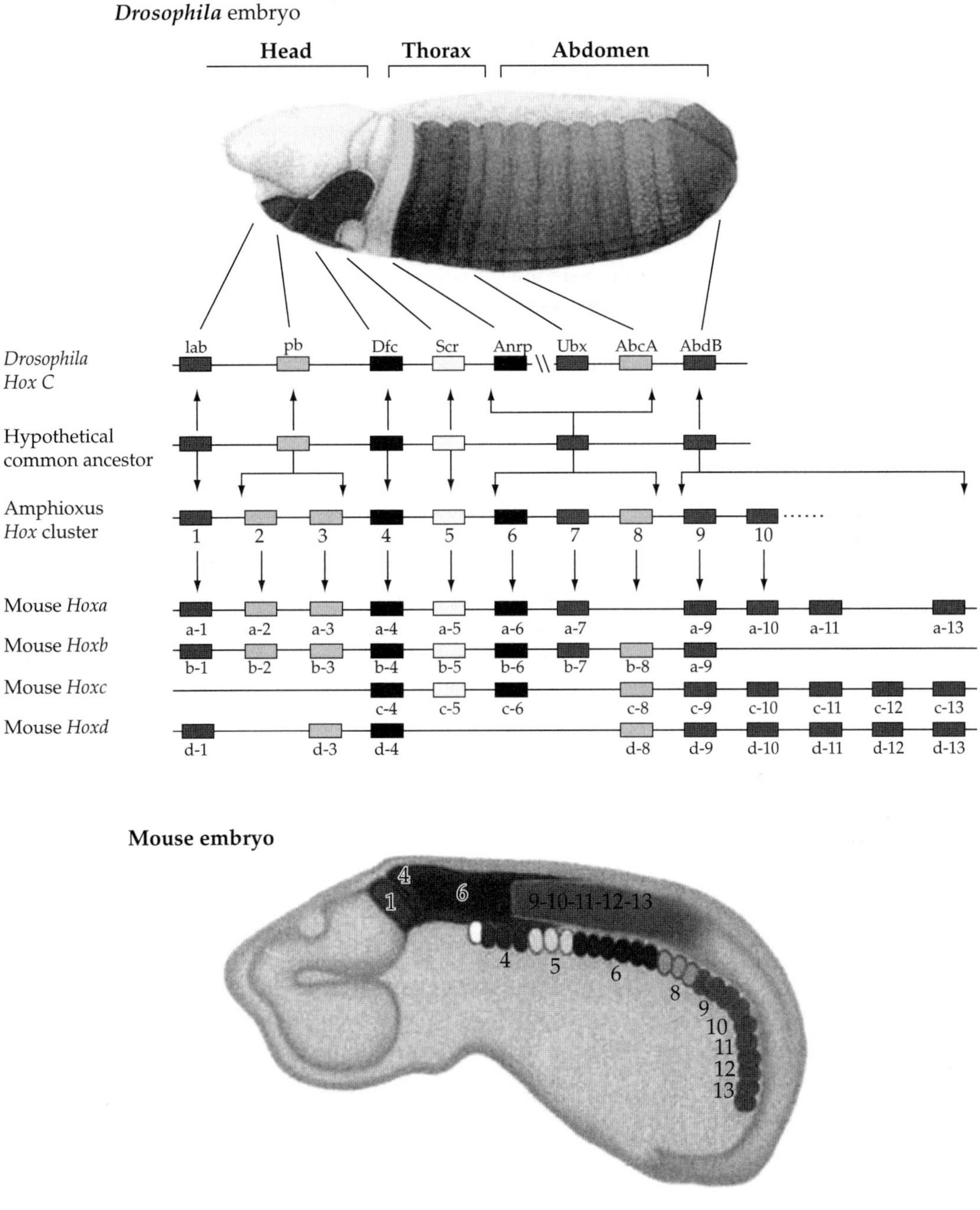

Fig. 2.3. Homologous *Hox* gene organization and expression. The anterior–posterior (A–P) body domains of Drosophila (top) and mouse (bottom) of *Hox* gene expression correspond to gene order within the *Hox* complexes. The middle of the figure depicts the homologous gene relationships (arrows) between Drosophila, Amphioxus, and mouse *Hox* clusters, and also the deduced *Hox* complex in the common ancestor of arthropods and chordates. From Carroll (1995) with permission from Nature Publishing Group (see also Plate 3).

genes (in cis, as designated by geneticists), frequently upstream of coding sequences (but sometimes downstream), allowing the Hox proteins to activate or repress those genes. These **cis-regulatory elements** (CREs) are just a few hundred nucleotides long and one gene may be flanked by several of them. Besides DNA binding proteins or **transcription factors**, as Hox and other master proteins are collectively named, there are other kinds of regulatory proteins. Some of them are signalling proteins that communicate among cells, others are cellular receptors bound by these signalling proteins, and others are hormones. The whole set constitutes a kind of tool-kit of proteins encoded by genes that affect (regulate) other genes in the kit by turning them on or off. This intricate suite of relationships builds complex GRNs whose evolutionary capabilities are enormous (see p. 55 of this chapter).

But before getting into the intricacies of GRNs, let's see how a gene knows where and when it must express or remain silent. The gene orientation process has been likened to the calculations that navigators must make in their ocean journeys or, in a current satellite-orientated version, the way that global positioning systems (GPS) work. For simplicity, let's concentrate on one dimension, namely longitude, along the fly embryo. The early developing embryo, about 100 cells long, is already flooded with morphogenes (a classic embryology designation for transcription factors) stratified along a few 15–25 cell bands that mark broad longitudinal regions. These bands fade away as another group of tool-kit genes turns on, generating a series of stripes separated by interstripes. These stripes also disappear, giving way to more stripes generated by new tool-kit gene expression. A total of 14 stripes are formed, most of them persisting throughout development. Then segmentation occurs and the *Hox* genes are activated (see Fig. 2.3. and Plate 3). The expression of specific *Hox* genes occurs in different sets of segments, spanning from two up to seven segments. The way these successive tiers of bands or stripes of gene activity are positioned results from the integration of the different activator or repressor proteins that are present in each longitude. Since a gene can host up to ten binding signatures, its expression is positioned at a longitude range where gene activation is not suppressed by any repressor binding proteins. Box 2.1 exemplifies

Box 2.1 The combinatorial power of genetic switches in the evolution of form or how a navigator cell finds its developing fate in a universe of switches

Provided that transcription factors are unevenly distributed along a body axis and that tool-kit genes are endowed with several DNA-binding signatures, the area of gene expression will be defined by the coordinates where combining all on and off signature inputs results in gene expression. Since this positioning logic can work simultaneously or in sequence for many genes along several coordinates in the embryo, it comes as no surprise that the combinatorial power of switches is enormous. For simplicity let's consider only one gene with four signatures, each one bound by one of four transcription factors, namely V, X, Y, Z proteins, whose differential expression extends along the embryo longitude as follows: V from 60° W to 10° E; X from 40° W to 40° E; Y from 30° W to 0° E; and Z from 20° W to 60 E. Figure A depicts how combining the gene activation (V, X) and repression (Y, Z) of proteins gives a net output of expression in a stripe 30–60 ° W long.

This logic works in *Hox* genes. The *Ultrabithorax* (*Ubx*) gene is responsible for differential shapes between hindwings and forewings in insects. In flies this difference is most striking: the forewings are flexible, venated, and used for flight, whereas the 'hindwings', named **halteres**, consist of two small, ball-shaped structures that serve to balance the flight. Both wings start to develop in the same way until the *Ubx* gene is expressed in all hindwing cells but not in the forewing cells. This differential expression suppresses a set of forewing genes in the hindwing and changes the usage in others (see Fig. B, where the letters stand for regulatory proteins; U stands for Ubx regulatory protein). No wonder that when the *Ubx* gene is impaired, the Drosophila Ubx mutant develops a pair of forewings in the third thoracic segment instead of halteres because of the lack of the Ubx functional protein. These changes in longitude expression exemplify a fundamental rule of differentiation along serially repeated modules along body axes.

continues

Box 2.1 *(continued)*

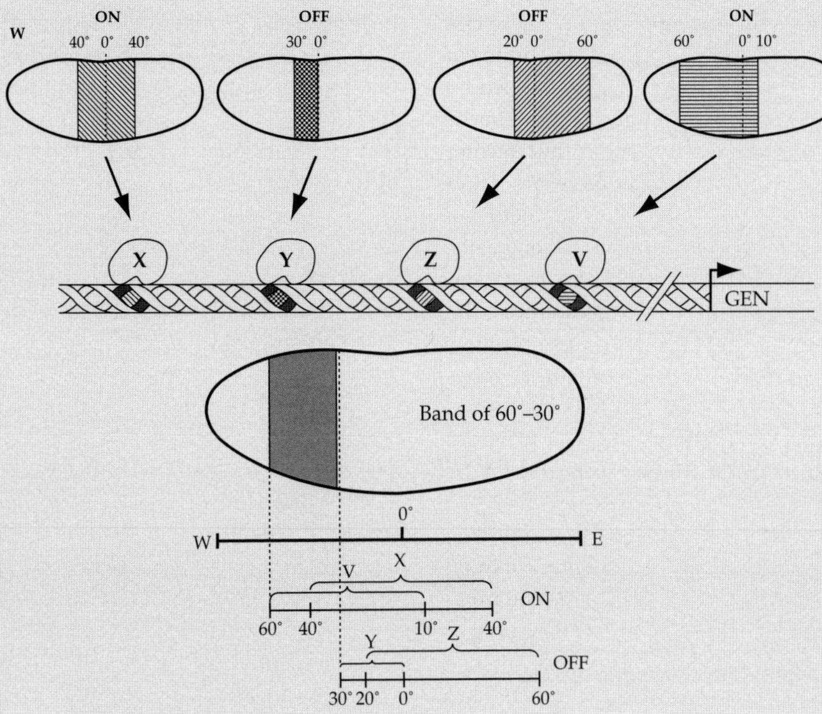

Fig. A. Schematic depiction of the way the transcription factors bind to gene signatures along the body axis to combine activation and repression of proteins. (See text for further details.) Drawing by Montserrat Peiró after a figure drawn by Joshua Klaiss from Carroll (2005), with permission from Sean B. Carroll.

Although this observation could suggest that wing evolution in insects was mediated solely by changes in *Ubx* longitudinal expression, other changes seem to have occurred. In butterflies and other insects with well-developed hindwings, one should expect naively that *Ubx* were not expressed in the third thoracic segment. That this is not true indicates that changes in the signatures of hindwing gene switches are likely responsible for the Ubx ability (or inability) to bind these **cis-regulatory elements**. In fact we know now that the Ubx protein modifies the hindwing pattern in ways characteristic to each insect group. Figure C shows that *Ubx* is expressed in the third thoracic segment of both dipterans and lepidopterans but affects a different set of genes in each group. The dramatic result is the formation of two radically distinct structures: a haltere or a hindwing. The lesson to take away from this observation is that the evolution of form can operate at two switch levels. One applies to changes in the *Hox* switches themselves that control at which coordinates they are expressed. The other level refers to gene signatures that are recognized by Hox proteins. As in the hindwing genes of many insect groups, changes in signatures of genes controlled by Hox proteins can modify the circuitries of development. Obviously, the presence of the same Hox proteins does not guarantee that they are going to work in the same way in all insects or other animal groups.

continues

Box 2.1 *(continued)*

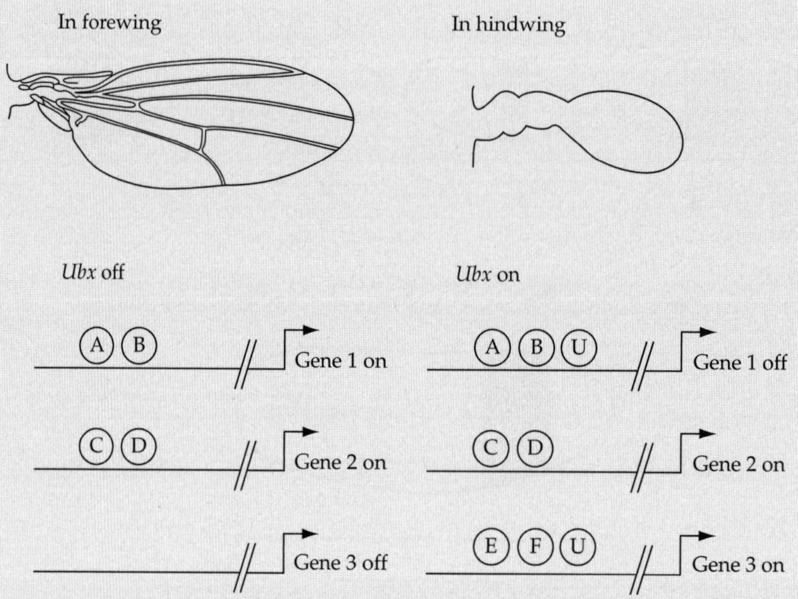

Fig. B. Scheme of the *Ultrabithorax* gene expression in hindwing cells but not in forewing cells of flies. See text for further details. From Carroll (2005), with permission from Sean B. Carroll. Drawing by Joshua Klaiss.

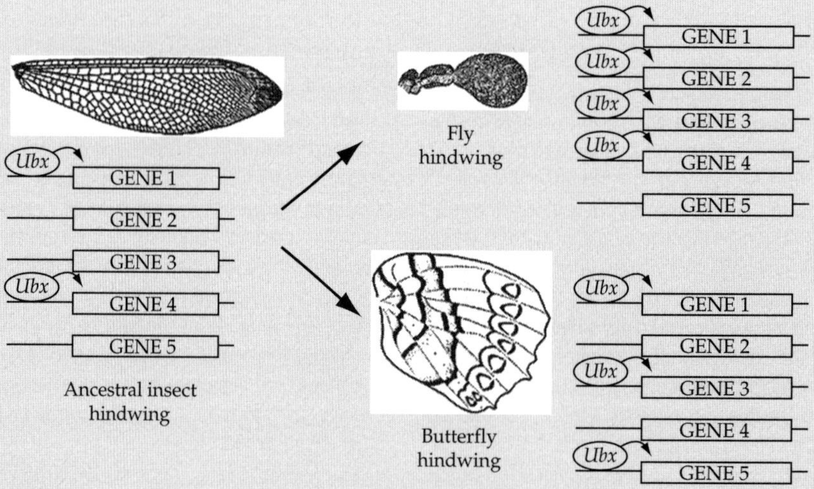

Fig. C. The figure depicts how changes in the signatures of hindwing gene switches allow the Ubx protein ability (or inability) to bind them and modify the hindwing pattern even when *Ubx* gene is expressed. See text for further details. From Carroll (2005), with permission from Sean B. Carroll. Drawing by Leanne Olds.

how this integration occurs along the embryo longitudinal axis.

Longitude is not the only coordinate needed for cell positioning. Simultaneously with longitude subdivision, the embryo is subdivided along its top-down axis (latitude) in broad cellular regions by latitude genes. These regions are, however, not serially striped, as in the longitude axis, but mark the future layers of the adult fly. Once longitudes and latitudes are finely established the cell field is set to place specific structures, i.e. organs, at particular coordinates. Groups of cells are recruited at these coordinates and a process of coordinate positioning, similar to that described for the whole embryo, starts anew at a smaller scale. This is how legs, wings, antennae, eyes, and other organs are built. For example, in the case of insect legs, the master gene *Dll*, an ancestral master gene involved in limb formation (see p. 40), is turned on just in the three thoracic segments, but not west or east of this area. To be more precise, *Dll* is activated at the southern end of each segment, exactly where a leg is supposed to develop. In a similar way, flies and all modern winged insects develop two pairs of wings, each positioned in the second and third thoracic segments. Another wing-related master gene is activated precisely in a coordinate north of where the legs are formed. We know that *Hox* genes are responsible for the master-gene activation in both leg and wing processes.

After emphasizing the prime role of coordinate orientation in development, one is tempted, evolutionarily speaking, to assert that most (if not all) major transitions in body forms are mediated throughout changes in latitudes and longitudes of gene expression. This appreciation is true, yet it is not the whole truth. Admittedly, comparative studies revealed that shifts in spatial boundaries of *Hox* gene expression along coordinates are largely responsible for large transitions between body plans. This is particularly true in the evolution of number and diversity of repeated units in arthropods, annelids, and vertebrates. But, when more closely related organisms are compared, other fine-tuning mechanisms downstream of *Hox* activity are discovered. The diversity of insect hindwings is an example of this (see Box 2.1).

Big explosions and small explosions: not so different after all

Now that we are starting to understand how cells orient themselves in the embryo geography to express the right tool-kit genes, it may be wise to go back to the Cambrian explosion paradox. Remember that the link between body plans was traditionally challenged mainly because revealing ancestral homologies was a hard task, futile even for great evolutionists like Mayr. With the discovery of *Hox* genes, and other tool-kit genes common to all phyla, the unity of kind seemed to be supported. Yet, the explosion of phyla was still not well understood because it required too many novelties in too short a time, geologically speaking. The long-held notion was that new genes are needed for new kinds of structures to arise. If time is short, the effect of these genetic changes must be large, much larger than that of the small mutations which natural selection is supposed to act upon. So, the spectrum of macromutations and hopeful monsters remained. The idea that new genes are necessary for novelties to evolve was so widespread among the scientific community that it was applied to *Hox* gene evolution. Edward B. Lewis (1978), a pioneer of *Hox* gene studies in Drosophila and Nobel Prize winner, suggested without hesitation that the insect *Hox* gene cluster evolved from a smaller set of ancestral arthropod *Hox* genes. This was an educated guess but was proved to be incorrect.

Once the detection of *Hox* genes had been completed for the overall phylogeny of phyla, it became clear that the major increase in tool-kit genes through duplication and divergence occurred deep in time, probably much earlier than Cambrian times (Fig. 2.4). Between the last common ancestor of diploblasts (cnidarians) and triploblasts, and the bilaterian radiation that originated most of the extant phyla, a lot of genetic and morphological complexity evolved. The radiation of bilaterians is not correlated with a corresponding increase and divergence in tool-kit genes. Some duplications, including at least two rounds of genome-wide duplications in early vertebrate evolution, have certainly occurred, but the basic tool-kit machinery has

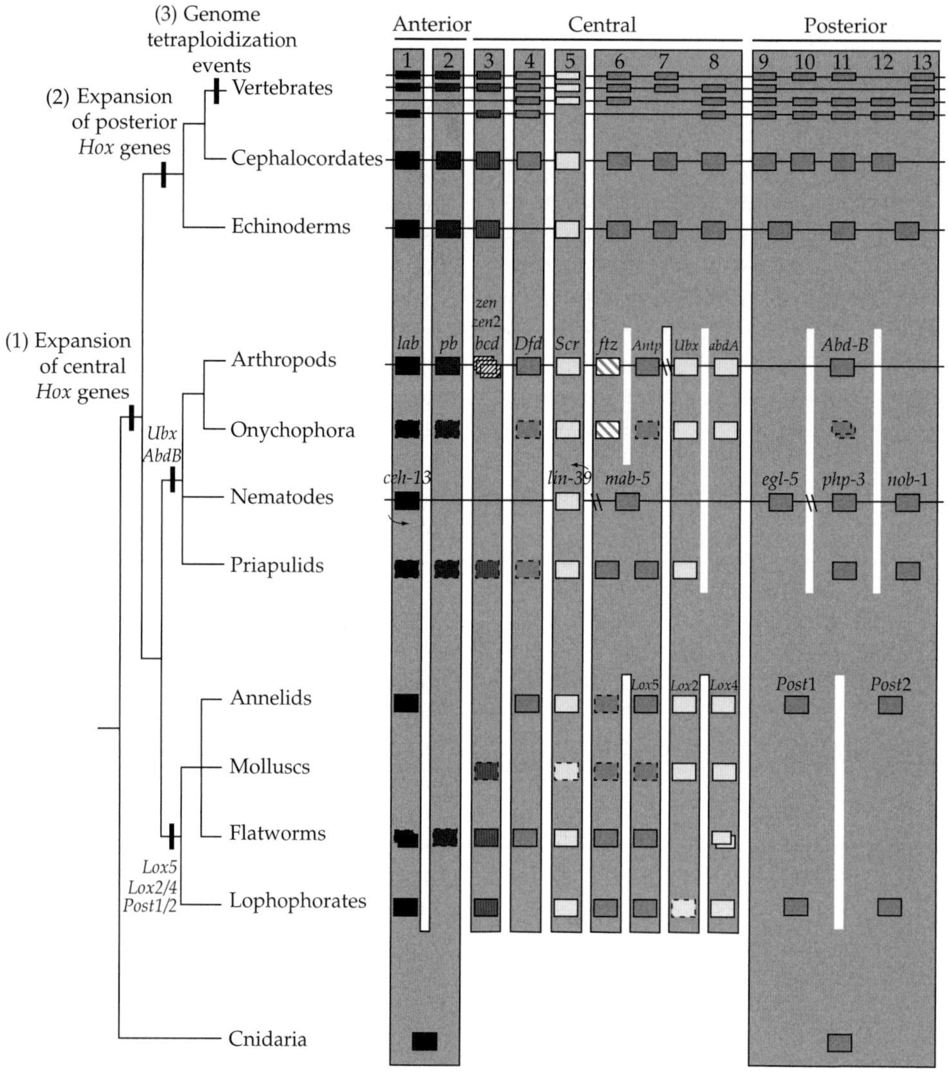

Fig. 2.4. *Hox* gene phylogeny for metazoan phyla. Right to each phylum the figure depicts the ordered gene complexes and their duplication degree. Homologous genes are vertically delineated in grey shaded boxes separated by white bars. The most ancestral pattern is that of Cnidarians (radial symmetric metazoans) that only shows two *Hox* genes located at both ends of the cluster. Many of the central genes expanded early at the base of the bilaterian (metazoans of bilateral symmetry) lineage (1), but the series *Hox*1–*Hox*5 was likely established much earlier. In the deuterostome lineage (2) the expansion towards the central and posterior gene clusters (*Hox*6–*Hox*13) is proceeding together with two rounds of genome duplication (3) in the vertebrates that gave rise to four complexes (see text). From Carroll *et al.* (2001), with permission from John Wiley & Sons Ltd.

been conserved through phyla across vast time periods that have spanned hundreds of millions of years. In particular, the ancestors of some well-studied groups, such as arthropods and vertebrates, were already deploying the same *Hox* gene batteries as their descendants. Then the question remains. If the same genes exist in ancestors and descendants, what makes the new forms so different? The educated guess was that the answer resides not in the number and divergence of genes, but in the way the genes are used in the body geography, namely where, when, and how are they expressed. This guess turned out to be correct.

How to go from crustacean ancestors to flies: a matter of Hox shifts

When discussing the Cambrian explosion I mentioned that lobopodians were considered the likely ancestors of arthropods. All Cambrian lobopodians have gone extinct, but animals with lobopods, unjointed limbs, are still around; they are the onychophorans. Arthropods have evolved a great diversity of appendages, ranging from legs, antennae, wings, and claws to highly specialized limbs for feeding (maxillipeds), swimming (pleopods), and other sophisticated activities such as defence (poison claws), copulation, and brooding. Actually, changes in segment number and appendage specialization are key to arthropod evolution. But if we are to understand how both the number and diversity of arthropod segments have evolved, it would be wise to find out what kind of developmental genes were present in their ancestors and how they have evolved in number and structure. This is exactly what Sean Carroll and his students thought. Their work focused on identifying all the *Hox* genes in Onychophora, the sister group of arthropods. To their surprise they found that, despite Onychophorans having only limited segment and appendage diversity, these comparatively less complex animals have the same *Hox* genes as are present in arthropods (Grenier *et al.* 1997). Obviously, the common ancestor to arthropods and onychoporans was already equipped with all the tool-kit genes needed for morphological evolution.

Further studies in *Hox* gene deployment along the arthropod body revealed that the key to segment diversification was in the differential geographic expression of the same *Hox* genes. Figure 2.5 depicts the expression domains of *Ubx* and *abd-A* along the segments in several arthropod orders. Remember that in insects the *Ubx* anterior boundary lies within the third thoracic segment (T3), where it is responsible for the haltere development. Here we see that *Ubx* and *abd-A* also express through most insect abdominal segments, both contributing to repress leg development. The anterior boundary of *Ubx* expression, however, changes with each arthropod body plan. For example, brine shrimps (a crustacean, genus Artemia) possess a simple thorax with very similar segments and appendages. Concordantly, the anterior boundary of *Ubx* has shifted to the first thoracic segment (T1), making all Artemia swimming appendages equal. Interestingly, other more derived crustaceans, such as lobsters, with specialized limbs in the anterior thoracic segments, show a backward shifting of the *Ubx* anterior boundary, so Ubx protein-free anterior thoracic segments (T1, T2, T3) can develop small appendages for feeding (maxillipeds). Looking at the Hox centipede (a myriapod) geography we notice that, like the brine shrimp, it shows a long trunk with identical segments and appendages where Hox and abd-A proteins are expressed all over, except in trunk segment 1 (Tr1), in which a poison claw develops. This specialized claw is reminiscent of the maxillipeds developed in anterior thoracic segments of crustaceans, where the *Ubx* gene is turned off. In sum, it seems that *Ubx* and *abd-A* expression shifts correlate with the evolution of specialized novelties in thoracic appendages.

Since expression of these *Hox* genes occurs in most of the non-specialized segments of arthropods, and because the onychophorans show a body plan with nearly identical segments, the guess would be that *Ubx* and *abd-A*, if present, would be turned on in most of the Onychophoran body length. Certainly these *Hox* genes exist in onychophorans, but their deployment is reduced to the tip of the embryo. This observation supports the notion that *Hox* genes are ancient and conserved in the arthropod ancestors but also indicates that during arthropod evolution, at least since onychophorans and arthropods split, their expression patterns have strikingly diverged by shifting their antero-posterior orientation. We cannot rewind Cambrian evolution but an educated extrapolation allows us to infer a similar story in the origin of this phylum. After all, fossil arthropods show much appendage specialization and body regionalization that suggests the presence of an ancestral conserved tool-kit of genes that evolved by shifting their *Hox* zone deployment through changes in cis-regulatory signatures.

Vertebrates as upside down arthropods (or vice versa)

Arthropods are not the only phylum in which this kind of expression shifting has been proved; our phylum, the chordates, has also used the same

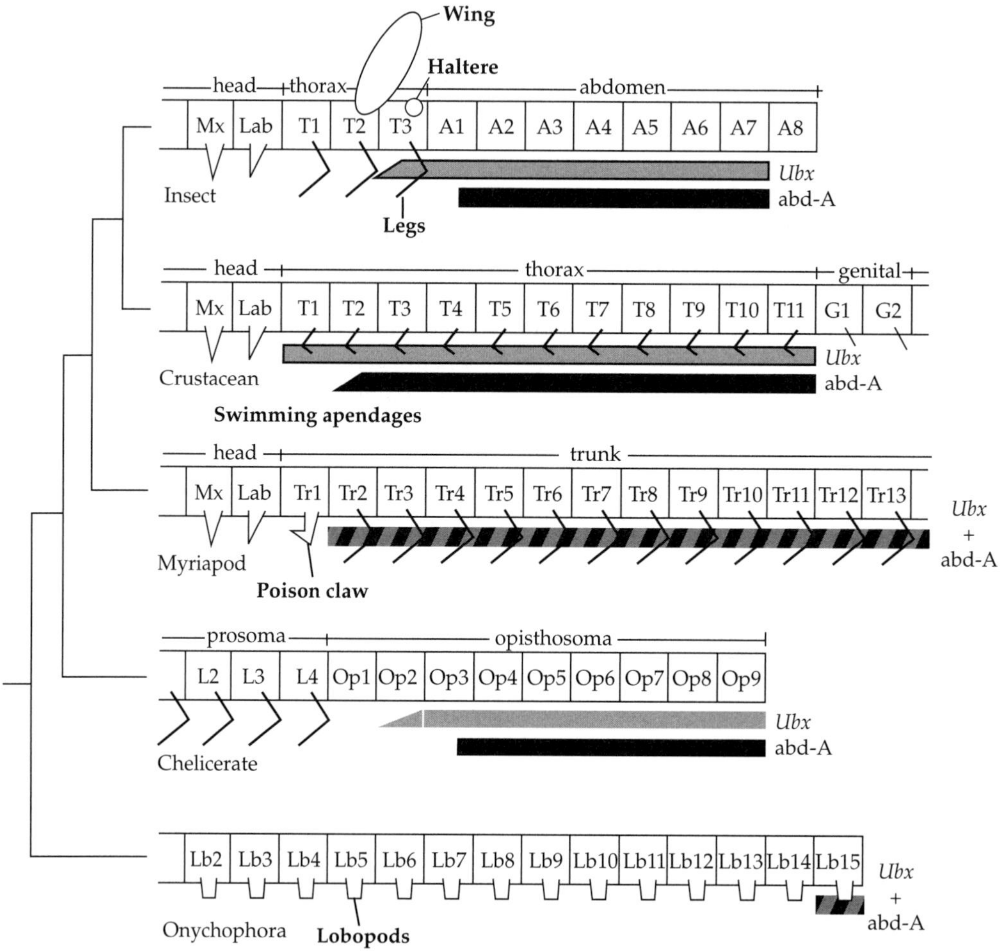

Fig. 2.5. Shifts in the anterior boundary of *Ubx* and *abd-A* gene expression in arthropods and onychophorans (arthropod ancestors). The expression domains of Ubx and abd-A proteins are depicted below the body plans of animals. Note how the anterior boundary of Ubx expression shifts in each group and marks a transition in segmental and appendage kind (see text for details). In onychophorans and myriapods the diagonal interspersed stripes denote the sum of the expression domains of both *Hox* genes. Segment annotations: A (abdominal), G (genital), Mx (maxillary), L (leg), Lab (labial), Lb (lobopod), Op (opisthosomal), T (thoracic), Tr (trunk). From Carroll *et al.* (2001), with permission from John Wiley & Sons Ltd.

evolutionary process in generating morphological diversity. As mentioned above, the lancelet is an extant invertebrate related to vertebrates that belong to a basal subgroup in the chordate phylum, the subphylum Cephalocordata. So it comes to no surprise that Garcia-Fernández and Holland (1994), two researchers at the University of Oxford, decided to study the lancelet *Hox* genes. They found one whole *Hox* battery of genes homologous to those present in vertebrates, but no trace of the four duplicated batteries present in the latter. The conclusion was, once more, that the primitive common ancestor of vertebrates and cephalocordates, and possibly of all chordates, already possessed the whole *Hox* basic tool-kit, suggesting that posterior diversifying evolution was likely to be dependent on changes in *Hox* expression zones. While this is true (see below), one cannot disregard the putative effect that further extensive, or whole, genome duplications might have had on vertebrate body plan evolution. In fact, at least two duplications have occurred in the lineage leading to vertebrates, one after the split from cephalochordates and the other some time before the origin of jawed fishes. Thus, it is tempting to advance that new duplicated *Hox* genes coupled with duplications of other regulatory genes in the same GRN could allow the speeding up of new adaptive body patterns without impairing the already fine-tuned gene interactions (see Fig. 1.7).

Despite their similarities with cephalocordates and other primitive chordates, including tunicates, vertebrates have evolved a wealth of striking novelties, including a head with a pair of eyes, jaws, and teeth; limbs in tetrapods; and other axial morphologies. In addition, one distinguishing characteristic of vertebrate complexity is the number of cell types. Higher vertebrates, including humans, possess different cell types that give rise to specialized tissues such as cartilage, bone, blood, nerve, and other sensory structures, not present in cephalocordates. This increasing cell type and tissue complexity has been accounted for by the advantage of extensive genome duplications comprising simultaneously the transcription factor gene and the gene regulated by this transcription factor. The logic of coupled duplication advantage means that if only the regulatory gene is duplicated, the two transcription factor copies should compete between them for binding to the unduplicated signatures, or CREs, of the regulated gene, preventing new mutations that could be selected to acquire a new regulatory function. Contrarily, if CREs are also duplicated, mutations arising in the new copies of regulatory genes would be freed from competition, favouring coevolution between new tool-kit genes, i.e. between regulator and regulated genes (see Chapter 1: Genome evolution by duplication).

There is no doubt that extensive duplication in the genome has contributed to the evolution of complexity, but an increase in the number of genes is neither a whole explanation of further vertebrate evolution, nor is it the original cause of new body patterns. Clearly, once the four *Hox* clusters were established a lot of important body changes in vertebrates occurred without more genome duplications. Comparing *Hox* gene expression along the body axes of different vertebrates has resulted in important shifts in *Hox* zones being detected, bringing to mind the situation described for arthropods. In brief, it has been shown that expression domains change according to vertebral anatomy (cervical, thoracic, lumbar, sacral, caudal), regardless of the number and extension of vertebra. Thus, as an example, the forward boundary of expression of *Hox6* gene is always located at the cervical/thoracic vertebral transition, be it in mice, chickens, or geese (Fig. 2.6), although the axial position of each transition, indicated by the building blocks of vertebrate bodies (somites), is different. Interestingly, in python snakes, where there is no clear cervical/thoracic transition, the *Hox6* forward boundary has shifted up to the head, expanding the number of thoracic vertebrae and eliminating its neck.

It is quite satisfying that in vertebrates and arthropods, two highly successful and diversified animal groups, the same shifting process of *Hox* gene expression along the longitudinal axis underlies their body building. This conservatism in developmental mechanisms gives further support to the unity of kind. Strikingly, a similar landscape of unity in expression applies to the latitudinal (dorsoventral, D–V) axis. Remember the Cuvier-Geoffroy Saint-Hilaire controversy and how Geoffroy Saint-Hilaire sustained the body plan unity of arthropods and vertebrates. He argued that a vertebrate was just an upside-down arthropod: what is dorsal in the former is ventral in the latter, and vice versa. Now we are in a position to test Geoffroy Saint-Hilaire's proposition by probing where the same tool-kit proteins express in embryos of both phyla. If a change of polarity in the dorsoventral axis has been produced since the chordate–arthropod lineage split, we must observe homologous developmental genes expressing in opposite axial extremes in each phylum. Figure 2.7 depicts the topographic expression patterns of the developmental genes in a schematic cross-section of vertebrates and arthropods. In an early embryo, dorsal and ventral cells have the potential to differentiate into nerve cells. Yet the reason why we, vertebrates, do not have nerve cords running down our bellies is because our embryo ventral cells secrete a protein called Bmp-4, which opposes the nerve tissue formation. However, in the dorsal cells the Bmp-4 protein is antagonized by a protein named Chordin that blocks its effects, allowing the formation of neurons and eventually the nerve cord. A homologous system to the pair Bmp/Chordin exists in arthropods, but in this case the Bmp-homologous

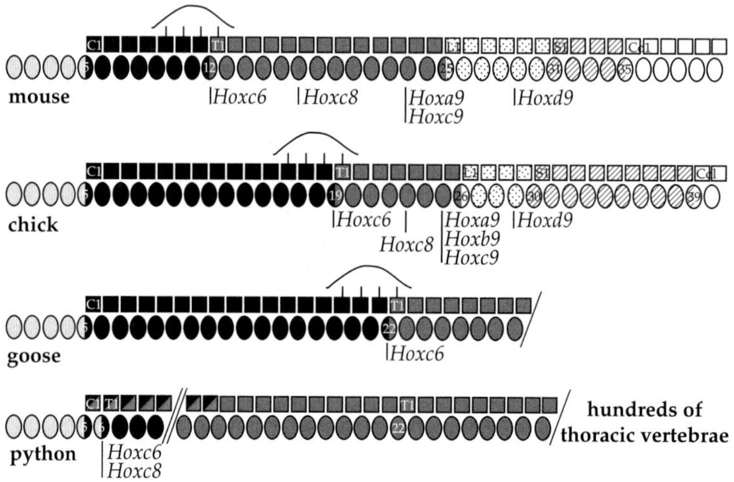

Fig. 2.6. The antero-posterior (A–P) axial organization of tetrapods depends on shifts in *Hox* gene expression domains. The anterior boundary of the *Hox6* gene expression marks the transition between neck (C somites, in black) and trunk (T somites, in grey). Shifts in this boundary differ in each organism, generating animals with short necks (mice), middle necks (chicken), long necks (geese), and no necks (pythons). Analogously, shifts in the anterior boundaries of other *Hox* gene expression along the A–P axis change the thoracic (T)–lumbar (L, dotted somites), the lumbar(L)–sacral(S, striped somites), and the sacral(S)–coxis(Co, open somites) transitions in different organisms. Vertebrae are depicted by squares with the same key design as in somites. From Carroll *et al.* (2001), with permission from John Wiley & Sons Ltd.

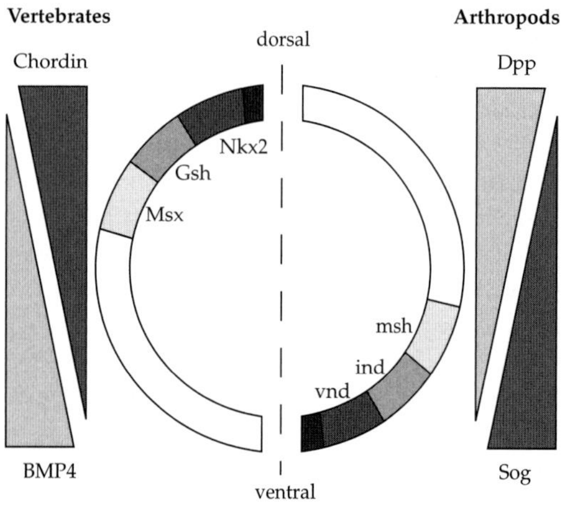

Fig. 2.7. Schematic patterns of developmental gene expression in a dorsoventral cross-section of vertebrates (left side) and arthropods (right side). The homologous gene pairs (*Chordin/Sog; BMP4/Dpp*) flip their expression in the dorsoventral axis between vertebrates and arthropods. Their secreted protein balance, depicted by triangles (grey for neurogenic inhibition; black for antagonistic action), induces the development of the neural cord, dorsally in vertebrates and ventrally in arthropods. Other sets of homologous homeobox genes show similar inverted gradients into medial (*vnd/Nkx2*), intermediate (*ind/Gsh*) and lateral (*msh/Msx*) neurogenic domains. From Lichtneckert and Reichert (2005) with permission from Nature Publishing Group.

protein Dpp is blocked ventrally instead of dorsally by the Chordin-homologous protein Sog, with the result that the nerve cord is formed along the ventral side. The degree of similarity between nerve-blocking genes (*Dpp* and *Bmp*) and between their antagonists (*Sog* and *Chordin*) is so high that they are exchangeable; for instance, if a *Sog* gene from Drosophila is inserted into frog embryo DNA a second nerve cord will form in the frog's belly. This flipping expression of the two homologous gene pairs *Dpp–Bmp* and *Sog–Cordin* in arthropods and vertebrates is paralleled by a battery of other homologous developmental genes that show the same expression pattern but in reverse latitudinal orientation. These observations support the D–V inversion polarity hypothesis, reinforcing the idea, held by the universal *Hox* expression patterns, that the basic mechanisms of development already existed in early evolutionary times, likely in the Precambrian, and have been conserved ever since.

Microevolution writ large

Current defenders of the idea that macroevolution and microevolution are decoupled argue that big transitions between body plans can be produced by large, rapid leaps instead of small, gradual changes. They state that Evo Devo supports the fact that changes in switches can likely be responsible for non-gradual, radical changes in body patterns. Admittedly, we lack a formal theory of developmental population genetics, probably because we still do not know all the intricacies of the complex GNRs that link the multiple effects of developmental toolkit genes. These genes have two basic properties. One is their **mosaic pleiotropism**, namely that they affect many other genes via the transcription factors they encode, and participate in building many different body structures. This is possible due to the **modularity of** CREs, which are switched on or off only in certain transcription factor-expressed modules or **compartments** (see below). This is their second basic property. In fact the two properties are the two sides of the same coin: the former refers to the cell-expression level, while the latter indicates that the modularity is a property of cis-regulatory regions at the gene level expression. Consideration of these complex properties justifies the fact that a population theory is still far from our reach. Yet some knowledge of modulations in *Hox* expression within conserved body plans may give some hints on gradual quantitative changes in morphology.

The origin of insect wings and their posterior evolution in shape and number may illustrate how structures evolve gradually. For decades insect evolution has been a highly contentious issue. Some argued that no primitive insects exist in the fossil record, others that it is not clear whether insect wings derived as independent outgrowths from wingless insects or from the gills of an aquatic arthropod. Both disputes are superseded by current evidence. Not only are there primitive insect fossils that look quite different from present insects, but the Evo Devo studies have given ample support to the gill-to-wing hypothesis. Currently, our knowledge on insect evolution has grown enough, and continues to grow, to be presented as an icon of how the merging of embryology, palaeontology, and developmental genetics can build a highly reliable picture of gradual evolution. Here is a brief summary of this work, carried out by many researchers, that has shed light on the idea that macroevolution is nothing less than microevolution writ large.

Hox genes underlie the appendage morphology patterning of arthropods in the way discussed above. Briefly, Ubx and abd-A proteins, by shifting their expression zones, turn on or off genes involved in limb development. Two of these genes, *apterous* and *nubbin*, are expressed in the respiratory lobe of crustacean limbs (the gill). Strikingly, the researchers (Averof and Cohen 1997) found that the Apterous and Nubbin proteins encoded by the same genes are essential for insect wing development. This coincidence was interpreted as a homology between gill and wing and gave strong support to the evolutionary origin of wings from gills. The gill-to-wing hypothesis is bolstered by much additional evidence. The most primitive insect fossils dating from the Carboniferous (about 300 mya) show wing-like structures on all segments, abdominal and thoracic alike. This fossil belongs to an aquatic animal and these structures are not wings but gills that resemble those found on present aquatic larvae of dragonflies and mayflies. The likely scenario suggested by this fossil evidence, and bolstered by developmental studies, is one in which early insects developed wings from larval gills. Interestingly, dragonflies and mayflies are the most primitive extant insects. Does this mean that they are the ancestors of all present insects? Not at all. They are just sister lineages of extant highly derived insects whose larvae may illustrate what a primitive insect-like arthropod did look like.

No apostle of macroevolution would argue that one of these aquatic shrimps could have gone ashore 300 mya and successfully started to fly using its wing-like appendages. This is most improbable. Rather, the evolution of wings from gills was likely a long journey of gradual steps of changes in developmental switches. At least two evolutionary processes must be explained to make insect evolution credible: the evolution of wing number and the diversification of wing morphology. In fact, the fossil record shows a gradual reduction in size of wing-like structures in the abdominal segments that eventually led to their disappearance (Fig. 2.8). In Drosophila, wings are repressed by *Hox* gene expression in specific segments: *Scr* in the first thoracic segment and *Ubx*, *abd-A*, and *Abd-B* in the abdominal segments. Reduction of wing size in the abdominal segments, however, was a gradual process mediated by

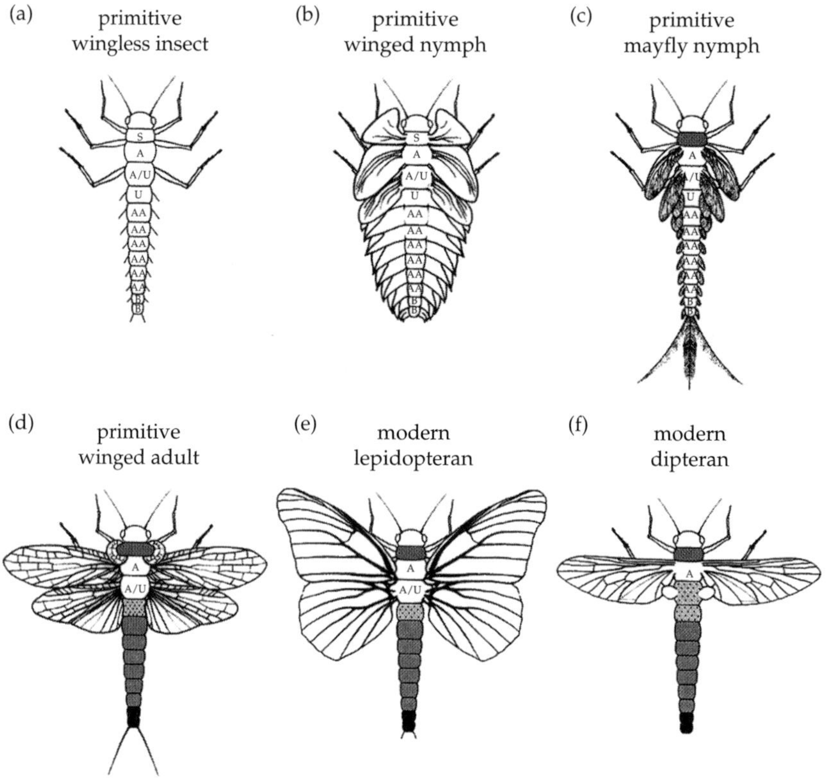

Fig. 2.8. How insect wings evolved from gill-like appendages and changed their number through time. Probably primitive insects were wingless (a) and the first insects with wing-like structures (gills) were extinct aquatic nymph forms (b) similar to mayfly nymphs (c). The evolutionary reduction of wing number was effected by the developmental repression mediated by *Hox* gene expression in the first thoracic segment (Scr: S, dark grey) and in the abdominal segments (A: Antp; U: Ubx, dotted; AA: abd-A, grey; B: abd-B, black). Note that the segmental domains of *Hox* gene expression were already present in primitive forms and changes occurred thereafter in gene switches (CREs). In dipterans (f) the action of Ubx is not only repressive but also regulatory of the size and shape of the haltere. From Carroll *et al.* (2001), with permission from John Wiley & Sons Ltd.

changes in *Hox* gene expression likely due to evolution of new CREs in target genes for wing formation. This is well documented in the diversification between forewings and hindwings in insects (Carroll *et al.* 1995). We discuss in Box 2.1 how hindwing-modified halteres in Drosophila can be accounted for by changes in *Ubx*-binding sites in CREs of target genes that encode proteins necessary for developing wing veins and sensory cells. Different evolutionary changes in CREs are responsible for *Ubx*-binding activation of target genes that build the hindwings of many other insects. Interestingly, similar changes in these selector genes (see below) are responsible for ancestral forelimb divergence in vertebrates mediated by the *Tbx5* transcription factor. This process allows the generation of a wealth of vertebrate forelimb morphologies, ranging from flight structures in bats and birds to our own arm.

This evidence points to gradual rather than to discontinuous evolution. In more detailed pattern modifications a correlation with *Hox* expression has been detected. The Drosophila second leg femur shows microhairs (trichomes) whose pattern differs among species. In *D. melanogaster* the absence of trichomes in the posterior region of the femur is associated with high *Ubx* expression. *D. simulans*, a sibling species, shows a larger area devoid of trichomes also associated with high *Ubx* expression. That *Ubx* expression is responsible for trichome repression was confirmed in *D. virilis*, in which very low levels of *Ubx* expression do not inhibit trichome formation. Thorough studies revealed that differences in *Ubx* expression depend on changes in CREs that evolved since the split of these Drosophila species (see Plate 16, Chapter 5). The lesson to take away from these results should be that if small changes of expression have evolved between closely related species producing slight changes in morphology, one could envisage that through long periods of time large evolutionary changes between highly diverged lineages would be quite likely.

Cambrian and other explosions revisited

After many years of studying the Cambrian fossil record and all the Evo Devo mechanisms, we can, merging both studies, propose a plausible scenario for the origin of phyla, in which classic population genetics and ecology are not alien characters in the drama. The early presence of most tool-kit genes of development underlies the Cambrian explosion, and also other posterior evolutionary radiations. But was it just the proliferation of *Hox*, or other selector genes, that really triggered the explosion of phyla? Paleontologists generally believe that building a minimum tool-kit gene network was necessary for the body plans to form. Yet other more ecologically orientated evolutionists think that deployment of a prime battery of developmental genes was not sufficient and other environmental stimuli were necessary to fuel the Cambrian disparity.

The first lesson from Evo Devo is that genes for building animal body plans are common to all and predate the Cambrian era. Two conclusions, at least, follow. One gives strong support to the Darwinian idea of descent with modification from one (or, in Darwin's words, a few) common ancestor (the unity of kind). The other champions the idea, also sponsored by Darwin, that new phyla (and new organs) did not evolve from scratch, but from earlier, more primitive, forms by modification of their evolutionary networks. The way that evolution acts is by tinkering with the original tool-kit genes to create new kinds of eyes, limbs, and other organs, which may fool the inexperienced, naked eye, giving to new forms the appearance of independent, irreducible designs. Yet nothing is farther from reality. As explained above, a close comparative genetic scrutiny reveals that under dramatically different body plans a common set of switching machinery is concealed.

Other deep ancestral homologies, revealed in developmental genes, are also bolstered by molecular phylogenetic reconstructions. Wray *et al.* (1996) and Blair and Hedges (2005), using protein sequences, found divergence times among metazoans dating between 800 and 1,200 mya. Some later studies corrected for the inconstancy of molecular clock. The most conservative studies dated the origin of phyla to at least 100 my earlier than the Cambrian. Thus these phylogenies agree with the idea supported by the Evo Devo that the early ancestor of all bilaterians (dubbed Urbilateria) must have existed several tens of millions of years before the Cambrian explosion. The last common ancestors of chordates and arthropods were most probably some inch-long, worm-like animals with mouth, gut, anus, muscles, and other simple organs similar to hearts, muscle, light-sensing organs, and maybe outgrowths in their body that are reminiscent of primordial appendages, the tracks of which were left in the Ediacaran sediments (Fig. 2.9). This speculative picture is based on the presence in this last common ancestor of a complete battery of at least six *Hox* genes plus a few hundred body-building genes that included *Pax-6*, *Distal-less*, and *tinman*, which all current bilaterians share. These soft-bodied animals could not leave body fossils. It was not until 600 mya, just before the Ediacaran fauna, that photosynthetic microbiotas, inhabiting the Earth for billions of years, were able to produce oxygen concentrations in the atmosphere high enough to develop complex metazoans (multicellular animals). Collagen, the basic cellular connective material, needs oxygen to be synthesized and aggregate the cells. Also, an oxygen concentration of 10% is essential to precipitate calcium to form skeletons. Both large cell aggregations and hard structures are requisites to build large, complex, and hard bodies, and also to leave fossil records.

The importance of oxygen levels notwithstanding, other ecological factors have also been advanced for the bilaterian explosion that produced 35 crown phyla

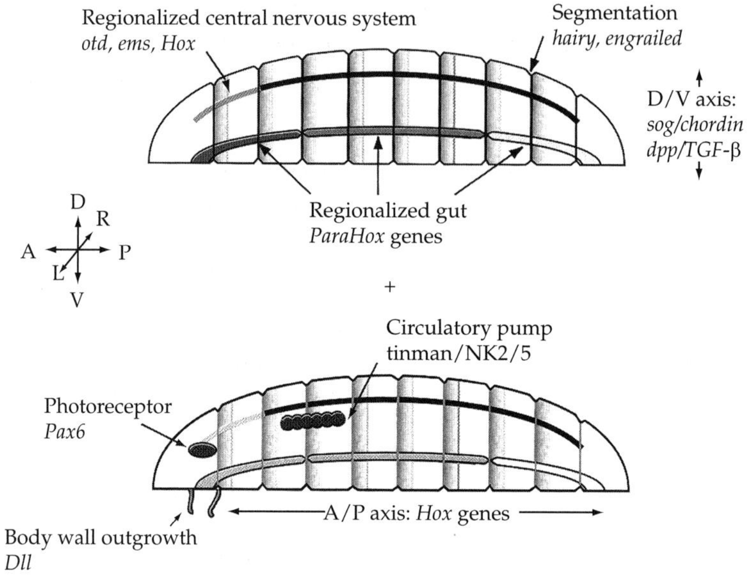

Fig. 2.9. Two views of a hypothetical urbilaterian body sketched from the conservation of genes and their developmental functions shared by extant bilaterians. The A/P axis was probably subdivided by nested, overlapping domains of *Hox* gene expression. The D/V axis was likely controlled by ancestral genes of the *sog/chordin* and TGF-*β* families. Some selector and master regulatory genes are depicted in their regions of expression. Also several tissues and organs are shown. The hypothetical size would not exceed 0,5 cm, as estimated from its fossilized tracks and burrows. (Top) Tissue layers were probably regionally patterned along the *A/P* axis, such as the gut (*ParaHox* genes) and the nervous system (*otd, ems, Hox*). Segmentation probably evolved under the regulation of ancestral *hairy* and *engrailed* genes. (Bottom) This scheme incorporates primitive organs with their ancestral regulatory genes (in parenthesis): a photoreceptor (*Pax6*), a circulatory pump (*tinman*), and body wall outhgrowths (*Dll*). Adapted from Carroll *et al.* (2001) with permission from Sean B. Carroll.

in less than 20 my. Among them, the continental drift that initiated the separation of land masses 540 mya may have influenced striking climatic changes. However, if environmental changes underlay the body plan formation, why have no new phyla, except the bryozoa, have been produced thereafter along the 500 my of climatic and planetary changes to the present? On the other hand, if the invention of tool-kit genes occurred prior to the Cambrian, what induced the Cambrian explosion? Certainly the reality of the Cambrian explosion is bolstered by recent thorough studies (Cartwright and Collins 2007), which adhere increasingly to the idea that it was an ecological event. Carroll (2005) suggests that once the cryptic tool-kit machinery of body building started to develop more obvious morphologies with complex structures, a runaway process of ecological invasions and competition was set off, unleashing further explosions. Current Evo Devo studies have revealed how the evolution of new body patterns divided into head, trunk, and tail parts, to allow movement and defence, and of new refined organs such as eyes, jointed appendages, hearts, and other structures of high competitive value, were made possible using old tools to cope with new challenges. No large genome structures needed to be invented anew; rather, new circuitries with old switches were evolved through the classic interplay of mutation and selection in regulatory elements. As stated by Carroll (2005, p.165): 'Genes in the tool kit are important actors in this picture, but the tool kit itself represents only possibilities, not destiny. The drama of the Cambrian was driven by ecology on a global scale.'

The evolutionary role of developmental constraints

Nowhere is the tendency to confound the patterns and the causes in evolution more obvious than in the explanation of the origin of form. Gould was likely the most active defender of the leading role of body pattern in the evolution of form. It is generally accepted that not all possible morphologies (patterns) are realized in the current universe (technically known as the morphospace). In fact, phyla occupy only a restricted morphospace area, but the crucial point is whether the external laws of ecology or the internal laws of development channel their evolutionary fate. The former constitute the basic mechanisms upon which natural selection acts (dubbed by Darwin as 'conditions of existence'); the latter are the genetic and developmental rules that underlie the phenotype structure (called 'laws of growth' by Darwin). It has become customary to designate the results from natural selection as 'adaptations' and those from development as 'constraints'. Whilst adaptation implies a positive meaning, constraint is often taken to include negative processes that limit selection. Here we want to emphasize the positive meaning of constraint, as defined and discussed in Box 2.2, to analyse its scope in channelling evolution. Much of this section tries to scrutinize an endeavour adopted by some contemporary developmental biologists that can be epitomized by Gould's words: 'We do need to reformulate, [. . .] the old notions of organic integrity, and structural determination from the "inside" of genetics and development, thus balancing our former functionalist faith in the full efficacy of adaptationism with positive concepts of internal and structural constraints' (Gould 2002, p. 1057). Amen.

Was Gould right? Deep homologies and parallel evolution

When Eric Davidson, a leader in current Evo Devo studies, was asked (see interview in *El Pais* 19 March 2006) whether Gould was right after all, he answered: 'Steve Gould did deal with no real mechanisms. His phenomenological intuitions did not influence the generation of our ideas.' Yet, Doug Erwin, a palaeontologist and co-author of a joint influential paper with Davidson (Davidson and Erwin 2006) added: 'Between our ideas and Gould's there is not a direct link, but a partial conceptual connexion.' But, what is this connection? I believe that the connection lies in the concept of developmental (or phylogenetic) constraints.

Homology is usually referred to as resulting from the derivation of characters in two species from a recent common ancestor, regardless of similarity in function or form. When this commonality applies to ancient common ancestors, we talk of deep homologies. As discussed above, Evo Devo has revealed many deep gene homologies. But the correlation between homologous morphological characters and homologous developmental genes is not always straightforward. In other words, homologous characters can show differences in some of their genetic bases. Although a commonness of gene machinery often underlies a common descent, and constitutes strong support for homology, the whole set of toolkit developmental genes can show significant differences between homologous characters. Wagner (2007) has raised the point that 'developmental variation in homologous characters is not randomly distributed, but affects some aspects of development more than others'. For instance, in Drosophila, genes that act in the early segmentation stages (discussed above), namely gap genes, which define longitudinal regions, and pair-rule genes, which generate stripes of alternating half segments, show higher interspecific variation. Moreover, some of them, like the gap gene *bicoid*, are present only in the higher diptera, and others like the pair-rule genes *fushi tarazu* and *even skipped*, do not have pair-rule function in some grasshoppers, but are expressed in the nervous tissue. Contrarily, genes that direct the later processes of segment formation (for example, segment-polarity genes) are invariant, at least among insects. This reminds one of the flexibility assigned to prephylotypic (pregastrulation) phases and the constraint that characterizes the phylotype (Box 2.2).

Since the phylotypic characters define the type identity, by analogy it seems sensible to assign a high level of conservatism to the GRN that specifies the character identity. These GRNs have been named ChINs (**character identity networks**) by Wagner (2007), to distinguish them from other GRNs that control the different shape and state of a character, i.e. its character states. There are several studied developmental genes that likely belong to a ChIN. We have already discussed one of them: the *Hox Ubx* gene. Insects have two pairs of wings, the forewings and the hindwings. Hindwing identity across species is determined by the specific *Ubx* expression in the third thoracic segment (T_3). Thus *Ubx* controls hindwing identity (the character), no matter what

Box 2.2 Are there constraints to natural selection? Or does the organism propose and the environment dispose?

Contrary to some misinterpretations, mutation is not random if by randomness we mean that all DNA changes have identical probability. We know that some genomic changes are more likely than others. In fact, Darwinian premises do not require this kind of random change. Precisely, what the natural selection mechanism requires is that mutations were isotropic towards adaptation. In other words, 'variation must be unrelated to the direction of evolutionary change', as stated by Gould (2002, p. 144). This undirected trend, together with its copiousness and smallness in extent, forms the three requirements for variation which natural selection acts upon. Darwin emphasized these variation properties to assert the leading role of natural selection in the direction of evolution. Yet, he recognized biased tendencies towards certain phenotypic states, but of lower strength than that of selection.

Darwin was not ignorant to the role of development, dubbed by him as 'the laws of growth' or 'the correlation of growth', in the evolutionary change. He was perfectly aware that the unity of kind imposes some phyletic restraint in the descendent phenotypes, constraining the operation of selection. But, contrary to the channelling role assigned to these internal constraints by many famed naturalists such as Owen, Geoffroy Saint-Hillaire, and many others, Darwin declared himself an advocate of 'the conditions of existence', his designation for ecological conditions, as the leading cue to adaptation through natural selection. In his own words: 'It is generally acknowledged that all organic beings have been formed on two great laws—Unity of Type, and the Conditions of Existence. [...] Hence, in fact, the law of the Conditions of Existence is the higher law; as it includes, through the inheritance of former adaptations, that of Unity of Type.' (Darwin 1859, p. 206).

Recent advances in developmental biology have filled the gap in the genetic mechanisms that build the body plans, as briefly discussed in this chapter, and given an opportunity, impossible in Darwin's time, to assess the putative role of deep homologies in evolutionary trends. Not all constraints imposed by descent are, however, of equal value, according to some authors (Gould 2002, pp.1029–37); some may be qualified as 'benign', or negative, for the Darwinian Theory, but a positive, leading role has been assigned to others. Among the former, the constraints imposed by the physical laws and the impossibility to optimize all phenotypic characters that often conflict (called 'trade offs') are the best known. These negative constraints have been accepted by all Darwinians because they do not challenge the leading role of natural selection. On the other hand, phyletic constraints imposed by developmental genetics have been qualified as positive by some authors, who propose that interactive genetic networks induce structures that channel the future evolutionary course (Gould and Lewontin 1979).

Positive constraints are those that must be understood if we are to decide between a structuralist (internalist) and a functionalist (externalist) theory of evolution. The long-held tenet, popularized by Haeckel, that the earliest stages of development are the most constrained (and the most conserved) is no longer acceptable. Recent work and reanalysis of older work (see Raff 1996 for a review) have shown that these early constrained phases, sensu Haeckel, are in fact intermediate in development (Fig. D. After Elinson 1987). The true early developmental phases (the embryo cleavage stages) are extremely plastic and variable among phyla. On the other hand, it is the intermediate phase, called the phylotype, which defines the phylum. The rationale that underlies this model is based on the number and kind of interactive regulatory events in the developmental networks. Though in the earliest phases interactions exist, they are few and act on a global scale (mainly defining the axis coordinates), allowing limited changes in morphogen distribution. Many organisms, from ascidians to sea urchins and to amphibians, have been able to modify their early development, suppressing even their larval stage. Changes in these prephylotypic phases, however, are not limited to larval development. In vertebrates the pharyngula, as it is designated the phyloptypic stage, is preceded by processes, such as diverse cleavages and cellular **transductions**, that strikingly differ among lineages. For example, the Bicoid protein that specifies the antero-posterior axis in the early Drosophila embryo is only present in some dipterans. This prephylotypic plasticity, however, is at odds with the rigidity of the phylotypic stage, in which interactions increase in number and complexity. These interactions act on a global scale and any change in one of them, even of a small magnitude, is likely to have dramatic effects on many other links and to eventually disrupt the body plan. In a way, the phylotype would represent a historical frozen stage that conditions the further evolution of the lineage and, as such, it has been considered by 'internalists' as a channelling force of evolution. After the phylotypic stage, new interactions occur but only within discrete modules, in a way that changes in one module have little or no effect in the others. Thus, developmental programmes of limbs, lungs, or hearts are complex and show large local interactions, but they do not communicate amongst them, which endows a greater

continues

Box 2.2 *(continued)*

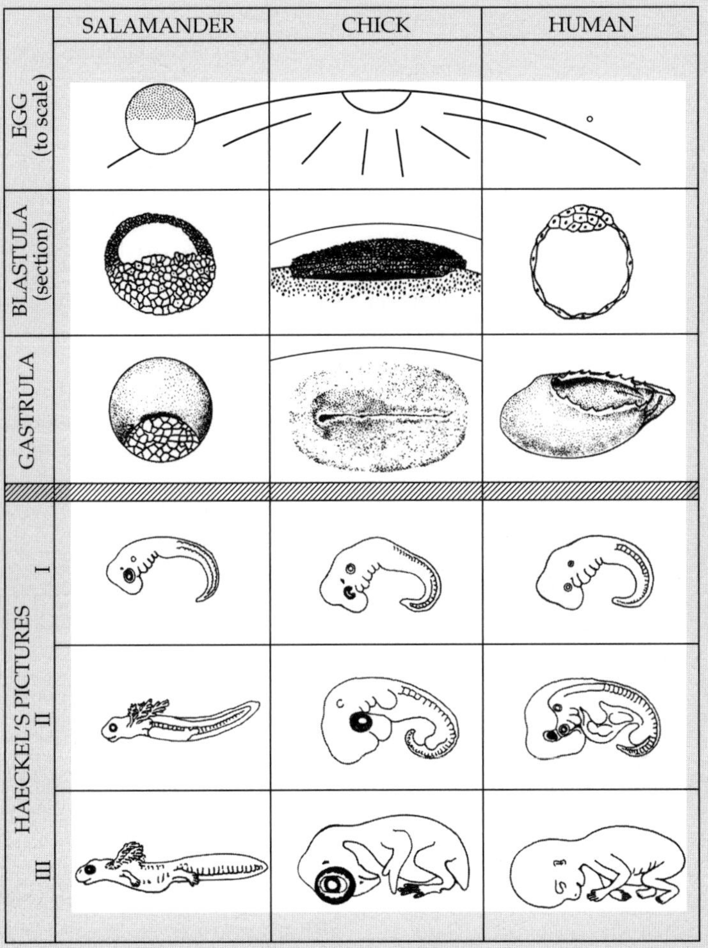

Fig. D. Schematic comparative view of early and intermediate embryonic phases of several vertebrates, showing that the former are extremely plastic and the latter (the intermediate) are constrained. (See text for further details) From Elinson (1987) with permission from John Wiley and Sons Ltd.

evolutionary plasticity to this phase. The sand clock model in Fig. E shows that all evolutionary trajectories cross the middle narrow point, which depicts a very conserved phase (the phylotype). (After Raff 1996.)

Much of the dispute on the leading role of natural selection versus development might be settled if the contenders would arrive at a common agreement on how natural selection operates. So many years since the formulation of the theory of natural selection this disagreement may seem ironic, but I believe it still exists. Contrary to the tenets of many internalists, the current teachings of Evo Devo are saying that the developmental regulatory networks, with their complex architecture in links, switches, and deployment of gene batteries, provide a wealth of flexibility through gradual changes. Admittedly, these genomic changes act largely on regulatory DNA stretches, a view likely ignored in the Modern Synthesis, although changes in structural DNA cannot be disregarded (Hoeckstra and Coyne 2007). These regulatory changes, the raw material upon which natural selection acts,

continues

Box 2.2 *(continued)*

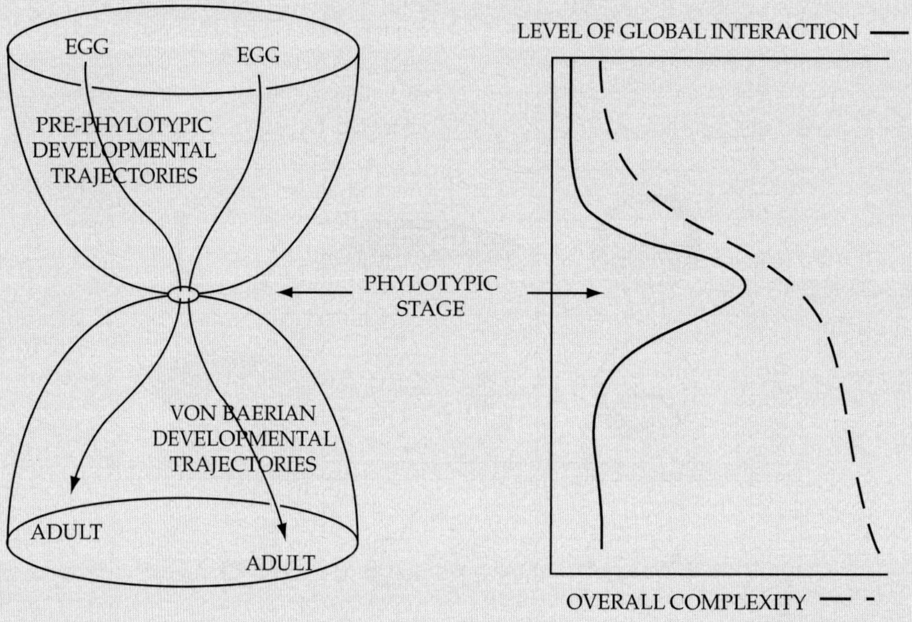

Fig. E. The sand clock model depicts the phylotypic stage as the point where all evolutionary trajectories cross and the degree of interaction is maximal. (See text for further details.) From Raff (1996) with permission from John Wiley and Sons Ltd.

are pervasive in the pre- and post-phylotypic stages. Compartmentalization in discrete modules facilitates the possibility that in each of them structures could evolve independently from others, thus avoiding the negative pleiotropic effects of regulatory genes. Who can deny that compartmentalization is the product of the opportunism of natural selection? Analogously, it is true that many master genes, such as *Hox* genes, have long been conserved, but this does not imply that building genes did not evolve in each specific organ nor that master genes were not conserved by natural selection. The ubiquitous control of eye development by the *Pax-6* gene has been argued as proof of a constraint that channels the eye independently from natural selection. Yet, *Pax-6* can be considered as a starting signal, analogous to the gun shot of any race contest. The signal may be the same but the elements, building genes in the eye and cars or athletes in the race, may be quite different in each case. Is this a channelling constraint or just a mechanism co-opted by natural selection? Or both?

The contention that natural selection has been unable to change developmental constraints is just a highly biased explanation to discredit the role of natural selection. Knowing that natural selection must take care that any new mutation does improve, though minimally, immediate adaptation, and its opportunistic tinkering of previous adaptive structures, it comes as no surprise that once a highly integrated developmental network is working any change would likely be discarded (or buffered). This applies mostly, but not exclusively, to the phylotypic stages. So, although the explosive origin of phyla could easily be accounted for by positive adaptation to the external global Precambrian conditions, their further conservatism spanning more than 500 my should not be simply attributed to phyletic constraint. It would be, however, more realistic to say that the organism proposes and natural selection disposes, but that organismal propositions are sometimes quite restrictive. In sum, selection is supreme in promoting conservatism or novelties, and there is no ground to invoke internal laws that channel the evolution of form. The pre-eminence of natural selection notwithstanding, the explanation of the evolution of form would be incomplete if we did not incorporate the constraints imposed by the unity of type, as Darwin himself recognized.

shape, blade or haltere, the character shows (the character state). Obviously, *Ubx* is likely invariant, its absence or impairment leads to an insect mutant phenotype that is qualified as unfit (an *Ubx* hopeless monster); however, it does not act alone and certainly belongs to a ChIN with other genes such as *Abd.*

Another example of GRN conservation has been described in more detail for the eye development in vertebrates and insects. We have already discussed the ubiquity and exchangeability of *Pax*-6 genes, which qualifies them as deep homologous. Yet, detailed studies strongly suggest that *Pax*-6 was separately incorporated into the ChINs for the image-forming of vertebrates and insects. As discussed above, anti-Darwinists have championed the implausibility of independent eye evolution in different phyla by Darwinian natural selection. Darwinists, on the other hand, have strained themselves to prove that independent evolution is possible by gradual convergent improvements upon a common basic structure. But, where homology ends and novelty starts has not been easy to decide. Only by probing into the genetic machinery that controls eye development has an ever-increasing overall picture emerged.

On the morphological side it is well established that all animal eyes consist of light-detecting cells, the photoreceptors, located adjacent to pigmented cells that shield light reaching the photoreceptors. Darwin's intuition described such a primitive device, stating that 'the simplest organ which can be called an eye consists of an optic nerve, surrounded by pigment-cells and covered by translucent skin, but without any lens or refractive body' (Darwin 1866, 4th edition, p. 216). This simple two-celled device would likely be present in the last common ancestor of all bilaterians, as it is found in some primitive organisms such as the larvae of the marine ragworm *Platynereis dumerili*, an annelid. Two photoreceptor types are recognized: rhabdomeric, which is typical of insects and other invertebrates, and ciliary, which is present in vertebrates. Both types are found in ragworms, the ciliary-type, expressed in the brain, and the rhabdomeric-type, expressed in the eyes. Again, the most parsimonious explanation is that both cell types were already present in the common bilaterian ancestor of vertebrates and invertebrates, and that each lineage used a different cell type for light detection. Figure 2.10 depicts a likely phylogeny of eye development, showing that vertebrates have also co-opted the rhabdomeric photoreceptors into ganglion cells which display a key role in image processing. This is a clear example of deep homology of eye development in cell type exaptations that demonstrates that eye morphologies have not been developed from scratch in each lineage. Moreover, deep homologies are not restricted to photoreceptors but to other structures such as target interneurons involved in light processing, and also to many transcription factors in the circuit batteries of genes. Interestingly, cnidarians, the sister group of bilaterians, also have a camera-type eye with a retina that expresses ciliary opsin proteins and a full set of genes typical of vertebrate phototransduction. Since rabdhomere opsins have not been detected in cnidarians, the rabdhomeric photoreceptor may be considered a bilaterian innovation. All this information suggests that the different eyes of jellyfishes, arthropods, octopuses, and vertebrates were not invented anew and they did not evolve by convergent evolution, but, contrarily, they are the evolutionary result of deep homologies which provide the raw materials, including cell types and developmental mechanisms, already present early in animal evolution. This evolutionary process, known as 'parallel evolution', has been interpreted by 'internalists' as an alternative to 'convergence' in that, instead of natural selection being the leading force of evolutionary change, as posited by convergent evolution, the shared long history of structures and mechanisms by different lineages channels the evolution of form. Thus, most, if not all, cases of convergent evolution should be interpreted in the light of parallelism, as Gould and the advocates of developmental constraints contend.

Are there structural novelties promoted by gene regulatory networks?

When the developmental interactions of master genes were studied in detail, it became clear that they were part of small networks that included other regulatory genes. These genes encode transcription factor proteins that interact among them. We discussed above how homologous master genes, like *Pax-6*, can induce the same structure (eyes) in different animal phyla. While this was rightly taken as proof of homology and conservatism, not all

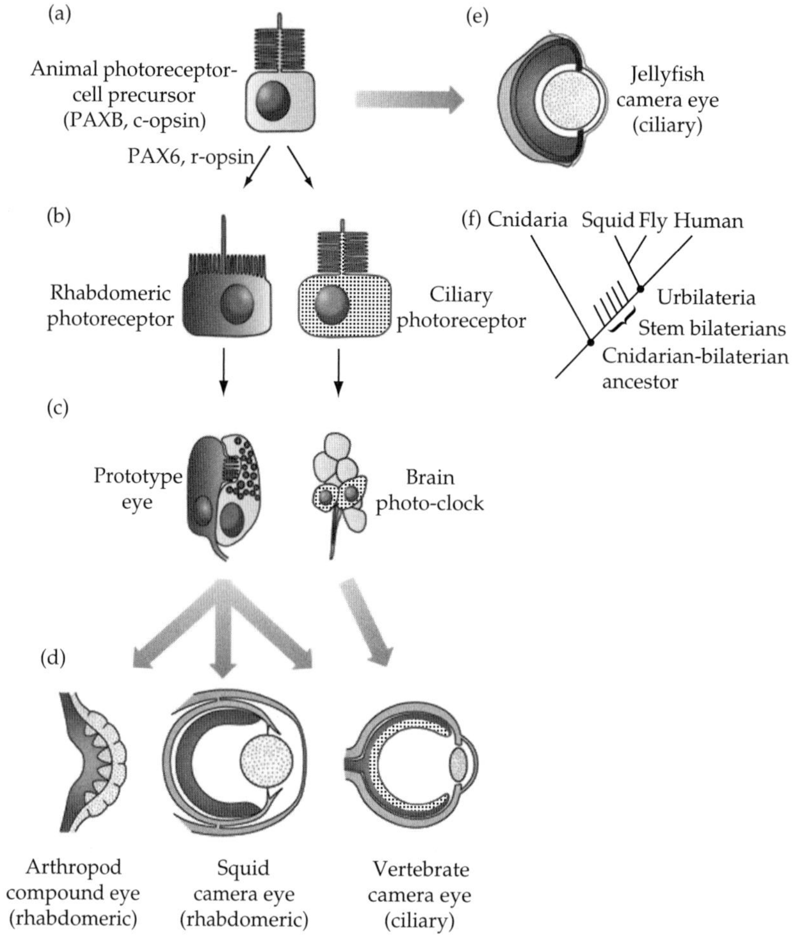

Fig. 2.10. Schematic proposed phylogeny of eye development evolution showing the deep homology of animal eyes. The cnidarian-bilaterian ancestor (a) had already photoreceptors that expressed c-opsin and PAXB transcription factors, the precursors of the r-opsin and PAX6 found in bilaterians. This ancestral photoreceptor cell diverged into two main kinds (b): the rhabdomeric (dark grey), used by insects and other invertebrates, and the ciliary (dotted), used by vertebrates (d). The path leading to this separation went through steps in which both types co-existed in the common bilaterian ancestor (c), and both are still found in polychaete annelids: with the ciliary-type opsin expressed in the brain and the rhabdomeric opsin expressed in the eye. Eventually, upon lineage separation, the rhabdomeric photoreceptor was adopted by invertebrates for light detection, but vertebrates incorporated both types of photoreceptors in the vertebrate eye (d), with the rhabdomeric photoreceptor cells (dark grey) being transformed into ganglion cells of high value for image processing. Jellyfish (cnidarians) evolved in parallel a ciliary camera-type eye (e). The tree (f) depicts the evolutionary relationships of taxa. From Shubin *et al.* (2009) with permission from Nature Publishing Group.

members of the master-gene regulatory networks are rightly homologous. Recent evidence is accumulating (summarized in Wagner 2007) that in eye morphogenesis the regulatory network differs between vertebrates and Drosophila (Fig. 2.11). For example, it is clear that in vertebrates *Pax-6* is regulated by *Rx*, but in Drosophila, the *Pax-6* homologous *eyeless* is under control of *toy*, a gene non-homologous of *Rx*. In fact, an *Rx* homologue is found in Drosophila, but it does not participate in insect eye development. This lack of correspondence in homology between elements of the eye ChINs is quite general for all the other genes.

There are two evolutionary explanations for this ChIN divergence between Drosophila and vertebrates. One hypothesizes that the gene network was already present in their most recent common ancestor and diverged later. The alternative hypothesis

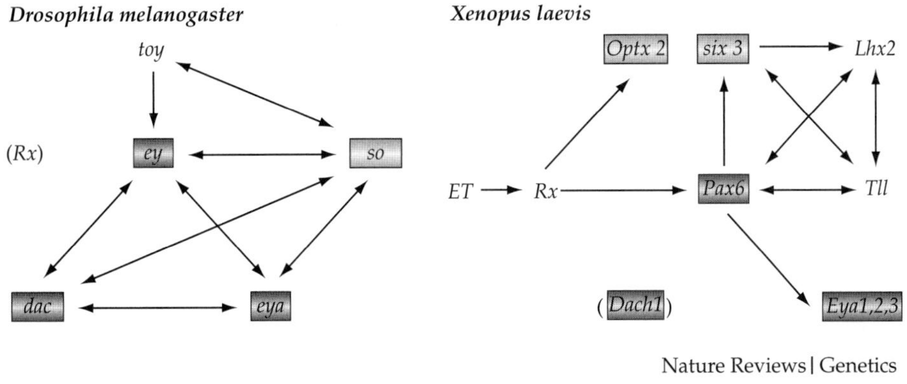

Fig. 2.11. Non-homologous GRNs control non-homologous eyes of insects and vertebrates. Note that the homeobox transcription factor gene (*Rx*), essential in vertebrate eye formation, is not involved in insect eye development. Similarly, *dac*, an essential gene in regulation of Drosophila eye, has a semi-orthologue (*Dach1*) present in vertebrates but absent in the GRN of Xenopus eye regulation. Many genes that have adopted similar functions in both GRNs, like *so* and *Optx*/*Six3*, are not **orthologous** but paralogous. The gene duplication that generated them occurred much earlier than the most recent common ancestor of insects and vertebrates. All this evidence bolsters the idea that the GRNs of insect and vertebrate eye development most likely derived independently. From Wagner (2007) with permission from Nature Publishing Group.

posits that a common gene network was not present in the common ancestor and that the vertebrate and insect lineages evolved different networks by recruiting several genes that existed in deeper time, perhaps earlier than the origin of multicellular animals. The latter hypothesis brings to mind the independent recruitment of parallel photoreceptor types that were already present in the common bilaterian ancestor (the Urbilateria). Recent phylogenetic studies with the *Six* family genes present in the eye-forming network, support the idea that **paralogous** members of the same family acting in different lineages predate the origin of multicellularity. Thus, the latter seems the most parsimonious hypothesis, which favours the idea that deeply homologous (paralogous) genes, often with diverse regulatory functions, have been independently recruited into different eye development lineages.

Most of the genes in ChINs belong to large gene families. This is particularly true in eye morphogenesis networks, but, interestingly, gene family members are also involved in the regulation of other structures, such as muscle and ear. This gene complexity implies a lot of gene duplication and divergence, but also requires an evolutionary explanation for the mechanism that keeps ChIN conserved. What maintains the network links tightly enough to specify the identity of the character? Remember how the presence of a hindwing depends on the *Ubx* expression of a specific transcription factor. This specificity has led perhaps to the false idea that transcription factors are functionally fixed, but recent evidence seems to contradict this view. Since the specificity of transcription factors does not entirely depend on DNA binding, but also on interactions with other transcription factors, it is clear that functional differences may be related to changes in parts not involved in DNA binding. Thus, it is amply documented that fly homeotic genes like *Ubx* and vertebrate *Hox* genes have usually not evolved specificities by changes in homeoboxes, but rather by changing those DNA stretches that encode protein amino acid sequences engaged in transcription factor interactions.

One experiment, among many others, that supports this view of functional non-equivalence concerns one of the most often widely qualified invariant hox genes: *Ubx*. When Grenier and Carroll (2000) compared the *Ubx* of an onychophoran species (*O-Ubx*) with that of Drosophila (*D-Ubx*), they found a high similarity (97%) in homeodomains and a great overall protein difference in sequence. Yet, although this sequence variation did not prevent transformation of antennas into legs or forewings into halteres when *O-Ubx* was expressed in flies, it could not repress the *Distal-less* in the leg rudiments as *D-Ubx* usually does. This was interpreted as the effect of differences in the non-homeodomain sequences that impede the adequate transcription factor interactions for leg development. Posterior experiments have identified protein repressor sequences. This and similar studies support the contention that transcription factors are not functionally invariant and likely evolve in a co-adapted way to be recruited in ChINs when novel characters appear.

We are still far away from understanding the large complexity of the GRNs in development. But we know some of them in certain detail. Davidson and Erwin (2006), in an attempt to define invariants in the GRNs, claim to have identified certain GRN components, coined '**kernels**', which, in their own words, 'because of their developmental role and their particular internal structure, are most impervious to change'. Kernels are similar, though non equivalent, to the ChINs, as explained by Wagner (2007). In brief, a kernel is '. . . [a] conserved subcircuit consisting of genes which interact with one another and which are dedicated to a specific developmental function', as defined by Davidson. On the other hand, Wagner defines ChINs as 'GRNs that subscribe the execution of a character-specific developmental programme, regardless how old they are'. In both cases the attempt to understand the invariance associated with some developmental programmes is basic to the whole endeavour. But, while kernels are specifically addressed to the conservation of phyletic body plans (phyla), ChINs are directed to understanding how tool-kit genes are wired and interact to specify new evolving characters.

Both views are related to 'positive constraints' and to their implications in the controversy between 'independent adaptations by convergence or similar solutions by constraints of parallelism', in Gould's words (Gould 2002, p. 1128). Both extreme positions, however, are likely to be untenable under the current wisdom provided by Evo Devo. Parallel evolution is probably a more acceptable proposition than strict convergence (Box 2.3), but the capabilities of natural selection to channel the evolutionary

Box 2.3 What is parallel evolution? Or are fish fins and tetrapod limbs homologous in their distal regions?

Homology, in the most inclusive definition, refers to a historical continuity (i.e. descent) in which characters in different species are derived from the same character in their most recent common ancestor (MRCA). Despite the fact that homologous characters often show similarities in morphology and function, this is by no means universal. Striking evolutionary changes in morphology and function have traditionally prevented recognition of homologous features. This has been rather common for morphological characters, as in vertebrate middle-ear ossicle evolution. In this case we say that old structures (jaws) have been co-opted to new adaptations (middle-ear ossicles), coined as 'exaptations'.

Recognition of difficult homologies has been greatly assisted by studying similarities in embryological features when adult characters differ strikingly. In the last 15 years, however, advances in Evo Devo have revealed a lot of unanticipated patterning mechanisms common to structures not considered homologous on any morphological grounds. Many of these mechanisms comprise more than specific tool-kit developmental genes; they involve complex regulatory circuits already present in ancient common ancestors. Carroll and his co-workers (Shubin *et al.* 1997) coined the term 'deep homology' to describe this sharing of ancient genetic regulatory apparatus.

One of the most challenging topics in evolution is to distinguish what is new from what is ancient. For centuries, limbs in four-legged vertebrates (tetrapods) have been considered a classic evolutionary novelty. Limbs of tetrapod species also exemplify homology because of their similarity in bone morphology and development. Yet, the comparison between tetrapod limbs and paired fins of fish, which are apparently related by descent, has always been a controversial issue. Limbs and fins show a similar pattern of proximal bones, but they differ strikingly in their distal region (the autopod). While unjointed dermal rays support the surface area of fins, limbs are supported by endochondral jointed elements like digits and ankle or wrist bones.

The classic view that autopod bones have no correlate in fins was bolstered by Evo Devo studies in the 1990s. These studies revealed that some developmental genes, like *Hoxd9–13* and *Shh*, show a late expression in the distal zone of tetrapod limbs. In fishes studied at the time this late phase

continues

Box 2.3 *(continued)*

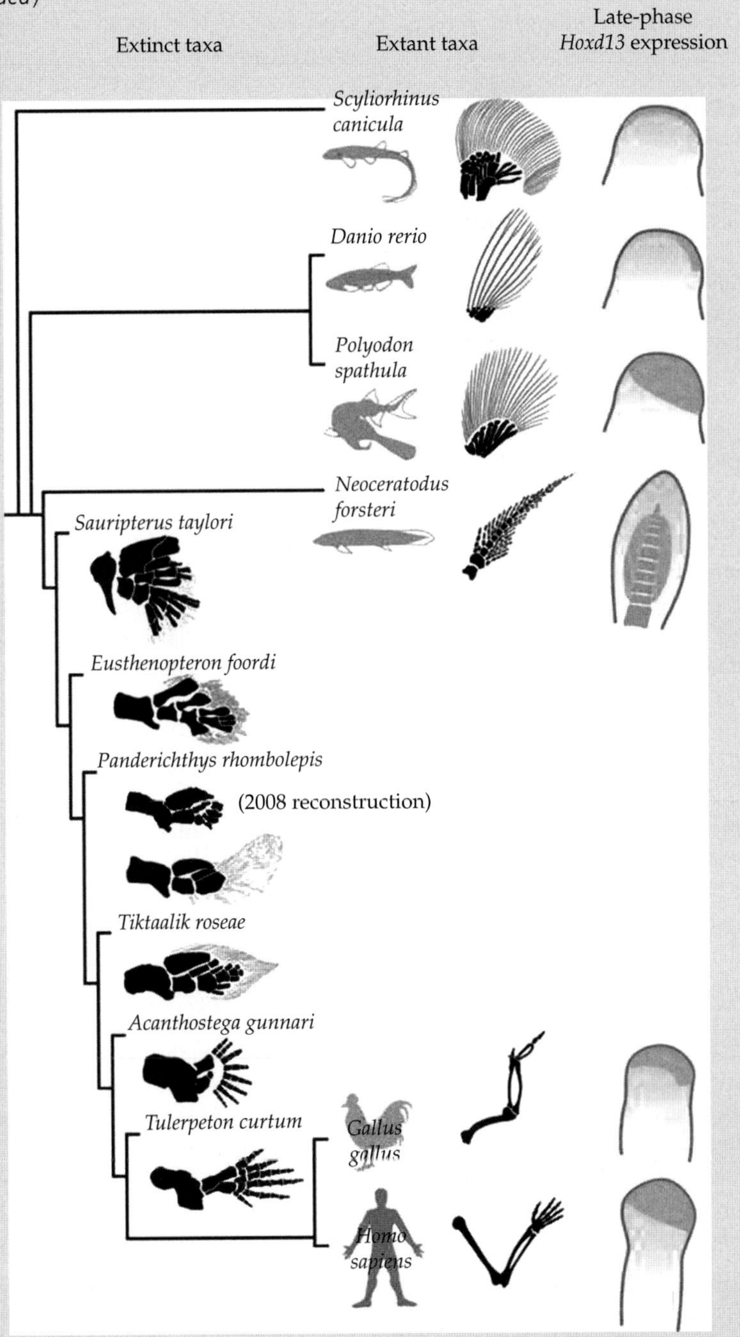

Fig. F. This phylogenetic tree shows the pectoral fins of extant and fossil fish taxa with known cases of late-phase expression of *Hoxd13* (in grey shading). In tetrapod autopods the same expression is also shown, which suggests deep homology between tetrapod limbs and fish fins. From Shubin *et al.* (2009) with permission from Nature Publishing Group.

continues

Box 2.3 *(continued)*

of expression was not found, since these genes were only expressed in the proximal zones of fins. The fact that the fossil sister group to extant tetrapods, the panderichthyids, lacked digit-like bones, supports the idea that the origin of the morphological novelty (the autopod) was paralleled by a genetic novelty. Yet, in the present decade, work on new fish species revealed that a late phase of expression is indeed present in the distal zone of fins of primitive fishes (Fig. F shows late-phase expression of *Hoxd13*. From Shubin *et al.* 2009). There are differences though. The most relevant is that late-phase expression in the fins spatially overlaps with earlier phases, something that does not occur in the limbs. But the key point to decide whether autopods are true novelties relies on the independent regulation of the tetrapod autopod. This is, as yet, unknown.

Interestingly, new fossil findings have led to the discovery of *Tiktaalik roseae*, the closest lobe-finned fish to tetrapods, whose pectoral fin shows a high reduction of its dermal skeleton and an expansion of its distal endoskeleton with incipient joints.

One way or the other, even if fin rays and autopods are not homologous, the more we probe into the regulatory patterns of expression, the more evidence we find of deep homologies in *Hox* gene networks and their targets. This may apply to autopod evolution as well as to other 'supposed' novelties like eyes (see text). Again, the basic tenet that natural selection never works from scratch is strongly reinforced by new studies of deep homologies. In this case it is not the ancient morphology that reveals the homologous ascendency but the gene regulatory circuits, long established since ancient forms, whose deployment underlies the evolutionary progress. This is what has been traditionally termed 'parallel evolution', presently uncovered by the power of developmental genetics.

trends cannot be wholly obscured by the constraints imposed by any historical developmental network. After all, it would be preposterous to forget that natural selection never works from scratch and always tinkers with the raw available materials to advance, no matter how little, the immediate adaptation. This applies to microevolution as well as macroevolution if we accept that the evolution of form has a long Precambrian history, in which many novel development circuits were built up by co-opting many deep homologous regulatory genes, as explained for *Pax-6* and *Six* gene-including networks, and that this process has been repeated ever since then.

Darwin redux

No sensible person would deny that the wealth of knowledge provided by the Evo Devo and other genomic studies accumulated in the past decades draws a pattern wholly consistent with the unity of type. Only those prejudiced by ignorance or intuitive non-rational ideas can still assert that this pattern is not compatible with the Darwinian theory of descent. But the pattern (unity of type) must be distinguished from the mechanism that produces it. Darwin's most celebrated innovation was to propose, and to demonstrate, a mechanism (natural selection) that bridged the gap between the unity of type, the cause of homology, and the conditions of existence, the cause of adaptation. How can organisms with similar body plans and structures adapt to different conditions of existence (habitats)? Should not these diversely adapted organisms show also diverse body plans and structures? These, and similar questions underlie many evolutionary debates in pre-Darwinian times. Darwin stated clearly this contentious issue in *The Origin of Species* (Darwin 1859, p. 434): 'What can be more curious than that the hand of a man, formed for grasping, that of a mole for digging, the leg of the horse, the paddle of the porpoise, and the wing of the bat, should all be constructed on the same pattern, and should include the same bones, in the same relative positions?'. But he immediately provided a solution to this riddle on the next page: 'The explanation is manifest on the theory of the natural selection of successive slight modifications—each modification being profitable in some way to the modified form, but often affecting by correlation of growth other parts of the organisation. In changes of this nature, there will be little or no tendency to modify the original pattern, or to transpose parts.' It is clear that Darwin had this explanation in mind almost two decades earlier when he wrote in his 1842 Essay the sentence quoted at the start of this chapter.

Before Darwin, the idea of organic adaptation, although on a local scale, was not alien to anatomists like Cuvier, and the primacy of adaptation versus that of unity of type to explain the organic order was a hot and contentious issue in scientific debates. Although Darwin's theory of descent with modification seemed to put an end to this opposing duality in favour of adaptation, the debate has never completely faded away. In fact, there is a long European tradition, popularized by adherents to the Goethe´s Natural Philosophy (*Naturphilosophie* in German) and followed by anatomists like Geoffroy Saint-Hilaire and Owen, that praised the primary role of the laws of form (laws of growth in Darwin's term) to explain the natural order. Yet, this primacy was negated by Darwin when he asserts that 'the law of the Conditions of Existence is the higher law; as it includes, through the inheritance of former adaptations, that of Unity of Type' (Darwin 1859, p. 206). Interestingly, Darwin never denied the role of the unity of type as a source of deep homologies, but he considers it as a law separate, and subsidiary, from the mechanism of natural selection. Gould clearly dislikes this secondary role of the law of form when he complains that 'modern constraint theorists, myself included, balk at Darwin's resolution because his argument demotes a large chunk of biology to a chink in a corner' (Gould 2002; see pp. 251–60 for a longer discussion).

The preceding sections have presented an overview of how the unity of type is bolstered by current knowledge in developmental biology and how this insight has helped us to understand the origin of large body plans in the light of the fossil record. Some developmental constraints, however, can be released experimentally, which gives us an insight into their putative adaptive value. While deep homologies are much more common than anticipated, and they contribute to the explanation of evolutionary constraints, the Darwinian ideas on the primacy of adaptation by natural selection versus evolution by the laws of form seem to be also supported by the creation of complexity at lower taxonomic levels. The evolution of fish pelvic fins and insect wing patterns are among the best worked processes at these levels. These topics are discussed in the following sections, but let me first discuss how development can also help to deconstrain the evolution of form.

How developmental constraints may deconstrain development

Perhaps the most acid criticism of the primacy of natural selection in the evolution of form is that natural selection has been unable to change the function of master genes. This may apply to the ubiquitous, exchangeable ability of *Pax-6* homologous genes to induce complex eyes in different phyla. Though this conservatism is not true for all master genes that are involved in eye, and other development networks (see above the ChIN concept), we also know that natural selection preserves gene structures that function well as much as it promotes novel genes. In selector gene networks once a gene, or subcircuit, regulates several other genes, natural selection's best strategy to open new developmental pathways may be to keep the upstream gene and change the regulatory sequences of regulated genes or to incorporate (co-opt) new genes.

Besides this regulatory evolution, largely documented, some authors have proposed that natural selection has also favoured other molecular and cell properties that facilitate the origin of developmental variability. Kirschner and Gerhart (2005) have coined the term 'facilitated variation' for this theory. Traditionally many evolutionary biologists have underestimated, according to these developmental biologists, the range of phenotypic variation (dubbed the reaction norm; see Box 5.2) from conserved processes (the genotype) as a significant factor to generate new variation of evolutionary value. They argue that an organism's capacity to evolve (i.e. their 'evolvability') depends on its ability to generate phenotypic variation in response to genotypic variation, including suppression of deleterious mutations as well as the quantity and quality of phenotypic change (see Chapter 5). Evolvability, thus defined, is largely based on the ability to link together conserved core processes that arose in evolutionary times, including, in historical order, basic mechanisms of DNA, RNA, and protein synthesis; functions of intracellular membranes and cytoskeleton in the eukaryote cell; functions of junctions and the extracellular matrix in metazoan; the role of the *Hox* genes in bilaterians; and developmental programmes such as limb formation in land vertebrates. Since the Cambrian era, most metazoan evolution has come from changes, through regulation, in the deployment of the unchanged core processes.

Three properties are considered fundamental to the theory of facilitated variation: weak linkage, exploratory behaviour, and compartmentation. Weak linkage refers to the fact that protein interactions are weak and indirect. Proteins are built by modules, some modules have regulatory properties and other modules are involved in functional activities. Regulatory modules can change without affecting the functional modules, and vice versa. Jacob and Monod pioneered the discovery of this dual protein activity in their model of bacterial enzymatic regulation; a property they dubbed 'allostery' (Box 2.4). But in multicellular organisms regulation must provide a specific set of proteins at a specific time of development for each of hundreds of cell types (more than 300 in humans) that build their body. The evolution of complex interactive regulatory processes, implying a wealth of regulatory factors, included in many intricate regulatory circuits, has been the workable solution to metazoan development.

However, contrary to the rigidity of prokaryotic regulation, in multicellular organisms the interactions between regulatory factors and DNA binding sites are less strict. The regulatory DNA sequence that surrounds a typical eukaryotic gene can stretch up to thousands of kilobases and include dozens of regulatory protein-binding sites (Fig. 2.12). This multiplicity of signals confers by itself a combinatorial flexibility (discussed above) that translates into time- and position-dependent changes in RNA transcripts. But the localization of regulatory factors in DNA binding sites is also much less precise than in bacteria, and some do not even have to touch the transcriptional machinery; they just interact by bringing enzymes that alter the structure of other regulatory proteins surrounding the gene. In sum, the use of alternative sites for regulation and for binding (the allostery concept) has allowed regulatory sites to evolve by weak linkage without touching the constrained and precise functional binding site. Moreover 'continual selection for the retention of weak linkage facilitates the generation of phenotypic variation and deconstrains the selection for new functions and new regulatory connections', quoting Kirschner and Gerhart (2005, p.133).

The second property of facilitated variation is exploratory behaviour. To illustrate this behaviour, think about the complexity of our neuronal web. How could the 24,000 genes of our genome encode millions of neurons and their fine connection pattern? Or, how could this complex neuronal web come to serve all the musculature in a complex organism? This is an example of the more general concern, raised by Darwin (and also by Paley) and

Box 2.4 The 'transvestite' life of regulatory proteins facilitates developmental evolution. Or what is allostery?

Prior to 1965, a protein was thought of as a molecule that had only a single kind of site to interact with other molecules for carrying out a chemical reaction. This site, dubbed the fixation site, reversibly unites to another molecule named the **ligand**. Ligands can be any kind of molecule, including other proteins. For instance, in the oxygen transportation by the blood haemoglobin the ligand is the oxygen and the fixation site is the hemo group in the haemoglobin that contains iron that binds to oxygen. Enzymes are a special class of proteins that transform (metabolize) other molecules, which are named substrates instead of ligands and their fixation site is known as the active site. Haemoglobin is always present in the blood to perform oxygen transportation, so the intuitive corollary suggests that enzymes should also be permanently present in the cell. However, this educated guess turns out to be astoundingly wrong—the cell only produces the appropriate enzyme when the specific substrate to metabolize is present.

Known by some bacterial biochemists since the early days of the twentieth century, this 'enzyme adaptation', as it was referred to by them, was unknown to Jacques Monod, a French molecular biologist at the Pasteur Institute in Paris, who was interested in the metabolic physiology of *Escherichia coli*, the fashionable model organism at the molecular biology labs of the time. He centred his research

continues

Box 2.4 (*continued*)

on the enzyme beta-galactosidase, dubbed lactase because it metabolizes the milk sugar lactose to allow bacterial growth. He observed that lactase synthesis was induced by the presence of lactose. Monod hypothesized that the cell may continuously synthesize an inhibitory protein, the repressor, which prevents lactase production by binding to the DNA region that encodes the enzyme. This bind would normally impede the lactase gene expression, but when lactose becomes available, his hypothesis followed, the sugar would bind to the repressor, nullifying its ability to bind DNA and releasing the inhibition of lactase production (Fig. G). The hypothesis was completely confirmed by Monod's experiments and opened up a new era in regulatory physiological biology that later had a strong influence on the understanding of the development of multicellular organisms.

Monod's hypothesis, known as the derepression model, posits that there must be proteins that exist in two alternative forms, mediated by ligands that change their molecular

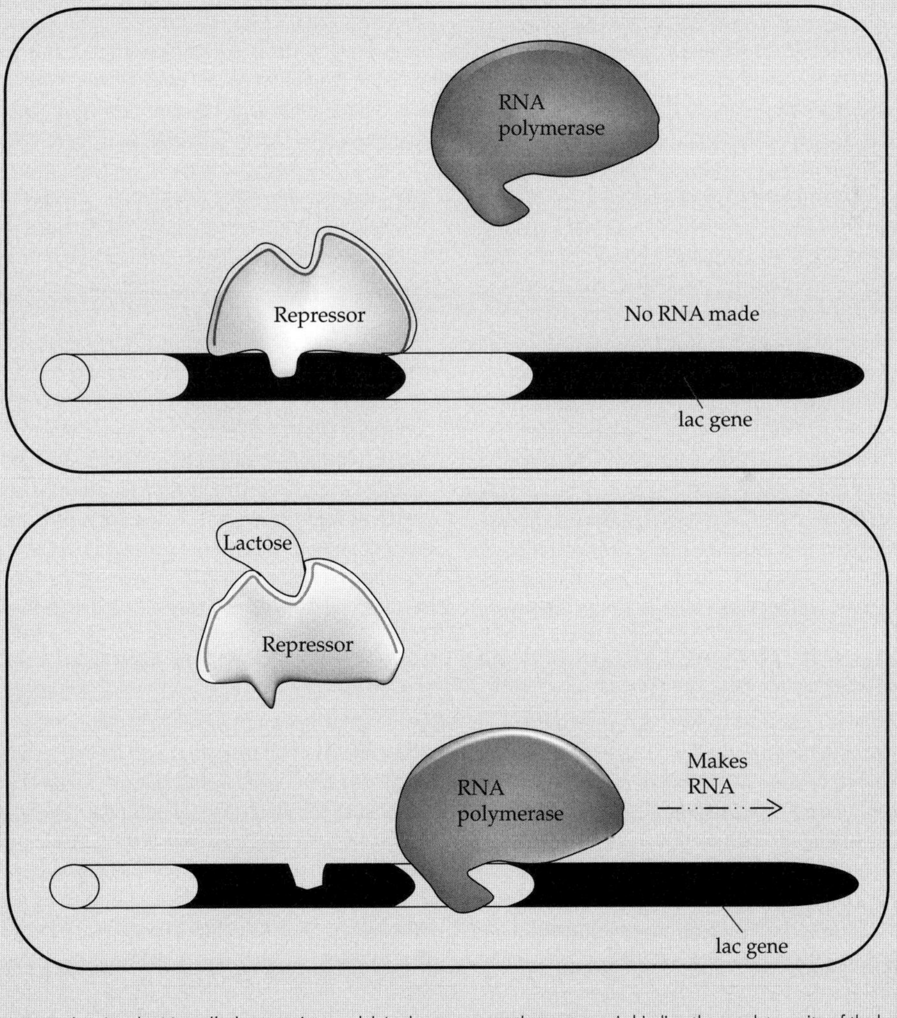

Fig. G. Diagram showing the Monod's derepression model. In the upper part the repressor is binding the regulatory site of the lactase gene, impeding lactase production. When lactose becomes available (bottom part) it then binds to the repressor, changes its form, and nullifies its ability to bind DNA, releasing the inhibition of lactase production by allowing the RNA polymerase binding and the ensuing lactose gene expression. Redrawn by Montserrat Peiró, adapted from Kirschner and Gerhart (2005), with permission from the illustrator, John Norton.

continues

Box 2.4 *(continued)*

conformation and their ensuing properties. Namely, the binding of one ligand (lactose in this case) in the regulatory site changes the binding capacity of another site to its ligand (the DNA). These double-form proteins are named 'allosteric' (from the Greek *allos* meaning other, and *stereos* meaning solid or form). These proteins can be not only repressors in the regulatory-free state, but may also be activators in this state, as François Jacob, Monod's colleague and co-discoverer, speculated. The work of both researchers elegantly solved the physiological adaptation to substrates in bacteria by introducing the concept of the genetic switch that turns the gene on and off through the regulatory binding of repressors or activators to non-coding DNA regions, in cis-flanking positions.

Jacob and Monod's insight further transcended metabolic regulation, delving into the intricate field of multicellular development. They were perfectly aware that regulatory sites can interact with any molecule, substrate or not, that impinges in any regulatory process. Freeing the regulatory site from the primary functional site (the catalytic site in enzymes and the DNA binding site in regulators) allows the independent evolution of the regulatory part of a protein without the need for evolving new functional proteins responding to internal or external conditions. Since in multicellular organisms the crucial process in development is a stepwise linked programme of cell expression signals from the single-celled zygote to the whole adult, the connection between steps could be regulated by changes in regulatory sites. Interestingly, complex organisms have evolved different development circuits (see GRNs in the text) that can also be easily connected because the regulatory proteins (including transcription factors) produced at each step are independent of what functional proteins (often also regulatory proteins) are produced at other steps. Moreover, these regulatory connections, due to the multiple network pathways and redundancy of circuits, generate a web of weak linkages that can easily be evolutionarily changed and redirected for other developmental operations.

An example of regulatory sequence versatility is provided by comparing cis-regulatory element (CRE) regions of the *even-skipped* gene, itself a transcription factor involved in segmentation, between closely related Drosophila species. Figure H shows in section (a) the schematic of the CRE involved in expression in the second stripe in the embryo, depicting binding sites for four transcription factors: Bicoid (Bcd), Hunchback (Hb), Giant (Gt), and Krüppel (Kr), the first two of which upregulate transcription, while the last two downregulate, or repress, transcription. Section (b) shows the conservation and evolution of the 17 transcription binding sites. Michel Ludwig, Martin Kreitman, and colleagues examined the sequence of these binding sites in five Drosophila species and found that, albeit that expression is always correct, only 3 of the 17 binding sites are conserved across species. They then constructed chimeric DNA sequences combining binding sites from different species, which were transferred to *D. melanogaster* flies. These transgenic flies did not develop the correct even-skipped stripe. Thus, these experiments suggest that the regulatory connections between transcription factors are quite fluid, because they are permissive of important changes in CRE DNA sequences without changing the time and place of expression of developmental genes. Moreover, mutational changes in CREs are not neutral because chimeric constructions did not work in transgenic flies, suggesting that there is strong selection inside each species to maintain the developmental function.

The easiness of conservatism in physiological function by the weak linkage behaviour of regulatory mechanisms has its counterpart in the facilitation of rapid evolution of regulation. Then, if selective conditions change, for instance through the appearance of new ecological niches, mutational changes in regulatory sequences could be positively selected, promoting new patterns of expression without large changes in the underlying complex and specific developmental circuits. The more the complexity of regulatory networks progresses, the higher is the facilitation of the evolution of form because the linkage between signals is weaker. As stated by Kirschner and Gerhart (2005, p. 142) 'ancient processes such as signal transduction and transcription are constrained to change, but they readily allow regulatory changes that alter the interaction of the conserved components with inputs and outputs'. 'This freedom to link inputs and outputs', they continue, 'is a significant form of deconstraint.' This reasoning, pioneered by Jacob and Monod's insight and pinpointed by recent empirical work, should convince, or at least induce reflection in, sceptics of the involvement of Darwinian **gradualism** in phenotypic evolution. In sum, there is no need for multiple large and complex (hopeful) mutations; only small regulatory changes are needed for the evolution of form.

continues

Box 2.4 *(continued)*

(a)

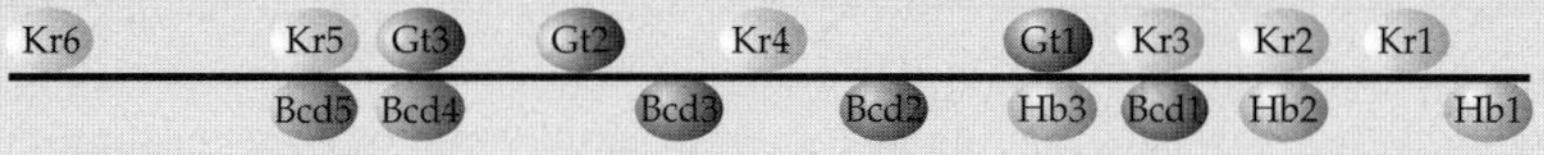

(b)

Conservation of Transcription Factor Binding Sites in Eve Stripe 2 Enhancer

Sites

Species	Kr 1	Kr 2	Kr 3	Kr 4	Kr 5	Kr 6	Bcd 1	Bcd 2	Bcd 3	Bcd 4	Bcd 5	Hb 1	Hb 2	Hb 3	Gt 1	Gt 2	Gt 3
D. simulans	P	P	P	P	P	P	P	P	S	S	P	S	P	P	P	P	S
D. yakuba	P	P	S	P	P	P	S	P	W	S	P	S	S	P	P	S	S
D. erecta	P	P	S	P	P	P	S	P	S	S	P	S	S	P	P	S	S
D. pseudobscura	S	S	S	S	P	P	S	S	A	S	P	A	P	S	S	W	S
D. picticornis	W	S	W	S	P	P	W	S	A	S	P	A	S	W	W	W	W

Conservation
P = perfect
S = strong (1–2 changes)
W = weak (3 or more changes)
A = absent

Fig. H. Regulatory sequence versatility in the evolution of the *even-skipped* gene in Drosophila. (a) Shows a scheme of the CRE involved in the second stripe bound by four transcription factors Bcd, Hb, Gt, and Kr. (b) Shows the conservation and evolution of the 17 binding sites among *Drosophila* species. The degree of sequence conservation within each site is indicated (P, S, W, A). Note that sites Kr5, Kr6, and Bcd5 are perfectly conserved, whereas other sites (Hb1 and Bcd3) are absent from some species. From Carroll *et al.* (2001) with permission from Sean B. Carroll.

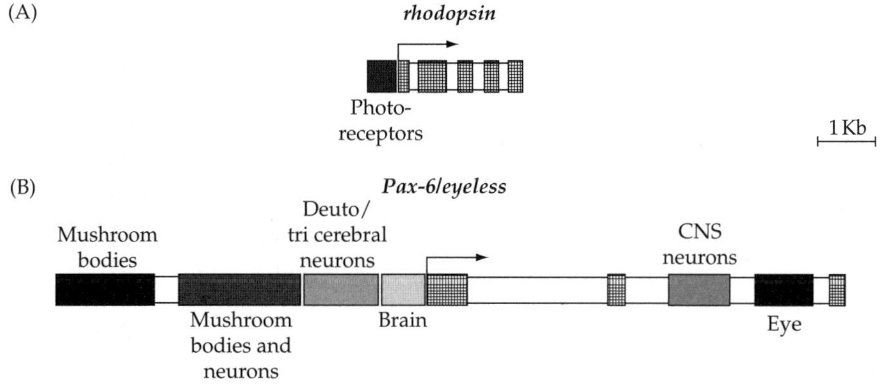

Fig. 2.12. Comparative cis-regulatory element (CRE) structure between locus encoding cell-type-specific proteins acting in physiological processes and a pleiotropic tool-kit locus. (A)The rhodopsin **locus** in Drosophila is typical of the former. It encodes a protein that is expressed in photoreceptor cell types. The cell-type expression specificity is regulated by a single CRE element. (B) The *Pax-6/Eyeless* (*Ey*) locus is representative of a tool-kit gene, which is required not only for eye development (it binds to rhodopsin CRE) but also for the development of the brain and the nervous system. In contrast to the rhodopsin single CRE, the *Ey* CRE architecture encompasses six CREs, totalling about 7 Kb in size, each one driving expression in a specific tissue or cell-type-eye, embryonic, larval, and adult brains, and the central nervous system. Exons are depicted as mesh rectangles and CREs as rectangles in distinct grey tones. From Carroll (2008) with permission from Nature Publishing Group.

many evolutionists, about the need for simultaneous changes in many systems to produce fitted novelties. Even the highly worked eye evolution discussed above requires that changes in muscle, vascular, nervous, and light-sensitive cell systems act in concert at each step towards evolutionary complexity. While large steps in evolution have been discarded due to their implausibility for producing viable phenotypes within the physiological (and morphological) adaptive range, even if the steps were small, how small should be and how many of them should be taken simultaneously in different cell systems to afford a stepwise increase in fitness? These requirements, though wholly possible to neo-Darwinians, have been argued as an anti-Darwinism stronghold by irreducible complexity advocates. Yet, we have seen that weak linkage and compartmentation are two properties that facilitate the evolution of complexity, which is explained perfectly by the gradual building of GRNs.

But what is the exploratory concept and how does it facilitate evolution? Going back to neurone development, we know that the cell bodies of neurons that control muscle activity are located in the spinal cord from where they send axons to the peripheral muscles. During embryo development, neurons explore, by means of the growing axon tips, every muscle cell they encounter. Initially several neurons contact each muscle cell, but eventually functional competition assures that each muscle cell is innervated by only one neuron. This is the way that the nervous system is wired, suggesting that this basic process does not need to be modified when new evolutionary novelties arise in the musculature to keep the organism adapted. A similar exploratory behaviour occurs in the development of the vascular system or the microtubules that shape cells, and any other system composed of parts that supply a structural or physiological component.

Exploratory behaviour is widely beneficial not only for adapting to genetic developmental changes but also for the normal range of phenotype variability. This can be shown in the muscular development of vertebrate limbs. Precursor vertebrate muscle cells migrate from the trunk close to the nerve cord outward into the neighbouring appendages. They explore and associate with cartilage and bone regardless of limb structure. Analogously, nerve cells explore and associate with muscles, and also migrating cells sent by the vascular system generate vessels whose number and calibre depend on muscle activity and oxygen consumption. All these concerted changes are controlled by signals produced in normal development, but they can be also elicited by artificial interventions like exercise training. Thus, in evolution of limbs, even if genetic changes only affect the bone structure, the nerve, muscular, and vascular systems can readily adapt by exploration without the need for simultaneous genetic changes in these systems. Once adaptation is assured there is time for subsequent evolutionary changes in these systems to occur. Exploratory processes facilitate evolutionary change because they deconstrain it from simultaneous occurrence in several systems, but also because their plasticity blunts the deleterious effects of mutation. Moreover, the ubiquity of the exploratory behaviour is likely favoured by natural selection because of its facilitation of evolvability.

Last but not least, compartmentation or modularity, the ability of multicellular organisms to express genes or gene circuits in some body compartments but not in others, is perhaps the most studied property of development. Its mechanism has already been delineated in previous sections. Here I want to emphasize its power to generate variation by bypassing the deleterious effects of pleiotropic random mutations in phenotypes.

Form evolution by compartment building

Traditionally two basic questions have undermined the explanatory hypotheses of development: what are the signals that tell each cell at each embryo position what type of cells it must differentiate into; and how can these enormously different kinds of signals be encoded by the limited number of genes in the genome? The **compartment** theory has recently significantly helped in answering both questions.

Curt Stern, one descendant of Morgan's Drosophila group, worked with the engrailed fly mutant that showed a number of morphological abnormalities, including changes in the bristle pattern that gives an engrailed thorax appearance. Most importantly, it also alters wing and leg morphology, changing the posterior parts to resemble the anterior parts as homeotic mutants do. By inducing patches of mutant cells in the developing leg, Stern showed that if the patch was on the anterior part of the developed leg then the bristle pattern was normal, but if the patch was on the posterior part the pattern was a mirror image of the anterior pattern. This was one of the first hints that the developing fly body could be divided by invisible compartments, involving specific expression patterns, underlying the visible detailed morphology.

The first significant breakthrough in the concept of developmental compartments, however, was made by Antonio Garcia-Bellido, a Spanish geneticist working in Madrid (Spain) at the Centro de Investigaciones Biológicas. In an attempt to understand the earliest episodes of pattern formation, he decided to work with Drosophila imaginal discs, small bags of larval cells that develop into adult structures like eyes, wings, or legs. By marking one of the 15 embryonic cells of the wing disc, he and his colleagues found that the descendant patch of adult cells occupied either the anterior or the posterior half of the wing, but they never crossed an invisible border separating both halves. This border, which does not coincide with any anatomical boundary, separated, as Garcia-Bellido suggested, two compartments of wing development in which the *engrailed* gene was differentially expressed only in the posterior compartment. This and later experiments with other mutants 'were the bases for the notion of selector genes' as stated by Garcia-Bellido (1998), adding that 'systemic transformations, like these homeotic ones, affecting individual cells, meant that the abstract specification of whole cell territories (as to segment or compartment) resided in developmental operations carried out by the individual cells'.

Remember that the Drosophila body is divided into 14 anatomical segments. Now we know that each segment has an anterior and a posterior compartment and that *engrailed* is expressed in the posterior compartment. *Engrailed* acts as a selector gene that differentiates each posterior compartment. In fact, segments often differentiate from one another by developing different structures, and this is achieved by specifying new smaller compartments where specific selector genes are expressed, as we have discussed above (see p. 42 and Fig. 2.5). The first hints to the identity and mapping of selector genes, later dubbed *Hox* genes, were found by Edward Lewis with homeotic mutants at the California Institute of Technology in Pasadena, California. Lewis (1978), a Nobel prize winner, fuelled the posterior molecular analysis of *Hox* genes in the two last decades of the twentieth century, sketched above. Now we know that differentiation starts by maternal substances in the egg that set the stage for future compartmentation through a programmed process of selector gene signalling. At the beginning, overall signalling is rather evolutionarily flexible for each phylum (see Box 2.2), but soon, after the embryo has divided into several thousand cells, compartments appear to generate a pattern that is highly conserved inside a phylum. This is the phylotypic stage that makes all embryos look alike, providing a true definition of type independent from the circular criteria of anatomical resemblance. After the compartments are established the evolution inside each compartment becomes flexible, generating the high diversity we observe inside each phylum.

Although the concept of compartment is rooted in the cell lineage concept and strictly applies to arthropods, and perhaps to some sections of the vertebrate architecture like the rhombomeres, the compartment-like patterning observed in the expression domains of selector genes across phyla allows us to establish evolutionary relationships among phyla: an endeavour long sought by Darwinians to counter the anti-Darwinian argument of independence among phyla. For instance, hemichordates are worm-like animals the anatomy of which, studied in the nineteenth century by Bateson and Morgan, tenuously resembles that of chordates: they have gill slits but lack a brain and central nervous systems. But now, more than a century later, the study of expression domains of selector genes in hemichordates reveals that they keep the same pattern as in chordates. Even the hemichordate anterior domains, which develop no brain, show a high similarity to the chordate forebrain and midbrain domains. Figure 2.13 sketches this close similarity and also a significant similarity to arthropods. Who could, after

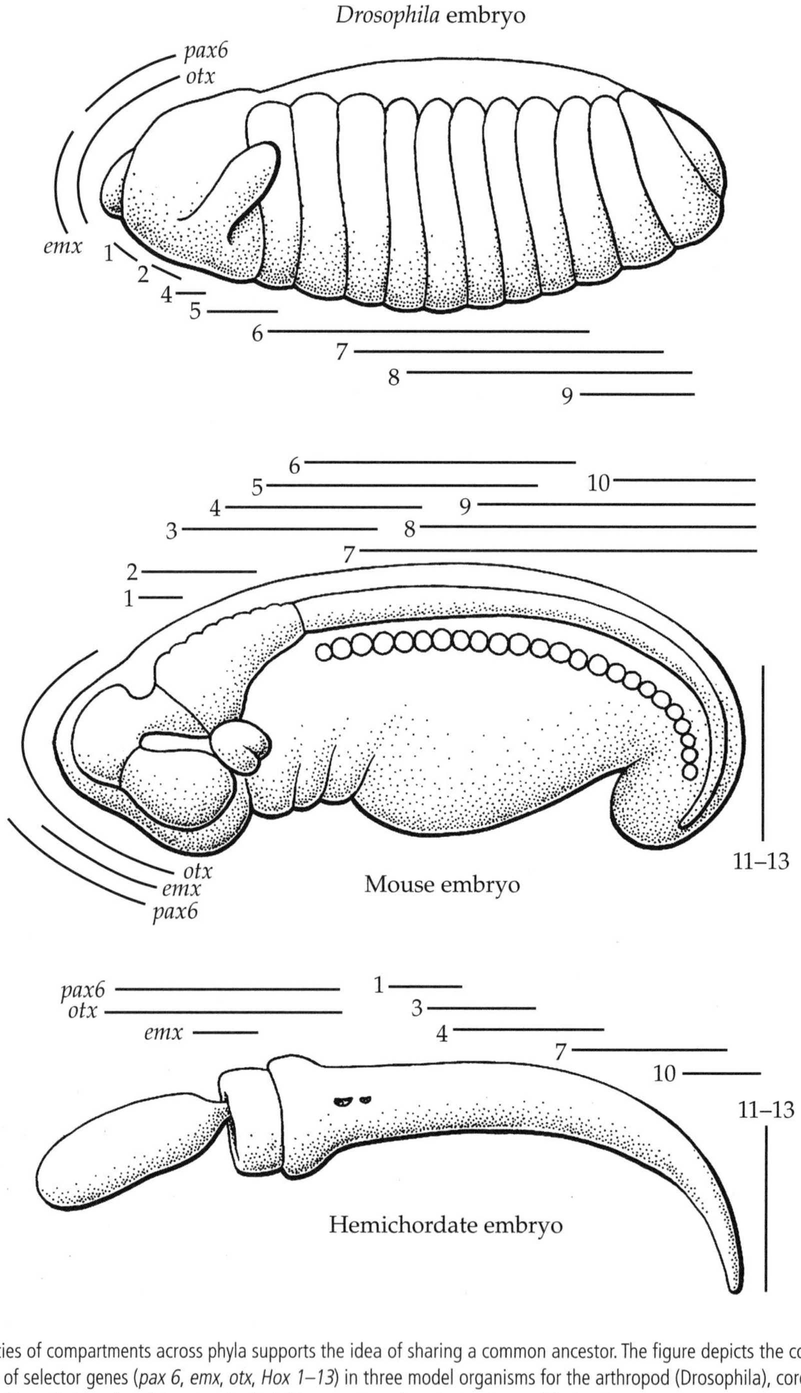

Fig. 2.13. Similarities of compartments across phyla supports the idea of sharing a common ancestor. The figure depicts the compartment map of expression domains of selector genes (*pax 6, emx, otx, Hox 1–13*) in three model organisms for the arthropod (Drosophila), cordate (mouse), and hemichordate phyla. Note that hemicordates, despite their bizarre external anatomy, share with chordates a conserved compartment body plan along the A–P axis. Even though hemichordates lack a head, they show a compartment pattern in their anterior body extremely similar to that of chordates. From Kirschner and Gerhart (2005), with permission from the illustrator, John Norton.

this evidence, negate that hemichordates and chordates had a common ancestor earlier than 500 mya? The same question could be posed for the common ancestor, though even more ancient, of chordates and arthropods.

Once the power of functional compartmentation to define the unity of type is well established, we may turn again to its implications for the processes that channel evolution. The extreme conservation of the phylotypic compartment map could be taken as proof that development constrains future evolutionary avenues, as Gould would like to argue. Yet, again, a deep understanding of how natural selection acts supports just the opposite view, namely that compartmentation facilitates the available variation for evolution. This is so because after the compartments have been established (the postphylotypic stage) each compartment shows large flexibility. Inside each compartment practically any kind of development can occur by using the same core processes in different combinations, amounts, and times. Compartments are deconstrained because within each one target genes can be used differently by the regulatory influence of compartment-specific selector genes. Besides, compartmentation allows an uncoupling among modules, which means that structures evolve in each module independently from others. This mitigates the **pleiotropy** problem, that is, the conflicting effect of those mutants whose action could be beneficial in one region and deleterious, or even lethal, in another. By subdividing the body into independent modules, mutations have only localized effects, eliminating this negative trade-off of overall-effect mutations in a non-subdivided embryo. In sum, developmental constraints imposed by compartmentation are not constraining future evolution; on the contrary, they deconstrain evolution by facilitating variation that can be used to generate new adaptations through natural selection. Kirschner and Gerhart (2005) capture the point when they assert that 'we should view the compartment body plan as maintained in evolution by selection'.

Darwinism revisited: neither chunk nor chink

We fruit fly evolutionists are often perceived as odd scientists by the general public, including educated people. Admittedly, working with an inconspicuous, small fly does not seem to have the same glamour as that associated with naturalists that study evolution in fancy organisms like lions, elephants, or gorillas. I am not going to enumerate here the obvious advantages that Drosophila exhibits for evolutionary studies, including easiness (and cheapness) of rearing in the laboratory and its extremely well known genetics. All these properties of the fly as a model organism are well known to scientists and are easy to adduce as proof of its scientific appeal. What many people, some scientists included, do not know, is that Drosophila is a genus that comprises more than 2,000 described species, many of them showing the most extraordinary morphologies which depart from the 'dull' *D. melanogaster* appearance (the fruit fly's scientific name).

Amongst Drosophila species, a group of over one hundred species inhabiting the Hawaiian archipelago, known as the 'picture wing', is famous mainly for the vivid wing pattern of its members. As a population geneticist working with *D. engyochracea*, a picture wing species, I had the happy opportunity to become fascinated by the colourful world of these flies. After spending several days collecting flies from dawn to dusk at the tropical forest patches (dubbed 'kipukas' in Hawaiian) in the volcanic Big Island, my previous image of a fruit fly changed dramatically. These flies have evolved a wealth of new adaptations in less than half a million years, including exquisite food specializations and a repertoire of territorial sexual behaviours highly unusual for a fruit fly. But their highly decorated wings are likely to be what capture most people's attention. The question is what evolutionary value, if any, can be assigned to these embellishments? Or are these beautiful wings just caprices of nature? Since the legendary nineteenth-century work of Henry Walter Bates in the Amazon with butterflies, we know that wing pigment patterns have a high adaptive value for escaping from predators, either by camouflage or by mimicking wing patterns of other distasteful forms. Darwin admired Bates' insight on the adaptive value of wing mimics, a phenomenon designated today as Batesian mimicry.

While in picture wing Drosophila we are not aware of any mimicry case, there is suggestive evidence that wing pattern fitness is poised between a fly's needs to hide from predators and to attract mates. The fly's escape strategy is to blend with the substrate instead of flying away. They do it by folding their wings back when resting on bark, leaves, or other feeding substrate, and even by mixing with the soil litter where they precipitate and play dead; a striking behaviour which I took advantage of when collecting by cautiously inserting my open vial under each upside-down resting fly. This wing-folded camouflage must, however, be substituted with male wing display during courtship. The work of Kaneshiro (1988) and his associates provides strong evidence that **sexual selection** has fuelled speciation, often by morphological change, in the Hawaiian flies. In fact, striking male-specific modifications of legs, bristles, antenna, and mouthparts characterize several species groups. Thus, fly wing pattern evolution, as exemplified by increasing complexity or changes in parts (Fig. 2.14), might also be triggered by sexual selection, as it is amply documented that the female chooses males with the most colourful displays in many species, including birds, fishes, and insects.

The 'picture wing' species group evolution, widely known thanks to the lifelong work of Hampton Carson and his followers, has revealed that its high diversification in morphology, including wing patterning, behaviour, and ecology is an example of an explosive radiation from a **founder** event that occurred approximately half a million years ago in the Big Island. Strikingly, this species radiation has entailed relatively little change in the DNA sequence. Yet although this scenario, molecularly and ecologically well defined, contains the right ingredients for a research programme towards a developmental study of wing pattern evolution, no such thing has been produced, as Edwards *et al.* (2007) bitterly complain. The reason for this may be that, due to their long generation time (2–3 months) and their complex life cycle, these flies demand high investment in space and labour for stock keeping. These limitations are not present, however, in other less fancy Drosophila that have been the main source of enlightenment for much of what we know of the wing development machinery.

Captured and gone with the wing

One of the most common wing patterns in Drosophila is a black spot at the tip. Melanin is the pigment deposited in the spot after being locally synthesized from a precursor molecule. This is a complex process involving many regulatory steps. We know that one step is mediated by up-regulating

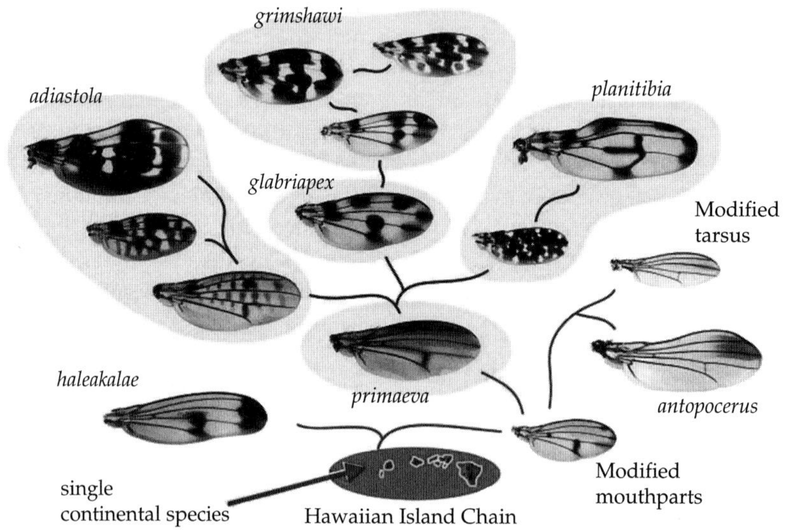

Fig. 2.14. Schematic view of Hawaiian Drosophila wing patterns in major species groups and picture wing subgroups. The figure depicts the phylogenetic relationships from the proposed single introduction in the archipelago, indicated by an arrow. Grey background denotes the picture wing subgroups, which show an increased wing pattern complexity towards the adiastola and grimshawi subgroups. From Edwards *et al.* (2007)

the *yellow* gene that normally encodes a protein required to pigment the cuticle. Yet the *yellow* gene is not alone in spot formation; other genes like *ebony* and *engrailed* participate in the regulatory circuit. As an example, the selector protein Engrailed, that controls the extent of the spot, has been co-opted in *D. biarmipes* (Gompel *et al.* 2005) by the evolution of *yellow* CREs (Fig. 2.15). So, the spot evolution is not a one-step process, but 'a multistep series of changes during which the intensity and the shape of the spot evolved', as Carroll (2006, p. 208) states. Moreover, wing spots can also be lost and gained repeatedly by independent co-options of distinct, ancestral CREs. This has been demonstrated in the *melanogaster* group species for the *yellow* gene, where two independent gains and five losses of wing pigmentation have been reported (Prud'homme *et al.* 2006), showing once more the tinkering nature of the evolutionary processes.

The multiplicity of paths available for natural selection to choose is bolstered by these results. The *yellow* gene is highly pleiotropic, which conforms to the idea (see above) that modularity allows cis-regulatory changes to drive a large flexibility in morphological evolution. On the other hand, evolutionary changes in coding sequences have been described in vertebrate genes that encode low level pleiotropic proteins involved in pigmentation,

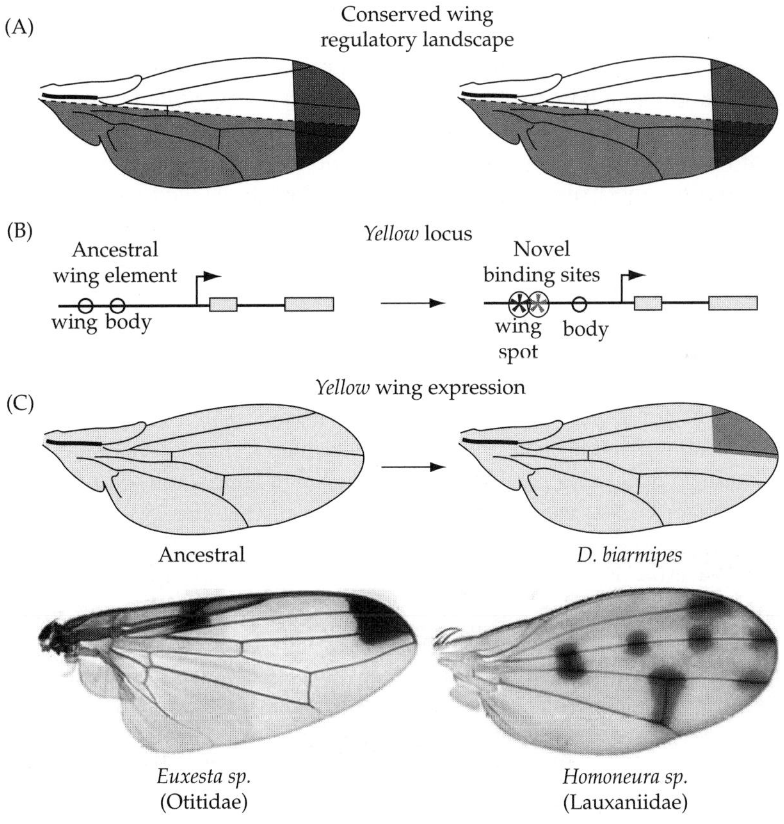

Fig. 2.15. Compartment cryptic pre-patterns foreshadow the evolution of novel gene expression patterns through the co-option of new cis-regulatory elements (CREs). (A) The shaded wing areas denote two wing compartment expression areas of transcriptional repressor (light grey) and activator (dark grey) regulators that pattern and shape the Drosophila wing. (B) The evolution of the *yellow* wing CRE through mutations in binding sites (encircled grey stars) may co-opt a set of those regulators (i.e. the engrailed protein) to modify the *yellow* expression. (C) This change, coupled with similar regulatory changes at other loci with distinct expression compartments, may result in new wing pigmentation: spotted patterns, ranging from one-spotted fly wings (*D. biarmipes*) to many-spotted fly wings (*Euxesta* sp.and *Homoneura* sp.). Adapted from Gompel *et al.* (2005) with permission from Nature Publishing Group.

confirming the value of modularity. All these suggestive opportunities for selection notwithstanding, and despite the fact that wing pattern has been repeatedly associated with sexual selection in Drosophila, the direct proof of a relationship between spot formation and environmental induction was lacking. Nor is it known how these fruit fly developmental rules could be applied to other organisms.

We have to turn to butterfly studies to get some insight on the putative action of natural selection in developmental patterns. Carroll and his colleagues decided to use the butterfly homologues of genes involved in fruit fly wing patterning to probe the mechanisms of wing eyespot formation. Most homologues were expressed in the same wing disc regions of both flies and butterflies. Yet *Distal-less*, a gene very familiar to us, involved in building arthropod limbs, had acquired a new role: the formation of eyespots in wing butterflies (Carroll *et al.* 1994). It turned out that, as previous cis-regulatory developmental examples anticipated, this new function evolved by the acquisition of a new *Distal-less* switch, namely a new CRE that responds in the precise eyespot coordinates to the selector tool-kit gene. The eyespots are complex structures made up of several rings that surround a central spot, each one with a distinct colour. It came as no surprise then that other tool-kit genes were found to be involved in eyespot building. Among them *Engrailed* and *Spalt*, two pleiotropic genes also participating in many developmental processes, were expressed in different eyespot rings, and their new roles were also accomplished by the evolution of new switches.

Butterfly students, amateurs and professionals alike, know that butterflies show an endless repertoire of wing pattern diversity, particularly in numbers, sizes, and colours of eyespots. Eyespots can even disappear in some species and also in some forms within the species. We now know that changes in the deployment of tool-kit genes underlie this phenotypic variability. But how much do we know about the causes that trigger the capture and release of wing pattern signalling? A significant breakthrough was made by Paul Brakefield and his students at Leiden University (The Netherlands) with *Bicyclus anynana*, an African butterfly species that has to endure drastic seasonal changes in the wild (see for a review Beldade and Brakefield 2002). Two morphs alternate their presence along the serial cyclical seasons; while in the wet season the wings show large highly visible eyespots (Plate 4b), the dry season morph bears no eyespots on its wings (Plate 4a). One explanatory account for this butterfly response to the environment could go like this. During the wet season, which also overlaps with the reproductive season, butterflies are very active, using their colourful wing display to attract mates and also to deflect predator attention from their main body to wing eyespots; butterflies with a wing chunk removed can still fly and reproduce. On the other hand, in the dry, cool season when the substrates are made of brown leaf litter and the butterflies are forcedly less active, a dull, eyespotless morph camouflaged in the litter seems to be the best strategy for avoiding predation. I can see the anti-Darwinians readily qualifying this story as panselectionist, but there is increasing evidence that selection, in the wild and in the laboratory as well, is a powerful actor in this drama.

When eyespotted butterflies were released in the dry season by Brakefield's team, they found that these wing-ornamented butterflies were predated much more often than the cryptic forms. In addition, artificial selection experiments performed by the same research group, starting from a variable population, showed that after only twenty generations two temperature-independent stable populations, one that developed large spots and the other that bore small eyespots, were established. These results are strong evidence that natural selection has the potential to change, although at a slower rate than in the lab, the wing eyespot appearance. Moreover, further experiments with *Distal-less* expression in caterpillars raised at different temperatures showed that at low temperatures expression is reduced to fewer eyespot cells than at higher temperatures. So, the *Distal-less* gene seems to have evolved an eyespot switch that responds to temperature, albeit that this response is likely mediated by season- and temperature-dependent hormone levels of expression. Most of these experiments were carried out in Byciclus, but looking at other species, including picture-winged Drosophila and butterfly mimicry, we see an extraordinary wealth in wing-spot diversity that presages an evolutionary story of adaptive speciation. Thus, the exact correspondence between *Distal-less* spots in wing discs and the number of wing eyespots in all species strongly suggests the evolution of signature sequences in the *Distal-less* gene triggered by environmental cues, like in Byciclus phenotypic adaptation.

Lessons from fishes

The butterfly wingspot account shows that the underlying evolutionary strategy of wing pattern formation relies mainly on changes and/or acquisitions of regulatory switches in tool-kit genes. These changes are elicited by environmental cues, including temperature shifts and predator pressures. There are many other examples of the natural selection effect on the melanic coat pattern in mammals, including cats, mice, and others. While these cases nicely illustrate how a novelty arises, they could be qualified as of minor significance when the extraordinary morphological changes experienced by the large body plans (i.e. phyla) are considered. Moreover, some would argue that wing pattern evolution is just a microevolutionary story decoupled from macroevolution. Although it may seem an ironic coincidence that the regulatory 'microevolution' strategies were the same as those found in the evolution of high body plans, other more 'macroevolutionary' changes can also be studied at the population level.

In preceding sections I have explained how evolution of form is often accomplished by differences in the number and form of serially repeated parts. These include segments in arthropods and vertebrae in vertebrates. We know the evolutionary process that underlies the origin, or disappearance, of appendages in both large groups. We could not be present one million years ago when this differentiation took place, but we can study similar processes in current populations. The threespine stickleback fish is found in two forms in lakes in North America: one shows reduced spines, the other is a full-spined form (Plate 14). Each form occupies a different habitat: the former occurs on the bottom of shallow waters and the latter in open waters. Both forms are adaptive. The bottom-dwelling form takes advantage of its reduced spines because bottom dragonfly larvae are voracious predators that grab the fishes by their spines. Contrarily, open water sticklebacks are protected from predators by their long spines. We know that both forms evolved from an ancestral marine species which was isolated in glacial lakes during the last glaciation. Since then, in less than 15,000 years, these isolated fish populations have experienced several episodes of allopatric and sympatric speciation (see Chapter 4) that led to the two stickleback forms (Rundle and Schluter 2004). Obviously, natural selection was responsible in great part for the spine changes, but how did it occur? What kind of genome changes have the fishes experienced? David Kingsley and Dolph Schulter with their collaborators (see Shapiro *et al.* 2004) were able to show that the reduction of the pelvis development fin bud is mostly due to the *Pitx*1 tool-kit gene. When they compared the Pitx1 proteins between both fish forms no difference was found, yet the gene regulatory region has been changed, as could be anticipated for a tool-kit gene (Fig. 2.16).

The fossil record tells us that the pelvic fin is the precursor of the tetrapod hindlimb. Thus, it is not surprising that the *Pitx1* gene was also involved in the vertebrate hindlimb development. For example, in the mouse it contributes to differentiate hindlimbs from their serial homolog forelimbs. As a typical tool-kit gene, the *Pitx1* gene participates in the making of many body structures, like the thymus and the lateral sensory organs of fishes, and also has homologs in many animal forms. By changing the CRE that controls its expression in the pelvic fin bud, the pelvis fin is reduced in sticklebacks without affecting other structures that are controlled by different CREs. Moreover, vertebrates have often experienced the reduction of hindlimbs, as in cetaceans and manatees. Thus, this is not an oddity restricted just to sticklebacks. Pelvic reduction by genetic switch evolution has been repeatedly produced in independent lineages when lifestyle transitions occurred, as in the aquatic mammals which evolved from land-dwelling ancestors.

Coda: deep homologies as strong pillars of tinkering selection or should Darwin be reconstructed or deconstructed?

We have made a long promenade in this chapter across many research fields on the understanding of

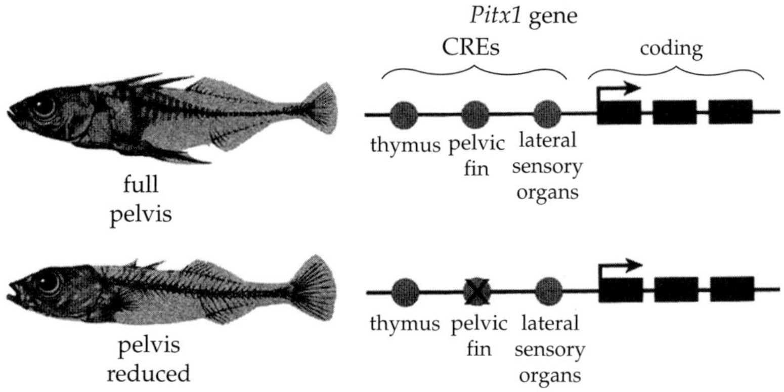

Fig. 2.16. Schematic evolution of the stickleback fish pelvic skeleton through a mutation change in a CRE. The reduction of the pelvis is due to a change in the CRE (X) that controls the expression of the *Pitx1* gene in the developing pelvic fin. From Carroll (2006), with permission from Sean B. Carroll.

the unity of type. This was a deeply cherished concept in Darwin's theory that is strongly supported by current advances in developmental biology. The unanticipated finding, that the genes responsible for body building are ancient, often predating the Cambrian era, gave a final blow to defenders of phylum independence. Tool-kit genes were present before organic diversification arose, they were there as a potentiality but they had to wait the specific environmental changes to act as drivers of adaptation.

Yet, the role of natural selection in the theory of 'descent with modification' is still debated in some academic circles. I believe that most of this controversy focuses on the contention, mainly held by the fathers of the Modern Synthesis, that small changes in coding gene sequences were the raw material upon which natural selection acts. That the evolution of form could be explained by the gradual, continuous selection of these small mutations was never accepted by some evolutionists, including some geneticists and many embryologists. We have explained (see p. 37) that embryology (i.e. development) was missing, and perhaps underrated, in the Modern Synthesis. Interestingly, embryology was highly prized by Darwin. To him 'embryology rises greatly in interest, when we thus look at the embryo as a picture, more or less obscured, of the common parent-form of each great class of animals' (*The Origin of Species*, p. 450, first edition), a strong tenet that is reinforced by his famous statement in a letter to Asa Gray (10 September, 1860): 'Embryology is to me by far the strongest single class of facts in favour of change of forms.'

But from Darwin's time until around 30 years ago the fields of embryology and genetics remained divorced. Almost nothing was known about how and which genes were involved in development, first because genetics was not fully incorporated into evolution until the late 1930s, but mostly because the molecular nature of the gene and its regulatory expression were not understood until the 1950s and 1960s, respectively. During this long period of obscurity many evolutionists proposed that morphological changes were not elicited by the small mutations responsible for population adaptations, but by 'macromutations' of large effect. Fortunately, as discussed in this chapter, the recent marriage between genetics and development has dismissed this spectre of 'hopeful monsters' that haunted many evolutionists. Now, we are in a privileged position to analyse the mutations that underlie the evolution of form due to the advances of Evo Devo. And this analysis is revealing that most, albeit not all, evolutionary changes in body plans are found in regulatory DNA sequences and not in the coding parts of genes. This introduces, to some evolutionists, a new view that deserves to be incorporated in a 'new evolutionary synthesis'.

While a reform of the 'synthesis' may look sensible, the question remains whether the long held tenet between 'conditions of existence' (adaptation) versus 'laws of growth' (development), in Darwin's terms, is still valid. In other words, is natural selection the leading force in evolution of form, or is evolution channelled by the internal rules of development? The more we know about the subtleties of development, the less obvious it is that development constrains adaptive evolution. On the contrary, take for instance the compartment theory. Compartmentation is a process that deconstrains development by allowing **mosaic pleiotropy**, that is, the modularity of tool-kit gene expression facilitates endless variation in each compartment (pp. 70–3). This universal pattern is not an evolutionary constraint but rather a strategy that merits conservation because it facilitates adaptation through evolving new structures, including appendages in chordates and arthropods or wing patterns in insects, or eliminating useless structures, such as pelvic fins or appendages, as illustrated above. But, again, natural selection is the supreme referee that decides what is going ahead among those available opportunities.

No doubt ancient regulatory networks and structures are fundamental building blocks of evolution of form. But we must remember that tinkering is an inherent property of natural selection. The conservation of networks and structures is an example, albeit on a large scale, of the gradual building of structures by preserving ancient circuits that worked efficiently for long spans of time. Natural selection not only innovates but also conserves. Take the *Hox* genes that have been preserved for more than 500 my. Does this mean that they have channelled all bilaterian evolution? Or, rather, that they have been preserved because they are included in large gene regulatory networks that have been universally co-opted by natural selection? I rather adhere myself to the latter proposition.

In sum, much is still to be learned from the explosive ongoing advances of Evo Devo. But, when looked at in perspective, the progress in the past two decades has taught us that the evolutionary changes in regulatory switches are not so different from the gradual changes proposed by Darwin. Admittedly these changes must be incorporated into the new synthesis, but does all the new knowledge in developmental genetics justify the search for a new 'Darwin', as many propose? Or, as stated by some post-modern critics, is it licit to deconstruct Darwinism? Ironically, the new advances, sketched here, have vindicated Darwin's intuitions and we are now closer to Darwin than before the molecular era. Unfortunately Darwin could not enjoy the genetic and development advances we can, but he would have been amply satisfied that his ideas still hold true. Maybe a wise position would be, instead of deconstructing, to reconstruct Darwin. Perhaps this reconstruction would position the large 'chunk' of development to where it deserves to be, far from a 'chink' in the corner, where Gould thought Darwin positioned development. Yet, I do not see reasons to relegate natural selection, as the prime channelling force of evolution, behind development.

CHAPTER 3

The genome is mobile

There is little doubt that genomes of some if not all organisms are fragile and that drastic changes may occur at rapid rates. These can lead to new genomic organizations and modified controls of type and time of gene expression. It is reasonable to believe that such genome shocks are responsible for the release of otherwise silent elements . . . Since the types of genome restructuring induced by such elements know few limits, their extensive release, followed by stabilization, could give rise to new species or even new genera.

—(McClintock, B. (1980) Modified gene expressions induced by transposable elements. In W.A. Scott, R. Werner, D.R. Joseph, and J. Schultz (eds.), *Miami Winter Symp.* **17**, 11–9. Academic Press, New York.)

As early as the mid-1940s the results of an experiment conducted in maize paved the way to the unexpected identification of movable pieces of DNA in the genome. It was Barbara McClintock, a plant cytogeneticist at the Carnegie Institution in Washington, who performed the seminal experiments with maize lines that led to the discovery of mutations as a result of insertions of stretches of DNA (named *Ds*) in otherwise normal genes. In 1950 she thought that 'the study of insertions of a *Ds* into known gene **loci** had progressed sufficiently . . . to warrant publication in a journal with a wide readership'. The paper (McClintock 1950) was published in the *Proceedings of the National Academy of Sciences of the USA*, a most prestigious journal, but 'it was clear that . . . the presented thesis . . . could not be accepted by the majority of geneticists or by other biologists', as McClintock recalls.

In a 1953 publication in *Genetics*, also a widely read scientific journal, McClintock reported again her findings about mutable loci. To her dismay she received only three reprint requests, at a time when, with no electronic communications, snail mail was the normal way to manifest interest for a published research. After this she concluded 'that no amount of published evidence would be effective' and since then most of her data were 'treated to an unpublished written account' and only 'the highlights of these studies were reported in the annual Year Books of the Carnegie Institution', as she bitterly explains (see the introduction of Moore, J.A. ed., 1987).

Reasons for this outstanding neglect must be attributed to conflicts with the genetics paradigm of the moment. It was a time when scientists were largely ignorant of the nature of genes and the structure of the genome. The genome was viewed as a stable array of genes, each one occupying a fixed place, coined 'a locus' by classic geneticists, in the chromosome. As McClintock states in retrospect, 'that genetic elements could move to new locations in the genome had no precedent and no place in these concepts.' Ironically, Watson and Crick reported their DNA structure model in *Nature* at the end of April 1953, the very same year of McClintock's neglected paper, inaugurating the molecular era in genetics that has made possible the understanding of genome dynamics.

McClintock was also a pioneer in emphasizing the primordial role of gene regulation in development, an aspect of mutation instability that attracted her attention from the early days of her career. But the field of genetics was not ready to accept the role of mobile elements in gene regulation. After all, it was not until the late 1950s, almost 15 years after McClintock's early experiments, that the elegant experiments of Jacob and Monod in bacterial gene regulation marked a turning point in gene expression theories (see Chapter 2). Although the movable genetic elements in maize, named 'controlling elements' by McClintock because of their regulatory role, and now known as **transposable elements (TEs)**, represent a quite different type of regulation from that discovered in bacteria, Jacob and Monod's ideas contributed to the acceptance of their importance in gene control.

Today, 50 years after these events took place, nobody would deny that genomes contain a wealth of DNA sequences able to move, usually referred to as transposition, from one genome site to another, using different mechanisms. But this consensus was difficult to reach because it represented a paradigm shift in theories of genome stability and control. Transposable elements are now incorporated into the contemporary concept of the genome as an entity with unsuspected dynamism and fluidity of far-reaching evolutionary consequences. It is the evolutionary impact of transposable elements that this chapter focuses on, emphasizing the ability of the genome to respond to challenges by transposable-mediated mechanisms ranging from pure control to 'domestication'. It is not by chance that McClintock discovered these elements when she became interested in the extraordinary response of the maize genome to the entrance of a single ruptured end of a chromosome into the nucleus. As she recalls in her Nobel acceptance lecture of 1983, 'it was this event that, basically, was responsible for activations of potentially transposable elements that are carried in a silent state in the maize genome' (McClintock 1984). Now we know that this observation fuelled a grand scientific endeavour that is far from complete.

How much of the genome is mobile?

Perhaps one of the most challenging discoveries of the 'genome era' is the assessment that mobile, or transposable, elements (TEs) are ubiquitous in all living beings. In particular, they constitute about half of the mammalian genome and up to 90% of some plant genomes (Table 1.2). Mobile elements deserve their name because they are made of DNA stretches able to move by transposition to new sites in the genome of origin. They may transpose by different mechanisms, autonomous or not, of which many details are still unknown. Mainly, two prevalent procedures are recognized. Class 1 TEs are generally designated as retroelements because the original element is transcribed 'in situ', without moving at all, and its RNA transcript is reverse transcribed into DNA that mobilizes and inserts into a new genomic site (a 'copy and paste' mechanism in short). On the other hand, class 2 TEs, usually named DNA **transposons**, excise from their original site and integrate into a new site, in most cases by a 'cut and paste' mechanism, mediated by an encoded transposase enzyme (Fig. 3.1).

Soon after their discovery by McClintock (1950) and their recognition as ubiquitous, the controversy concerning the influence TEs exert on the genome took place. First, these elements were recognized as parasitic elements and as such considered a burden that the host genome should keep controlled if it were to survive. Yet not all scientists viewed TEs as negative and some, among them McClintock, viewed TEs as DNA sequences coding for functions that could be used for the benefit of the genome. The parasitic selfish behaviour of TEs stems from their ability to self-replicate and to invade the genome with as many dispersed copies as possible. To illustrate this parasitic nature of TEs, the parallelism between class 1 elements and retroviruses is most convincing. **LTR** retrotransposons, named after their Long Terminal Repeats at both ends, are a kind of class 1 element similar to retroviruses. Structurally, LTR retrotransposons and retroviruses both share genes that code for the same viral enzymatic functions, namely a capside protein (*gag*), and a polyprotein (*pol*), containing domains for protease (P), reverse transcriptase (RT), ribonuclease H (RH), and integrase (INT), to make cDNA from transcribed RNA and to insert it in a new genomic site. Retroviruses differ from LTR retrotransposons in that they possess an envelope gene (*env*) that codes for proteins to facilitate their infection from cell to cell. However, this difference is only partial in many instances because some LTR retrotransposon lineages have independently acquired *env*-like genes (Eickbush and Malik 2002) and in a few of them, like the *Gypsy* element of *D. melanogaster*, they are functional (Fig. 3.1).

This analogy between TEs and genome parasites prompted some early researchers (e.g. Doolittle and Sapienza 1980; Orgel and Crick 1980) to consider the large amounts of these selfish elements found in the genome as 'junk DNA'. This idea was reinforced when initial genome sequencing uncovered enormous amounts of non-coding DNA in many organisms, especially in plants and higher vertebrates (see Chapter 1 for a longer treatment). As an example, in the human genome only about 1.5% is protein-coding DNA. This extreme abundance of non-coding DNA is still a mystery, although in recent years a wealth of new information has been produced in an effort to understand its prevalence. In the 1980s the most conventional explanation was to view most of this 'excess' DNA as the result of an evolutionary equilibrium between selfish DNA invasion by non-phenotypic positive selection on the side of **selfish**

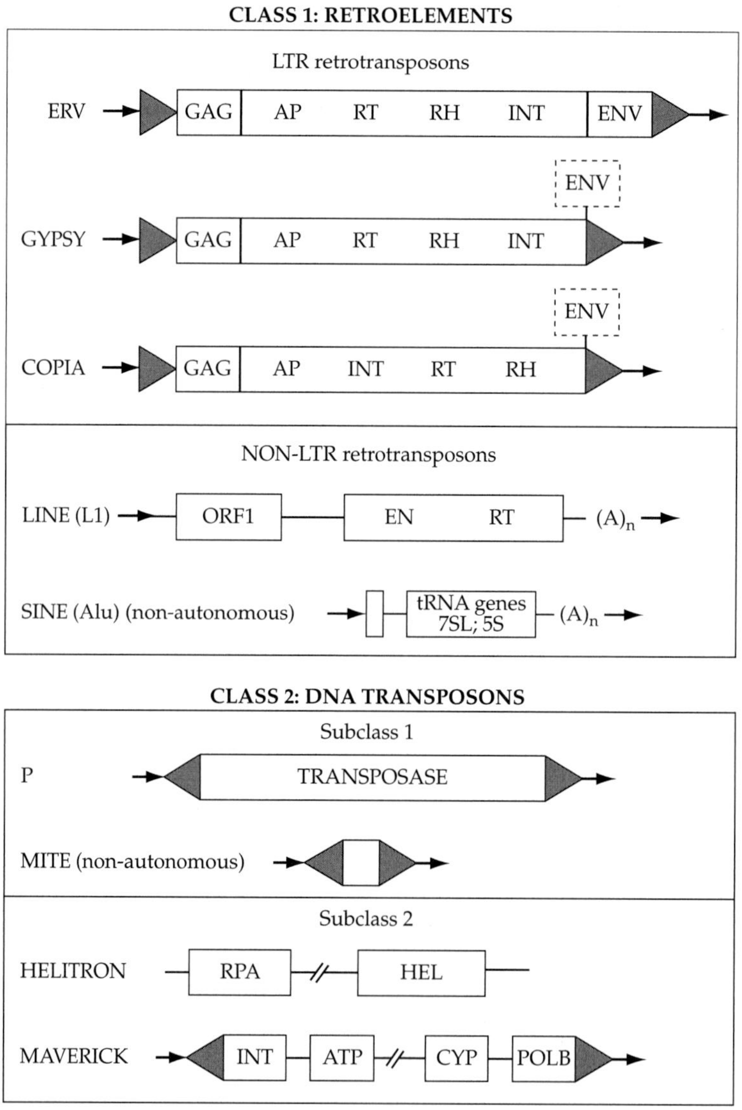

Fig. 3.1. A simplified view of transposable element types and structures. Retroelements are divided into two subclasses on the presence of two long terminal DNA repeats (LTR retrotransposons) or on their absence (non-LTR retrotransposons). The former resemble retroviruses in both their

continues

DNA and negative selection at the phenotypic level on the organism (Charlesworth and Langley 1989). Since then many new insights into the structure of the genome have produced a new perspective on the impact of TEs on the host-genome evolution, quite different from the selfish paradigm. This is not to say that all this 'excess' DNA can be explained by TE beneficial activity, rather that a significant fraction may play a role in genome dynamics, influenced by TE evolution. Most importantly, it is likely that the opportunistic condition of natural selection has been largely effective in shaping novelties, structural and regulatory, by co-opting TE functions, selfish at the beginning, in a process of domestication. The recent mastering of new 'in silico' (bioinformatic) and 'in vitro' (molecular) techniques has been instrumental in detecting TE signatures in a large number of genomic sequences ranging from regulatory to structural gene elements. This insight, growing steadily, is resolving the picture of a genome landscape that contradicts the 'junk-selfish' view prevalent a quarter of century ago.

The mobile genome landscape

A picture of the statics and the dynamics of the mobile genome for many model organisms is beginning to emerge. In Chapter 1, I explained some mechanisms that are likely to underlie the TE abundance and distribution in different organisms. Far from being a complete summary, these mechanisms, ranging from diverse insertion and excision rates, to explosive invasive bursts due to genomic and environmental stresses, and also to horizontal TE transfer, provide an updated view of TE dynamics in genomes, the result of which gives a varied TE type distribution in different organisms. Thus, low TE frequency in yeast is achieved by a small number of retrotransposon types, each one containing less than 100 elements. In contrast, the Drosophila genome harbours many different types of retrotransposons, but each one at low frequency. On the other hand, humans and mice contain a low number of retrotransposon types, each one with a large number of copies (over 1 million for *Alu* elements alone).

Mobile genome features are the end result of evolutionary processes, namely insertion and removal of TEs. Random insertion is a mutagenic process that may occur in a coding or regulatory site. Consequently, the overall phenotype-altering effect of TEs should vary across organisms depending on their coding DNA proportion. In many organisms, such as humans and mice, with less than 10% of

Fig. 3.1. Continued

structure and transposition mechanism. Compare the structure of the *gypsy* and *copia* superfamilies to the endogenous retroviruses (ERV), a superfamily of retroviruses that have lost their capacity of extracellular infection. All of them share two genes: the *gag* gene, encoding the capsid-like protein, and the *pol* gene, encoding a polyprotein that contains a protease (AP), a reverse transcriptase (RT), an RNase H (RH), and an integrase (INT) that are crucial for the element transposition. Yet the ERVs contain a specific envelope gene (*ENV*), albeit inactive, that encodes a transmembrane host receptor-binding protein that is responsible for transmission of active retroviruses. Interestingly, the *ENV* gene can be acquired by some *gypsy* and *copia* elements (depicted by dotted rectangles), and at least in the case of *gypsy* the generated virus-like particles have been shown to be infective. Non-LTR retrotransposons are subdivided into LINEs (long interspersed elements) and SINEs (short interspersed elements). L1, a LINE representative superfamily, contains an ORF1 (encoding an RNA binding protein) and another gene encoding an endonuclease (EN) and a reverse transcriptase. However, SINEs are non-autonomous and rely on the LINE machinery for transposition. *Alu*, a well studied SINE and the most common TE in the human genome, derives from 7SLRNA, but other SINEs derive from tRNA genes. Both LINEs and SINEs only share a terminal simple sequence repeat, usually poly(A). DNA transposons are divided into two subclasses. Subclass 1 comprises a number of superfamilies that share a transposase gene and two terminal inverted repeats (TIRs). The P element is a well-known example famous for its invading capacity (see text and Box 3.4). MITEs (miniature inverted-repeat transposable elements) are non-autonomous defective DNA elements, often consisting of two TIRs that include a short non-coding DNA sequence (not larger then a few hundred bp in size) that originated from very different DNA transposon families likely by internal deletions. Subclass 2 includes DNA TEs that transpose without double-strand cleavage as in subclass 1 transposons. Thus Helitrons seem to replicate via a rolling-circle mechanism that produces only one strand cut. The autonomous Helitron encodes a replication protein RPA (only in plants), a helicase-like domain (HEL) active in the rolling-circle replication, and often may contain host genome gene fragments that have been captured. Finally, Mavericks encode various proteins but do not encode RT, in conformity with their transposition without RNA intermediates. They are large TEs (10–20 Kb), bordered by long TIRs, found in eukaryotes, except in plants. Although some superfamilies are typically non-autonomous (i.e. SINEs and MITEs), most superfamilies comprise non-autonomous along with autonomous copies. See Wicker *et al.* (2007) for a complete TE classification. Shaded triangles at both ends of TEs denote LTR in class 1 TEs and TIR in class 2 TEs. Arrows represent short direct repeats that are generated on both flanks of a TE upon insertion (named target site duplications: TSD). Drawn by Montserrat Peiró.

coding and regulatory DNA, phenotypic insertion mutations are relatively less frequent compared to other organisms with a lower proportion of non-coding DNA. This latter case includes Drosophila, where more than 70% of morphological and regulatory mutants are insertional.

Moreover, the removal of TEs, by excision and/or ectopic recombination (Fig. 1.10, Chapter 1), also plays a significant role in TE distribution and frequency. These mechanisms, and other organism-dependent mechanisms, may explain some exceptions, like maize (see Chapter 1, pp. 23–5), in which large genomes with a high proportion of non-coding DNA show many phenotypic mutants. In fact the discovery of TEs was made in maize by studying pigmentation mutants (see McClintock 1980) (Fig. 3.2 and Plate 5). Humans and mice show, however, few phenotypic insertional mutants in accordance with their low mobility and low coding DNA proportion. This is likely to explain why TEs were discovered in these organisms only after advanced molecular techniques were available.

Removal of TEs is not the only mechanism to counteract the invasive behaviour of TEs, nor is the presence of transposition-inactivating mutations; other host-mediated control mechanisms of TE mobility exist. Otherwise how could we explain that some intact autonomous TEs are not expressed in host genomes? These 'cryptic elements' retain their DNA coding potential to mobilize but they are unable to produce the necessary proteins to transpose. These elements are silenced by a set of 'epigenetic' mechanisms that defend against the potential harmful effect of TEs to the genome. Among these epigenetic controls, DNA cytosine methylation is one of the best documented. Methylation is known to inhibit transcription at specific loci in a programmed genome, and is related to basic functions such as genome **imprinting**, i.e.

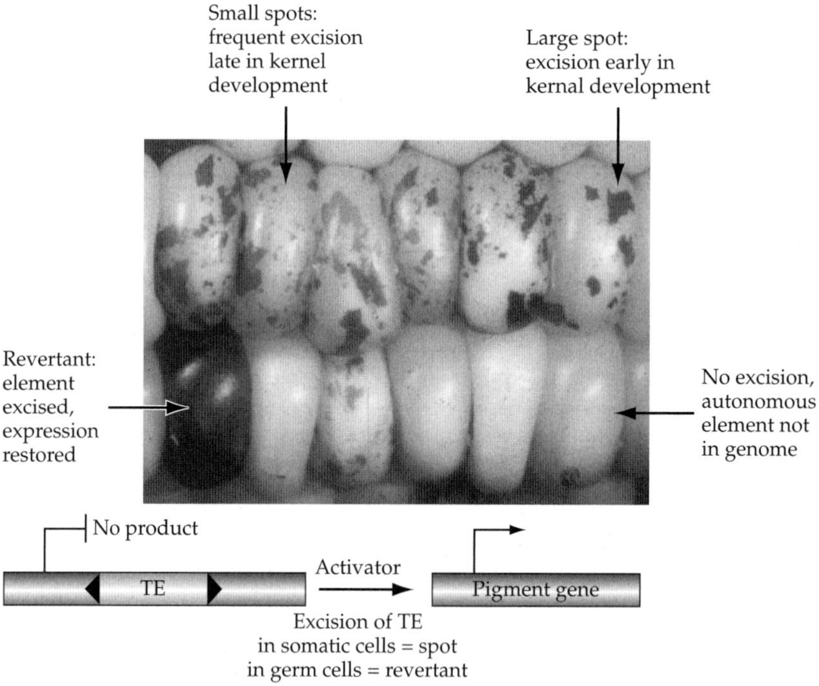

Fig. 3.2. Using kernel phenotypes to study transposon behaviour. Kernels on a maize ear show unstable phenotypes due to the interplay between a transposable element insertion in an activator gene and an activated gene that encodes an enzyme in the anthocyanin (pigment) biosynthetic pathway. Sectors of revertant (pigmented) aleurone tissue result from the excision of the TE from the activator gene that restores its encoding expression in a single cell. The size of the sector reflects the time in kernel development at which excision occurred, i.e. the larger the size the earlier the excision. An understanding of the genetic basis of this and similar mutant phenotypes led to the discovery of TEs. From Feschotte *et al.* (2002) with permission from Nature Publishing Group (see also Plate 5).

the differential expression of maternal and paternal **alleles** (see below).

TE silencing is also related to **RNA interference (RNAi)** (Box 3.1), a widely documented mechanism of genome regulation (see below). This relationship has been strongly demonstrated for a variety of cases including *Ty1* retrotransposition in yeast, Drosophila non-LTR retrotransposon *I*, transposon-related DNA sequences in Tetrahymena, and *Tc-1* transposition in *C. elegans*—probably the best documented case. The discovery of RNAi mechanisms has led to the development of a new, albeit still changing, view on genome regulation (see a review in Slotkin and Martienssen 2007). This is most extraordinary, in that a mechanism whose original function was likely a defence against invasive sequences—mobile elements among them—has been co-opted to service genome regulation.

In the paragraphs below I will try to review what we do, and do not, know about the evolution and the role of the mobile fraction of the genome. It suffices to say that the present mobile genome landscape is the result of many processes, some of them very primordial and others more recently derived but far from simple. Many of these mechanisms, like RNAi silencing, are still poorly understood, and are changing our perspective not only about the regulation of gene coding DNA but also the concept of 'junk DNA'. We continue to learn new functions for the non-coding fraction of the genome, of which TEs constitute an important part, and to detect many TE-related sequences in coding DNA. Thus, we can describe mobile element machinery as builders of genomes, as we look at their effects after thousands of millions of years of evolution.

Mobile elements as an ancient conflict to be domesticated

The recognition of foreign, invasive DNA by host genomes is of paramount importance in maintaining their integrity through mechanisms that disrupt

Box 3.1 Post-transcriptional RNA interference gene silencing

Understanding gene silencing as a co-opted function from a host defence mechanism needs a sound, general basis of the host recognition of self and non-self DNA sequences. Basically, recognition of foreign DNA is associated with the presence of double-stranded RNA (dsRNA) molecules, a common structure of many RNA viruses, which can be produced when repetitive elements, such as TEs or transgenic copies, enter the genome.

In the early 1990s, plant molecular biologists tried to increase gene expression of the enzyme responsible for purple flowers in petunias by introducing extra copies of the gene in an attempt to produce a more vivid flower colour (Krol *et al.* 1990; Napoli *et al.* 1990). To their dismay, they found that the transgenic petunias produced white flowers (Plate 8). This unexpected phenomenon by which extra copies of a gene suppress the expression of both the endogenous and the introduced gene has been called 'co-suppression' and turned out to be a general phenomenon. Later, it was discovered that gene silencing was

Fig. A. A sample of Petunia phenotypes due to RNA interference. White areas in flowers are generated by silencing a pigment gene through RNAi mechanisms. From Grosshans and Filipovicz (2008), with permission from Nature Publishing Group. See also Plate 8.

continues

Box 3.1 *(continued)*

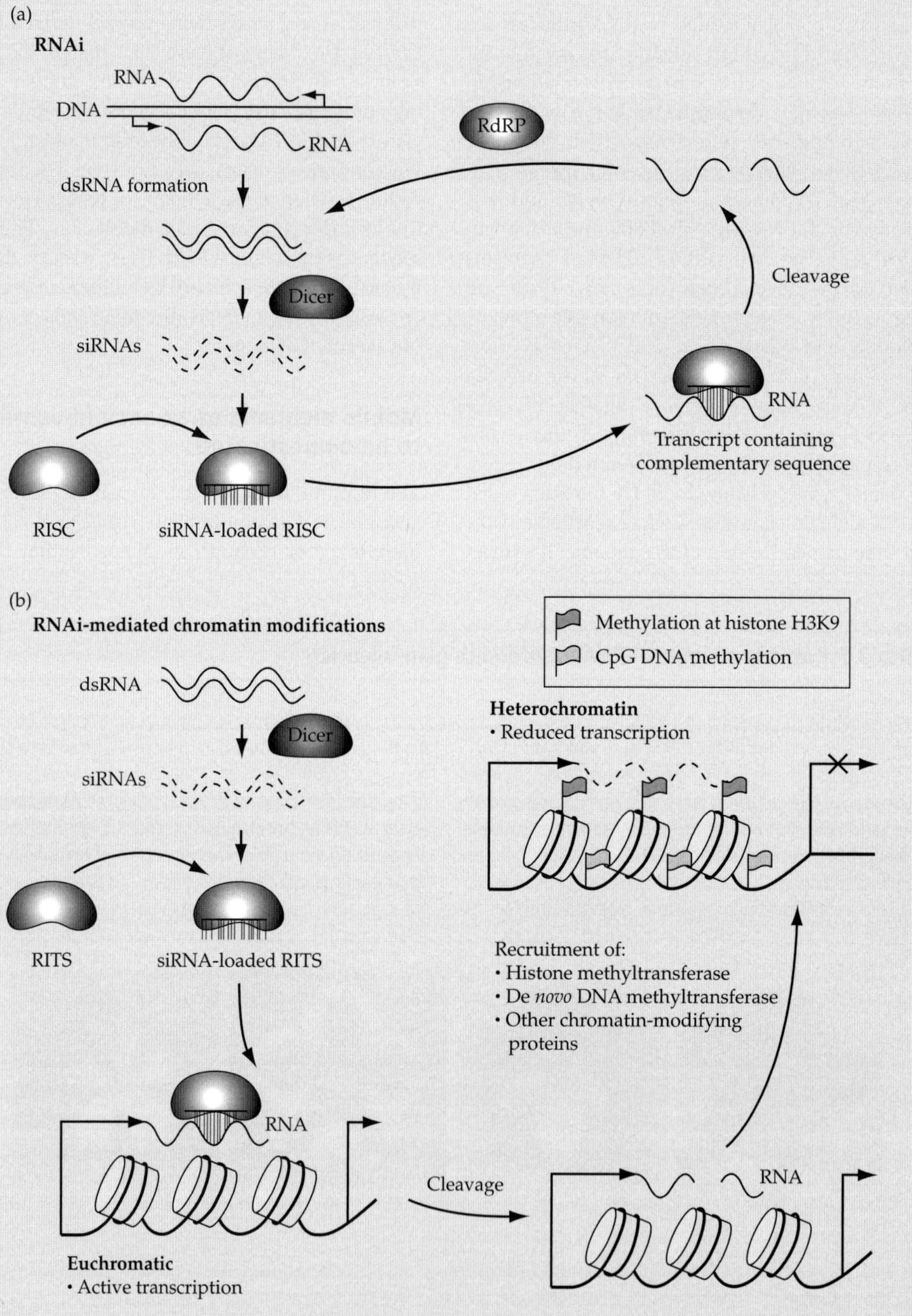

Fig. B. Post-transcriptional silencing by RNAi. From Slotkin and Martienssen (2007) with permission from Nature Publishing group. See also Plate 9.

continues

Box 3.1 *(continued)*
produced by injecting just mRNA transcribed from the gene to be silenced. We now know that co-suppression is due to post-transcriptional gene silencing (PTGS) in which transcription of the silenced gene is normal but the mRNA is destroyed in the cytoplasm prior to translation. Although not understood at the time of its discovery, the presence of contaminating double-stranded RNA (dsRNA) in the transgenic preparations, and the demonstration that this dsRNA was necessary for gene silencing, gave the first clues to the cause of co-suppression. This phenomenon is now known as RNA interference (RNAi).

Two mechanisms for RNAi are depicted in Fig. B (see Plate 9). In brief, RNAi comprises a set of silencing pathways that share processing of the dsRNA by a member of the dicer protein family into short 21–23-nucleotide **small interfering RNAs (siRNAs)**, whose antisense strand is targeted to its complementary mRNA. In the PTGS pathway (a) the siRNAs are incorporated into a transcript-cleavage complex (RISC), which contains an argonaute-family protein as the catalytic component. The siRNA-loaded RISC complex pairs mRNA transcripts complementary to the siRNA sequence and cleaves them. RdRP (RNA-dependent RNA polymerase) can additionally produce dsRNA from the cleaved RNA products. Another pathway (b) involves the use of a different transcript-cleavage complex (RITS in the fission yeast *Schizosaccharomyces pombe*) containing an argonaute-family protein. The siRNA-loaded RITS complex targets nascent complementary transcripts, still attached to the DNA strand. Cleavage of these transcripts targets the attached DNA region for chromatin modification by the recruitment of methyltransferases and other proteins (see text).

these parasitic sequences. These defence mechanisms have been termed the genome's immune system (Plasterk 2002) in view of their parallelism to the analogous system in vertebrates (Box 3.2). Viruses and TEs are the most important classes of foreign DNA sequences to be controlled by the host genome. The conflict of interests between parasitic elements and their hosts can be equated to an arms race responsible for the evolution of a wealth of coordinated attack–defence mechanisms in all living beings. Some of these defence mechanisms, such as methylation, are also present in bacteria as a defence against phage infection, indicating their ancient origins. The present consensus is that the evolutionary scope of host defence mechanisms transcends their protective influence and is considered an evolutionary trigger of control mechanisms of gene expression. Thus, an ancient conflict has not only induced an evolutionary arms race but is also responsible for important regulatory mechanisms, elicited by host 'domestication' of functions meant for other purposes, namely the host genome defence from DNA parasites. This process is analogous to the co-option mechanism discussed in Chapter 2, p. 34.

Several strategies have been advanced to suppress TEs. One comprises epigenetic global mechanisms that regulate gene expression without changes in nucleotide sequence. These mechanisms, reversible and short-term heritable, include selective DNA cytosine methylation and changes in chromatin configuration for TE sequences that suppress TE transcription. Both methylation and chromatin alteration are intimately connected (Matzke *et al.* 2000). Another silencing strategy, discovered only at the end of the twentieth century, is based on an ancient, conserved RNA silencing mechanism that has been independently discovered in different organisms, from plants to viruses and animals. Initially presented under different descriptions, this RNA silencing process is now recognized as the same mechanism: RNAi.

What is epigenetics?

Epigenetics has a long history. Waddington (1942), a famed geneticist in Edinburgh (UK), coined the term epigenetics to explain 'the interactions of genes with their environment which bring the phenotype into being'. Barely understood during his time, these interactions would have to wait until just recently to be deciphered. It has been known since the 1990s that the mouse locus responsible for the brown coat colour in mice (the *agouti* locus), has, besides the normal *agouti* allele (*A*) and its recessive (*a*), at least four alleles (A^{vy}) that carry a retrotransposon inserted upstream of the gene. Transcription originating in a cryptic promoter of this retrotransposon causes the anomalous expression of agouti protein, which results in yellow fur plus other abnormalities

Box 3.2 The immune system: a transposon-domesticated adaptive trait

About 450 mya, after the divergence between jawless (lampreys) and jawed vertebrates, a crucial transposition occurred in the jawed vertebrate genome that was likely to transform the way of life of all vertebrate descendants. This was the invention of the immune system, a mechanism that most probably contributed to the invasion of new ecological niches by enabling primitive vertebrates to combat parasites and infections. Remarkably, we have solid evidence that the immune system, with all its complexity, evolved from the insertion of a transposon into a primordial

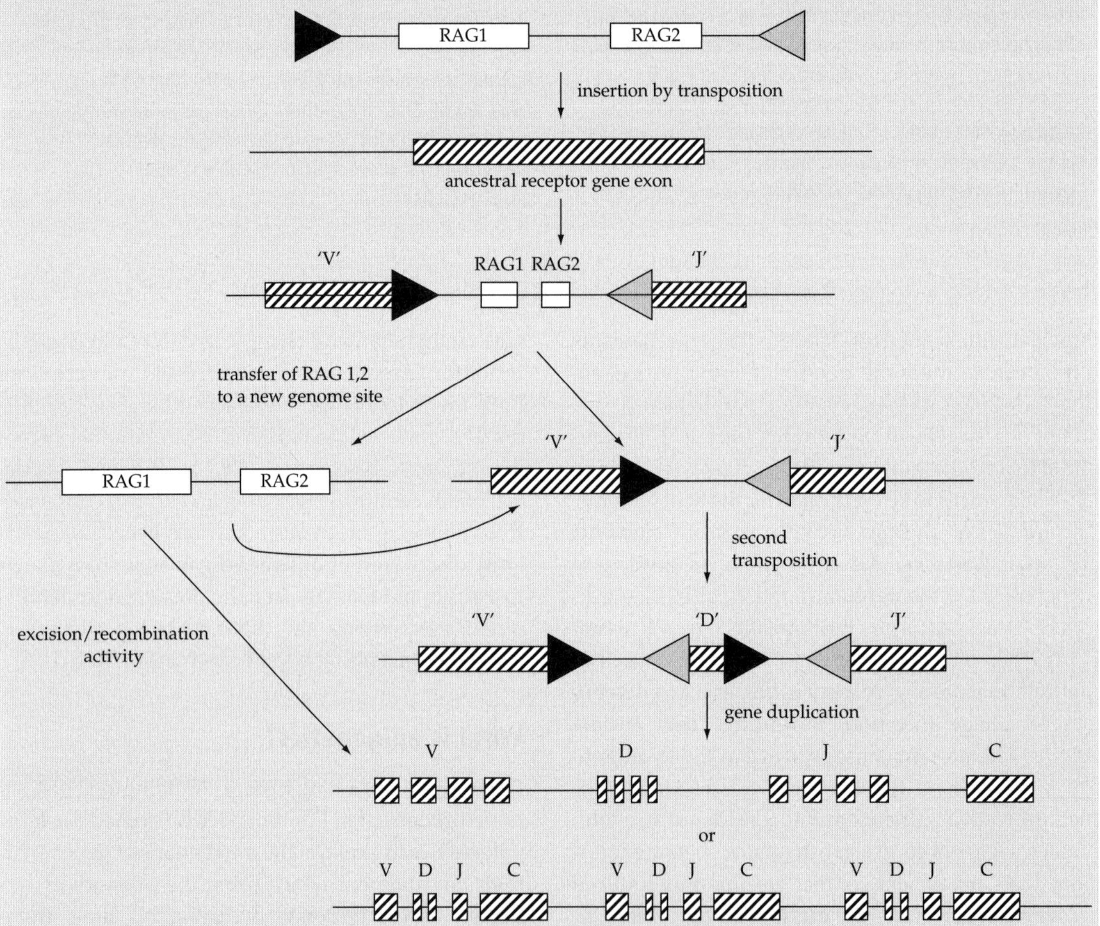

Fig. C. Diagram depiction of an evolutionary hypothesis for the origin of the vertebrate immune system. Initially an ancestral RAG-containing transposon (top) inserted into an ancestral receptor gene (striped bar), which become split into different elements flanked by RSS. The RAG-encoded proteins recognize the terminal sequences of the element (depicted by triangles), excise it from donor sequences and then catalyse a transposition reaction with striking similarities to those catalysed by other transposases. Then *RAG1* and 2 genes excised from split genes to become independent in function and in location. This process of transposition and loss is repeated to generate the "V", "D", "J"-type split genes found today followed by several rounds of duplication that led to the modern clusters of V. D, and J elements (bottom). Today these gene clusters are inactive unless they mature into RAG-driven recombination V-D-J assembles that code antibody-variable regions. The constant region exons are depicted as C bars. Interestingly, *RAG* genes recognize the RSS flanking each V, D, J gene to perform the somatic recombination in a way reminiscent of their ancestral role in transposition and excision. Redrawn from Agrawal *et al.* (1998) by Montserrat Peiró, with permission from Nature Publishing Group.

continues

Box 3.2 *(continued)*

vertebrate gene. Thereafter, a cascade of splits, duplications, and recombination events led to a complex genome architecture that codes for a wealth of proteins (antibodies) able to recognize the myriad alien molecules (antigens) that invade organisms. In order for an organism to be protected, all of these antigens must be recognized and thus each organism must produce of the order of hundreds of millions antibodies. Immunoglobulins (Igs) and T cell receptors (TCRs), produced by B and T lymphocytes respectively, are the major antigen-specific proteins. Since the number of antibody genes is limited (there are only three Ig genes and four TCR genes in the human genome) how can each person generate the huge antibody diversity necessary to build a defence against the massive antigen invasion?

This riddle, the province of 'Lamarckian' speculations for some time, was solved when the 'split' architecture of the Ig and TCR genes was discovered along with the need to assemble its parts by somatic recombination before they can be expressed. The molecules of antibodies are composed of heavy (H) and light (L) polypeptide chains, whose most external end constitutes the variable region responsible for specific-antigen recognition (antigen receptor). The rest of the polypeptide is constant and serves other functions. The variable region is encoded by genes composed of three types of gene segments: V (variable), J (joining), and D (diversity) segments. In the germ line cells there are multiple copies of each gene segment type, thus in the human IgH gene that encodes the variable region of the heavy chain, there are about 125 V segments, 25 D segments, and 9 J segments. These genes are non-functional and are only expressed in B and T cells due to a specific V-D-J somatic recombination. In each lymphocyte a unique specific combination of one of each V, D, and J segments, or of a V and J segment, creates a V(D)J exon-like active assembly that expresses functional products. The total number of V(D)J assemblies is huge (in the order of thousands) for each kind of chain (light or heavy) and when combined with similar numbers for the other chain (heavy or light) it amounts to millions of antibody molecules. Moreover, in vertebrates, somatic hypermutation, gene conversion, and specific recombination with different types of constant units increase antibody diversity. In sum, individual B (and T) lymphocytes produce unique Igs (and TCRs). Every individual is, therefore, a mosaic of millions of lymphocytes (of the order of 10^{12} in humans), each one with a characteristic antibody. This solves the riddle of the immune response.

The V(D)J assembly process is effected by an endonuclease protein (RAG), encoded by *RAG*1 and *RAG*2 genes, that recognizes recombination signal sequences (RSS) flanking every gene segment. The assembly process involves DNA double-strand breaks, repairs, and rejoining, in a manner reminiscent of the initiation of the 'cut-and-paste' mechanism of DNA transposons. Intriguingly, the *RAG* genes lack introns and *RAG1* and *RAG2* form a transposase that can induce in vitro transposition of DNA sequences. The researchers studying this process (Agrawal *et al.* 1998) also found that insertions generate short duplications flanking the transposed segment. These observations led them to propose that *RAG*1, *RAG*2, and the RSS were part of a transposon that entered the jawed-vertebrate genome, and inserted into an ancestral V domain. This event initiated a series of splits, transpositions, and duplications that culminated in the complex immune-response machinery of today (Fig. C). This well-supported hypothesis illustrates nicely how the selfish behaviour of a TE can be co-opted for the benefit of the host. The fact that linkage between *RAG* genes and the inverted repeats they recognize is broken shows that the originally selfish *RAG* genes are not working for their own benefit any more. Rather, they are under host control to act on the V(D)J assembly at the host-required place (the lymphocytes) and time (antigen invasions).

including obesity, diabetes, and susceptibility to tumours. However, this yellow phenotype displays variable expressivity ranging from full yellow, through variegated, to full agouti (pseudoagouti phenotype). Interestingly, these phenotypes correlate with methylation of the inserted retrotransposon: the more methyl groups are added, the browner the colour intensity produced. A yellow dam would be expected to give only yellow pups if the brown phenotype was only due to methylation. But yellow dam progenies contain some brown agouti pups. This was attributed to an effect of maternal metabolism on the phenotype of offspring, rather than the persistence of epigenetic modifications through generations. In the late 1990s Emma Whitelaw and her team at the University of Sydney (Australia) designed experiments to distinguish the two possibilities and they showed that, contrary to the then current thought, the erasure of epigenetic methylation was incomplete during the gametogenesis. Since methylation modifies a transposon-inserted promoter region that affects transcription, the authors

(Morgan *et al.* 1999) anticipated that 'because retrotransposons are abundant in mammalian genomes, this type of inheritance may be common'. This was an important landmark in understanding the role of epigenetics in heredity, but the origin of methyl groups remained obscure. There were some hints that dietary methyl supplementation of normal dams changed the offspring colour distribution towards the agouti phenotype, but no controlled study had yet been carried out.

A major breakthrough came in 2003 when a group of researchers at Duke University (USA) fed yellow, fat pregnant mice with vitamin supplements, as pregnant women are treated today. Although these mice were inbred for the A^{vy} gene responsible for their fat body and yellow coat, this group of vitamin-supplemented females unexpectedly gave birth to thin, brown pups. This came as a big surprise but was tentatively explained by some kind of random, unidentified mutation. Yet, when the babies were genetically examined, they were found to have the same DNA sequence in the A^{vy} allele as their mothers and thus should have been yellow. So what was the explanation for such a coat colour change? The solution to this mystery was uncovered when the DNA sequence for the *agouti* gene was found to be modified by the addition of methyl groups in the thin, brown babies (Waterland and Jirtle 2003). Methylation had malformed the normal expression of the agouti gene to low levels without changing the DNA nucleotide composition (Plate 6). Actually, the vitamin supplement included compounds (vitamin B_{12}, folic cid, betaine, and choline) that form the methyl groups that were responsible for this gene inactivation. The importance of this finding can be summarized in Jirtle's words: 'For the first time ever, we have shown precisely how nutritional supplementation to the mother can permanently alter gene expression on her offspring without altering the genes themselves.'

Two additional results increased the relevance of this experiment. First, methyl groups were not bound randomly along the gene; rather, they were found at a site of the gene occupied by a retrotransposon. This implies that TEs may be attractors of methyl groups and suggests that they take an active part in genome regulation. Second, it has been shown that in subsequent generations, yellow dams produced fewer brown offspring than did brown dams. Initially this was attributed to a maternal effect on metabolic differentiation, but a recent body of work suggests that this is caused by incomplete erasure of epigenetic marks in the female germ line (Morgan *et al.* 1999; Cropley *et al.* 2006). This is not surprising if you recall that the egg that formed your zygote was generated in your mother's ovaries while she was in your grandmother's womb. Thus, it is likely that when your grandmother passed epigenetic signals to your mother, those signals were passed to the egg she contributed to your zygote. In sum, epigenetic signals can imprint genomes across generations providing a complex, transgenerational mechanism of adaptive evolution.

Are your children the product of what you eat during pregnancy?

The link between the mother's nutrition (or other environmental exposures) during pregnancy and the child's future health is common knowledge. What is, then, the epigenetic contribution to this idea? Usually health problems in babies are qualified as birth defects attributed to accidents during gestation. While this is the likely explanation in many cases, it is by no means valid for all occasions. Many experiments similar to that described above with vitamin-supplemented pregnant mice, in which rats, sheep, and other animals were fed a low-protein diet during their early days of pregnancy, have shown that their progenies were prone to health problems, such as high blood pressure and thickened arteries because of abnormal fat storage. Surprisingly, this only happened when these babies were fed with a normal diet and not when they were undernourished. In view of this diet-conditioned response, the likely explanation is the acquisition of an adaptive response to optimize food availability in a resource-poor environment. Most probably this is also due to an epigenetic modification of the DNA that generates a kind of 'thrifty' phenotype that works efficiently under frugal conditions but gets out of control in abundance.

Obesity is a plague in children of our time. In the United States one-third of children, that's about 25 million, are overweight or obese. This figure is similar or steadily approached in other economically developed countries, including new affluent societies like Spain. It is no surprise that this problem is catching the attention of health managers and the general public because obesity is a major cause of heart diseases due to high blood pressure and high levels of cholesterol, triglycerides, and sugar. Traditionally, the culprits of this epidemic were identified as a diet composed mainly of sugary and fatty products, combined with many hours of sedentary TV watching and video game playing. There is certainly ample support for this, but recent research suggests that other agents may be important as well. Among them, DNA epigenetic modifications induced by what these children's parents, especially mothers in early pregnancy, are exposed to, like food quality and cigarette smoke, could be of paramount importance.

The link between poor foetal nutrition and later obesity was suggested 20 years ago by David Barker, a British medical professor. This hypothesis (Hales and Barker 2001), named the 'thrifty phenotype' hypothesis (or the Barker hypothesis), states that foetuses that sense a poor nutritional environment develop 'thrifty' metabolisms that are very efficient at sparing energy. This process was likely very effective in the prehistoric nutrient-poor environment where it evolved, but is very unfit in the present nutrient-rich environment because it induces obesity. For quite some time, researchers have noticed the capacity of organisms to produce offspring that seem to be modelled according to their mother's environment during pregnancy. These observations include organisms as diverse as locusts, voles, freshwater fleas (Daphnia, a crustacean), and lizards. For instance, baby voles are born with either a thick fur or a thin fur depending on the intensity of light the mother experiences during pregnancy (an indication of the year's season); Daphnia mothers give birth to offspring with a larger helmet and spines if the environment is crowded with predators; at least one species of lizard is born larger if the mother smelled a snake during pregnancy, as if big bodies were more protective against predators; and, most surprisingly, when locusts find themselves crowded because of a period of food scarcity that follows a period of abundance, baby locusts are born with more colour and with gregarious behaviour, which protects them from predators. In all these cases there are no DNA changes, rather epigenetic changes during pregnancy that control foetal development. In mammals, including humans, epigenetic effects in response to poor nutrition during pregnancy provide newborn babies with an initial bonus to cope with a harsh nutritional environment early after birth, akin to that a baby would likely find prior to the agricultural era 10,000 years ago. But foetuses that developed a thrifty metabolism are likely to become fat if they are born to our current, high-calorie (but often nutritionally poor) diet.

As in the case of epigenetically modified thin, brown baby mice, there are many observations in humans suggestive of epigenetic changes that passed through several generations. One of the most studied cases is known as the 'Hunger Winter' of 1944–45 in Holland. Due to a combination of a harsh winter and a cruel embargo by the Nazis during the Second World War, the population of Holland experienced a major famine. Women in their first six months of pregnancy during this winter gave birth to small children who were later more prone to obesity, heart disease, and even to cancer. Moreover, the children of women born during this famine, who grew up in a well nourished environment, were also born small. We still do not know whether this transgenerational effect is due to an incomplete erasure of methyl markers that contributed to this growth phenotype, but it is a likely possibility.

Is your child's behaviour the result of your parental grooming?

Michael Meany, a researcher at McGill University, Canada, and his team published in 2004 another epigenetic blockbuster. So far I have described cases of epigenetic changes before birth, but Meany and his colleagues showed that methylation changes to DNA can also occur after birth. Moreover, they showed in rats that these changes can be induced by the interaction between mothers and their babies.

The experiment (Weaver *et al.* 2004) was carried out with two groups of rats. One consisted of mothers that licked and groomed (LG) and arched-back nursed (ABN) their babies with high intensity; mothers in the other group were much less attentive to their offspring. The offspring of 'high-LG-ABN' mothers grew into less fearful rats and could handle stress better than the offspring of 'low-LG-ABN' mothers. While this could be explained by purely genetic differences, the researchers performed cross-fostering studies that showed otherwise. They provided babies from 'low loving' mothers to 'high loving' mothers, and vice versa. Those pups that were licked and groomed while nursed developed into calm and stress-managed adults regardless of their natural mother's behaviour (and vice versa). Now, again, this could be explained by purely non-genetic transmission of maternal behaviour. So, the nature vs nurture argument remained unsolved.

The riddle was solved when analysis of a glucocorticoid receptor (*GR*) gene, involved in brain development, showed that its promoter was less methylated in highly groomed pups than in pups that received less attention from their mothers, biological or adopted (Fig. 3.3). It seems as if the mother's care had triggered the removal of methyl radicals that would have impeded the development of a part of their babies' brain. These differences appeared over the first week of life, were reversed by cross-fostering, and persisted into adulthood. Although these results were met with scepticism when first published, they have increasingly been accepted since then.

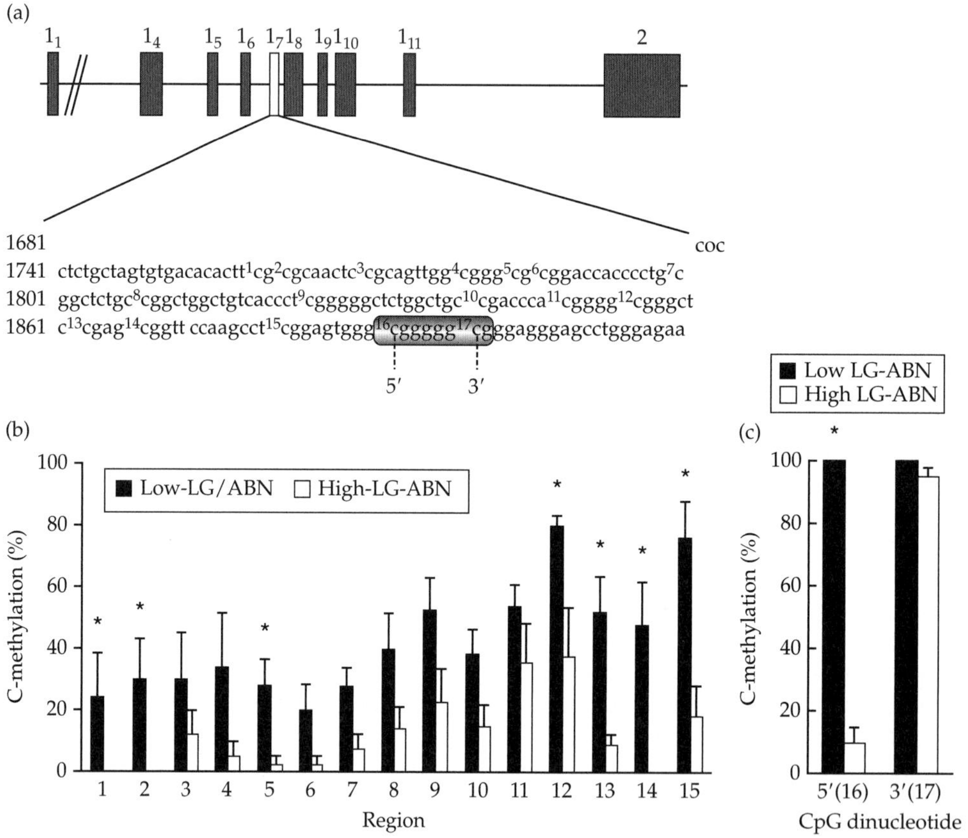

Fig. 3.3. (a) Scheme of the GR gene showing the non-coding exon 1 region. This region includes a promoter region (exon 1_7) that contains a binding site for a transcription factor (nerve growth factor-inducible protein A: NGFI-A). The figure depicts part of the exon 1_7 sequence showing the 17 CpG dinucleotides (depicted by superscript numbers) and the NGFI-A binding region (encircled). (b and c) Bar diagrams showing that all the 17 CpG dinucleotides are less methylated (open bars vs full bars) in offspring of rat mothers that provided increased pup licking and grooming (LG) and arched-back nursing (ABN). * indicates statistically highly significant differences. From Weaver *et al.* (2004), with permission from Nature Publishing Group.

Epigenetics: the ultimate human genome frontier

We don't know whether parental care for human babies has the same influence on human brain development through DNA methylation as it does in rats. But, we do know that a loving, relaxed parental environment tends to make children happier and healthier; and the reverse is also true. Why then, if cognitive and physical development in humans are more correlated than in most other animals, shouldn't humans be prone to similar epigenetic changes in childhood? Epigenetics as a field is in its infancy, but the more we discover, the more we uncover connections between methylation and regulatory pathways in humans. Scientists have reported connections between cancer and gene methylation in such different kinds as breast, colon, prostate, and esophageal cancers. These incipient clues notwithstanding, we have a long way to go before we can decipher all of the intricacies of epigenetic pathways. What seems to be very clear is that now that we have a well-advanced Human Genome Project it is time to engage in the Human Epigenome Project if we want to fully understand genome dynamics. This project, underway since 2003 and started by a group of European scientists, aims at assigning a map of every place in the human genome where methyl markers can change the expression of a gene.

Epigenetics and TE silencing

The relevance of epigenetic mechanisms for adjusting an organism to environmental stresses is paralleled by their importance as a defence against genome invasions. In fact, both processes may represent two sides of the same coin. McClintock (1984) pioneered the idea of 'the significance of responses of the genome to challenge', as she presented her Nobel lecture with this title. The more we know about the non-coding fraction of the genome (about 98% of the transcriptional output in humans) the greater our understanding is of the regulatory potential of this fraction, especially in the form of small RNA transcripts (see Chapter 1). Since a major component of this non-coding fraction is made of TEs, it comes as no surprise that TEs are likely to be important participants in the regulatory pathways of the genome. Moreover, since TEs act as parasitic elements to the genome, they must be the subject of strong genomic control in order to attain a stable coexistence in this ancient conflict between them and the genome. Although the regulation of TEs may be exerted via different pathways, RNAi-epigenetic mechanisms are likely to be some of the most prevalent.

Among the best documented silencing-epigenetic mechanisms, methylation—the binding of methyl groups—appears at the top of the list. Most of the methylation in eukaryotes is found in TEs and in centromeric DNA, mainly composed of TE sequences and other repeats (Yoder *et al.* 1997). This association, and the use of methylation by bacteria as a defence against phages, as already mentioned, suggests that methylation is a primary, ancient host defence mechanism against foreign DNA sequences, namely TEs. That methylation is an ancestral mechanism is also supported by its presence in all phylogenetic groups. Even the fruit fly (*Drosophila melanogaster*), long considered the epitome of a non-methylated organism, has shown a low degree of methylation in all life stages except in early embryos, where methylation is higher. However, the target nucleotide sequence for the binding of methyl groups is different from the classical CpG dinucleotides (Lyko 2001). These peculiarities are likely to be responsible for the historical difficulties in detecting methylation in this species. Recently, methylation has been detected in many Drosophila species and a wide range of other insects (Marhold *et al.* 2004), although it may have evolved to perform various epigenetic functions different from those in vertebrates.

TE silencing and chromatin modification

Methylation is usually targeted to cytosine residues of CpG nucleotides, but not exclusively (see Plate 7 for a summary of epigenetic modifications). Methyl groups can also be found on histone (H3) at lysine 9 residues in inactivated chromatin associated with TEs. In fact, TE inactivation via chromatin modification is mediated by cytosine and histone methylation, but proteins that modify chromatin structure are also required, as has been demonstrated in the plant *Arabidopsis thaliana* (Lippman *et al.* 2004). In a number of eukaryotes, RNAi-mediated chromatin modifications have also been shown. In *Schizosacharomyces pombe*, the fission yeast, pericentromeric heterochromatin (the portion of chromatin that remains densely stained and condensed after cell division) is modified by an RNAi-mediated mechanism that parallels the PTGS (post-transcriptional gene silencing) but in this case it works on nascent RNAs (see Box 3.1). These RNAs, still bound to the chromosome, are cleaved by the RITS (RNA Induced Transcriptional Silencing) complex, which contains an argonaute-family protein. This cleavage recruits methyltransferases, the enzymes that incorporate methyl groups in both histones and DNA, and other chromatin-modifying proteins to this region.

Centromeres and telomeres, two chromosomal features essential for chromosome and genome functions, are defined by a heterochromatic structure. In most eukaryotes, centromeres consist of inner tandem arrays of short-sequence repeats (satellite DNA) flanked by an outer pericentromeric region that contains mostly TEs. In humans, the inner centromeric region is generally free of TEs, but the pericentromeric region is composed of long stretches of non-LTR retrotransposons. The DNA composition of centromeres differs widely between species (Dawe and Henikoff 2006). For example, *A. thaliana*, a highly studied species for RNAi-silencing mechanisms, shows a satellite region interspersed with retrotransposons and a pericentromeric region full of DNA transposons. In this species RNAi genes have duplicated and diversified extensively to code for several different chromatin-modifying proteins, each with a specific function. Thus, two distinct siRNAs, one long and another short, are generated, depending on the specific dicer protein involved. The long siRNA size class operates in RNA-dependent DNA methylation, whereas the short size class participates in PTGS. Since these siRNAs (see below) are derived predominantly from TEs and tandem repeats, it is intriguing that TEs that are silenced in heterochromatin can give rise to RNA transcripts that are subsequently cleaved into smaller RNAs. In *A. thaliana*, the plant-specific RNA polymerase IV is responsible for this process (Herr *et al.* 2005), but in mammals the mechanisms participating in RNAi-mediated chromatin modification are still poorly understood.

Regardless of the variety of heterochromatin modification mechanisms, the involvement of TEs in shaping chromosome form and function cannot be denied. Centromeres are important structures that assure the symmetrical chromosome segregation in mitosis and meiosis. When epigenetically silenced TEs and satellite repeats are reactivated, chromosome segregation is impaired, accompanied by losses of centromere condensation and sister chromatid cohesion (Lippman and Martienssen 2004). Analogously, inactivation of dicer proteins in mammals results in accumulation of centromeric transcripts, most probably because cleavage into siRNA is impaired (Fukagawa *et al.* 2004). This evidence bolsters the prime role of TEs in heterochromatin function and formation. Even centromeric satellite repeats might be the outcome of TE dynamics, as their sequence homology to some TEs suggests.

TEs are also responsible for telomere form and function. Telomere structures resemble that of centromeres. This distal part of the chromosome consists of short tandem repeats attached to proximal stretches of heterochromatin, in a subtelomeric position, which is reminiscent of the pericentromeric heterochromatin as it too is composed of TEs. In many species, repeats are added by telomerase, a reverse transcriptase, which uses an RNA template analogous to non-LTR retrotransposon insertion. In other species that lack telomerase, like Drosophila species, addition is carried out by insertion of specific non-LTR retrotransposons (named *HeT-A* and *TART*). These retrotransposons are epigenetically regulated in the same way as described above for other TEs. Although RNAi-mediated chromatin modification is not as well deciphered in mammals, some experiments relate methylation with artificially inserted siRNA, which makes it a likely mechanism of silencing.

RNA silencing on centromeric repeat transcripts is well established in plants and animals and, at least in the fission yeast *Schizosaccharomyces pombe*, serves to support centromere-specific cohesion in sister-chromatid mitosis segregation. The suggestion that outer centromeric repeats and, possibly, satellite sequences within centromeres are derived from or show similarities to TEs, is based on resemblances between TE terminal inverted repeats and DNA protein-binding motifs in human centromeric satellite repeats. To date, much experimental evidence shows that small RNA-directed chromatin modifications and gene silencing is present in centromeres of fungi, metazoans, and plants (see Almeida and Allshire 2005 for a review).

The mysterious origin of double-stranded RNAs (dsRNAs) and their role as clues to foreign invasions of the genome

Perhaps the least documented step of the RNAi mechanism is the origin of dsRNA from TEs and other sources. The origin of dsRNAs is likely to be multiple (Fig. 3.4). It may originate from exogenous sources, such as RNA viruses, or it can be induced by foreign DNA, like TEs. In organisms such as *C. elegans* and plants, at least three processes for the derivation of dsRNAs have been advanced. First, DNA transposons with terminal inverted repeats (TIR) can be transcribed to form a hairpin dsRNA structure by homologous pairing of these terminal sequences (Plasterk 2002). Second, multiple random insertions of an element increases the chance transcription from both ends, generating two complementary RNA strands to form dsRNAs; one, the sense strand, starting from its own flanking promoter, the other, the

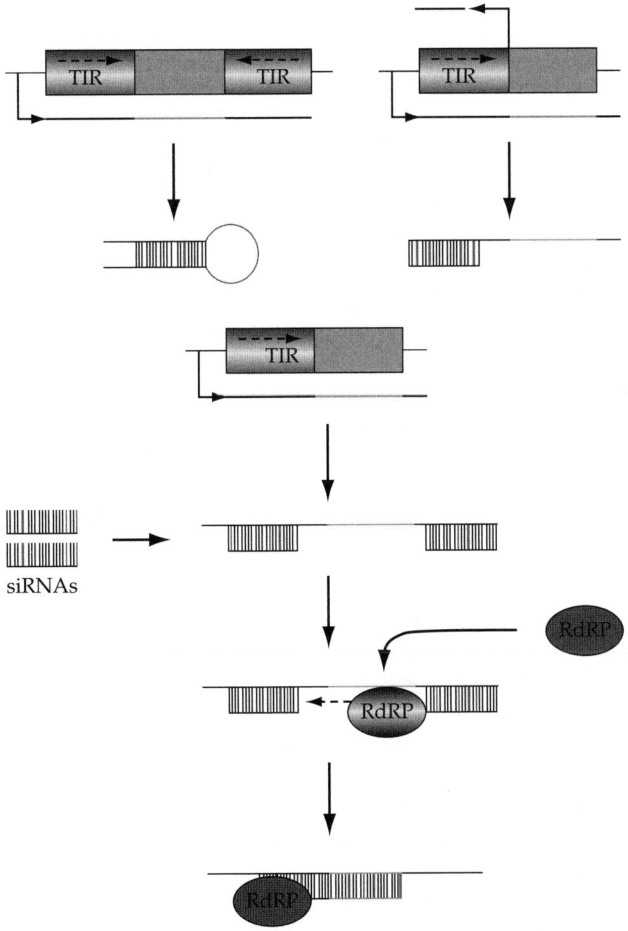

Fig. 3.4. How dsRNAs are generated. First, read-through transcription of the entire DNA transposons might produce a hairpin due to their terminal inverted-repeat (TIR) structure (top left). Second, dsRNA could also be formed by two overlapping antiparallel transcripts (top right). Finally (bottom), dsRNA could be produced by RNA-dependent RNA polymerases (RdRPs) that copy ssRNA into dsRNA. How RdRP select transcripts as foreign is not well understood (see text for details). From Slotkin and Martienssen (2007) with permission from Oxford University Press.

antisense strand, initiated by a flanking promoter accidentally located close to the insertion site. This mechanism acts as a sensor for TE spreading in the genome; the more copies are transposed the greater the probability of finding a genome promoter in the vicinity of the element. The third mechanism relies on the ability of the genome to sense the foreign DNA sequences. In general, when an overproduction of RNA transcripts exists, it is believed that an RNA-dependent RNA polymerase (RdRP) can utilize directly these single-stranded RNA (ssRNA) transcripts as templates to synthesize dsRNAs that enter the RNAi cycle. There is quite a bit of controversy over how these RNA transcripts are recognized as foreign by RdRP, the standard explanation being that these are 'aberrant' ssRNAs produced by several defective processes such as premature termination of transcription, inappropriate splicing, lack of poly(A)$^+$, or a failure to be translated. Aberrant ssRNA increases its probability with transcript overproduction, yet this model is still under experimental tests (see Zamore 2002 for a review).

From host genome defence to imprinting and gene control

The recruitment of host defence mechanisms for gene control is dependent, following the above RNAi model, on gene sequence recognition as foreign DNA. Imprinting, the differential expression of alleles according to their maternal or paternal origin, may be co-opted from the host defence mechanisms following TE invasion (McDonald *et al.* 2005). Methylation of cytosines is the primary mark of imprinting (Lie *et al.* 1993). Cytosine positions are found in megabase stretches of CpG dinucleotides (CpG islands) that are located in the promoter regions of many genes, including housekeeping genes and X-chromosome inactivated genes, in addition to those found in TEs. Most, if not all, imprinted genes contain tandemly repeated sequences located near CpG islands, suggesting that they aid in the methylation process. In fact, all imprinted genes are methylated in regions containing, or adjacent to, direct repeats. In addition, many imprinted genes, at least in mammals, transcribe non-coding RNAs and overlapping sense and antisense sequences. All these features (DNA repeats, 'de novo' methylation, sense/antisense RNA transcripts) are reminiscent of the RNAi mechanism and suggest that imprinted genes may be perceived as alien DNA by the host.

As is the case for X-chromosome inactivation, the silencing of most genes in one X chromosome in female mammals, genomic imprinting is typically found in clusters of genes usually located in large domains in mammals (Lee 2003). X-chromosome inactivation is dependent on sense (*Xist* RNA) and antisense (*Tsix* RNA) transcription of the *Xist* locus (Luikenhuis *et al.* 2001). It is postulated that *Xist* RNA may progress along the X chromosome and associate with promoter regions that become targets of de novo methylation. Originally proposed by Lyon (1998), the non-random chromosomal distribution of long interspersed nuclear elements (LINE1 retrotransposons) supports this model (Bailey *et al.* 2000). Not only are LINE1 sequences more abundant in the human X chromosome than on the autosomes, but, most importantly, these elements cluster within the X inactivation centre (Xic) of inactivated genes and are significantly reduced in numbers in the non-inactivated genes. The fact that the Xic-specific sequence is present in the *Xist* RNA strongly suggests that inactivated genes show enhanced sequence target homology for *Xist*-directed methylation, a condition absent from the LINE-depleted promoters of non-inactivated genes.

As stated above, much of the RNA silencing through RNAi mechanisms is targeted to centromeric repeats. However, many repetitive sequences are dispersed throughout the genome and we thus must ask whether a similar mechanism can be assigned to gene regulation in other non-centromeric regions. As mentioned above, many gene mutations are due to TE insertions. Among them, those that influence the expression levels of neighbouring genes due to their effect on the transcriptional activity are most relevant to RNA-directed gene silencing. Some evidence points to the formation of silent chromatin at TE insertions that influences changes in gene expression. A comprehensive survey of the complete *S. pombe* genome showed that most of the 300 Tf1/2 LTRs are dispersed as solo elements and only a few are included in full-length retrotransposons. At least in one case, repression of neighbouring genes was related to RNA silencing by an LTR subjected to RNA-mediated histone methylation and heterochromatin protein binding (see for a review Almeida and Allshire 2005). In plants, methylation is circumscribed mainly to TE sequences and promoters of silenced genes. Also, in mammalian somatic cells, satellite DNA and TEs show extensive DNA and histone methylation and histone deacetylation. In the mouse genome, chromatin studies show that modifications associated with TEs and LTRs are very active during differentiation of embryonic stem cells, but it is not clear whether transcriptional repression is the process involved in cell differentiation, although it is a likely possibility. In sum, the relevance of RNA-directed gene silencing is clear and evident in processes such as chromatin structure, imprinting, X-chromosome inactivation, and gene regulation. Most importantly, on the evolutionary side, is that all of these mechanisms may be co-opted from an ancient mechanism of host defence against the genome invasion by foreign nucleic acid sequences, such as viruses and mobile elements.

The evolutionary force of junk DNA

The preceding paragraphs provide insights on the putative, co-opted role of TEs as regulatory elements in the genome. Although McClintock's visionary tenets, dated from more than half a century ago, were bolstered by numerous observations of the impact of TEs in gene function and structure (see Kidwell and Lisch 2000), the real evolutionary force of TEs was underappreciated until the current genomics era. Thus, whole genome sequencing has revealed that, in spite of the long-held view that TE insertions in coding or regulatory regions mainly inactivate genes, many active genes contain TEs. In humans, for example, more than 200,000 *Alu*s (a non-LTR retrotransposon) are in genes, and more than 20% of our genes have TEs or TE-derived sequences in non-coding flanking regions, many of them identified as gene promoters (Jordan *et al.* 2003). Moreover, the contribution of MITEs to the building of regulatory regions seems likely in many plants ranging from maize to rice and sugarcane (see Burt and Trivers 2006, pp. 272–7 for a review).

Despite the fact that their magnitude may seem staggering, these figures are probably a low estimate of the real impact of TE insertions in the genome. Current whole genome comparisons are revealing that some conserved non-coding elements (CNEs), i.e those that have evolved through strong purifying selection, derive from TEs (Feschotte 2008). Since most of the insertions detected are very old, many pre-dating the mammalian radiation or much earlier, the accumulation of many mutations makes it difficult to trace them back to TEs. This handicap notwithstanding, a few cases of recent insertions in Drosophila have shown the selective value of TE sequences co-opted to regulate gene expression (see Box 3.3). Moreover, other historical episodes of population bursts of transposition (see below) to gene-rich regions of the genome may have contributed to adaptation to new foreign environments. These observations give support for an adaptive value for most, if not all, CNEs whose derivation from TE cannot be unambiguously traced.

In view of their ubiquity TEs have the potential to provide the genome with large amounts of variability that can be co-opted for novel functions. Yet insertion variability is not the only mechanism by which TEs contribute to genome evolution. It has long been known that TEs can reorganize the genome by mobilizing stretches of DNA ranging from a few hundred nucleotides to long chromosomal segments. These mechanisms have been documented many times and deserve close scrutiny to understand the evolutionary force of TEs in total.

Recombination and exonization: the double life of *Alus*

The great majority of TEs in our genome belong to two non-LTR retrotransposon families, namely LINEs and SINEs (see Fig. 3.1). Although LINEs can transpose autonomously, SINEs are not autonomous and depend on the transposition machinery of LINEs. In spite of its dependency, the *Alu* element, a member of the SINE family, has invaded our genome with more than 1 million copies, which represents about 13% of our total genomic content. Since more than 200,000 *Alu* copies are found within genes, the obvious question is how our genome can stand such a negative insertional mutational load. In fact, more than 20 human diseases can be explained by *Alu* insertions, which illustrates their mutagenic potential.

Box 3.3 Insecticide resistance by transposition

Insecticide resistance is a prime example of how evolution can be observed with our own eyes in a short time span. Because of its rapid establishment, the easy identification of its selective agent, and its economic importance, a lot of effort has been devoted to decipher the underlying adaptive mechanism of insecticide resistance. In *Drosophila melanogaster* DDT resistance is due to overexpression of the cytochrome *Cyp6g1* gene. Daborn and his coworkers at the University of Melbourne (Australia) have shown that up-regulation of this gene is correlated with the presence of a terminal direct repeat of an *Accord* transposable element inserted upstream from the transcription start site of this gene (see Fig. D). This insertion is polymorphic in *D. melanogaster*: present and widespread in resistant populations, but absent in susceptible populations. Interestingly, all the resistant *Drosophila* strains analysed not only contain the *Accord* insertion but they also show an identical nucleotide *Cyp6g1* gene sequence. In contrast, the susceptible strains revealed a set of diverse sequences. Population geneticists interpreted this sequence distribution as the result of selection pressure in favour of the widespread unique gene sequence, i.e. the allele or **haplotype** with the insertion. Although the association between resistance and TE insertion was very tight, in this early study (Daborn *et al.* 2002) the causal relationship between TE and overexpression was not proven. In other words, it could be that *Cyp6g1* was up-regulated by another closely linked regulatory sequence that hitchhiked on the *Accord* insertion. Fortunately, the Daborn team, using transgenic flies, later (Chung *et al.* 2007) showed that the *Accord* LTR hosts regulatory enhancer sequences responsible for the overexpression of *Cyp6g1* in tissues involved in detoxification, such as the midgut, Malpighian tubules, and body fat.

Another group of researchers has provided an extraordinary case of parallel evolution of insecticide resistance. Schlenke and Begun (2004) were interested in detecting genomic regions that showed reduced variability because they are putative markers of beneficial mutation sites. When a beneficial mutation appears, selection spreads the mutation and its adjacent DNA as a block, at such a speed that it impedes recombination along the flanking region of the mutation. Called selective sweep or hitchhiking, this phenomenon generates a species-wide invasion by a unique long stretch of DNA devoid of variability. Thus, the almost worldwide sequence identity of the *Accord*-resistant *Cyp6g1* region is explained by a selective sweep. Schlenke and Begun were not, in principle, interested in insecticide-resistant mutation sites, but their 3 Mb genomic survey in *D. simulans* led them to discover a 100 Kb region devoid of variability. To their surprise, this region hosted the same *D. melanogaster Cyp6g1* gene with an upstream TE insertion, albeit a different one. This insertion (*Doc*) occurs at very high frequency in all California strains, but it is absent in original populations in Africa, all in accordance with a recent selective sweep. Moreover, though it was a different element, the insertion also caused *Cyp6g1* over-transcription, as in the *Accord*-inserted strains.

This remarkable parallelism between the two species for insertion-mediated population and regulatory evolution prompted the researchers to investigate the putative association between the *Doc* insertion and DDT resistance. Although the *Doc* insertion lines showed an average resistance

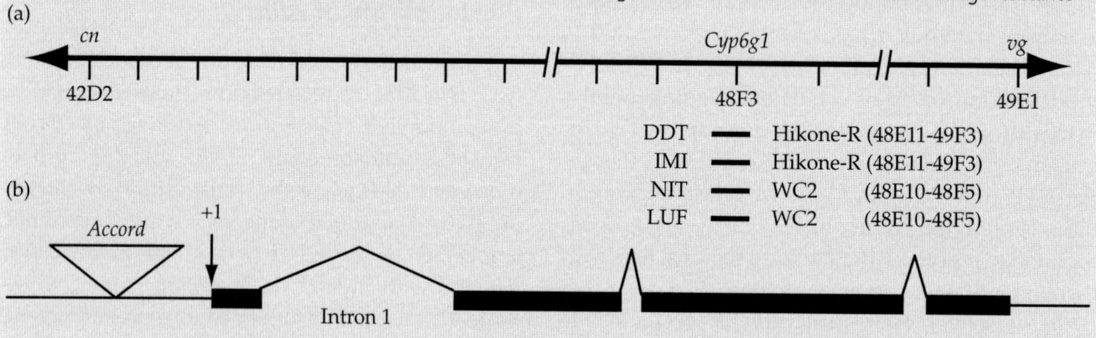

Fig. D. (a) Genetic map using visible mutants (*cn*, cinnabar eyes, and *vg*, vestigial wings), *P*-element insertions and other markers show that resistance to DDT and other insecticides (IMI, NIT, and LUF) map to the region encompassing *Cyp6g1* (location estimations depicted as bars followed by the name of the resistant allele and its predicted cytological interval position, in parenthesis). (b) Detailed map of the genomic structure of the *Cyp6g1* locus. Exons are depicted by solid bars. Note the *Accord* insertion site in 5′ position of the locus. From Daborn *et al.* (2002), with permission from the American Association for the Advancement of Science.

continues

Box 3.3 *(continued)*

slightly higher than non-*Doc* lines, this difference was only attributable to one line of extremely lower mortality. Most probably, the researchers conclude, 'other types of selection pressures at Cyp*6g1* remain highly plausible'. Since DDT has been banned for more than 30 years in California, yet continues to be used in tse-tse fly control in Africa, 'it is difficult to explain', the researchers follow, 'why the Doc insertion haplotype continues to persist at high frequencies in California and has not occurred in Africa'. The likely explanation is that the selective agent in *D. simulans* is not DDT but other insecticide, a natural toxin, or even an environmental contaminant present in California but not in Africa.

Another example of the *Doc* TE family versatility in providing adaptive resistance has been produced by the Petrov's group at Stanford University. Yael Aminetzach, Petrov's graduate student, observed that *Doc*1420, one member of the *Doc* family, was present in 80% of *D. melanogaster* world-wide populations (Aminetzach *et al.* 2005), suggesting that this insertion may play a selective role. Briefly, the coding gene, into the second exon of which the TE inserted, resembles those involved in choline metabolism, which is affected by pesticides. The researchers showed that *Doc*-carrier strains survive an organophosphate insecticide (AZM) better than those lacking the insertion. Since the insertion generates several truncated transcripts, it was hypothesized that no regulatory changes were likely to be at work and rather a new protein product was produced that induced insecticide resistance.

All in all, these works are compatible with the idea that TE insertions underlie pesticide resistance evolution and that adaptive selection is rapid and opportunistic often, but not always, by co-opting former regulatory functions already present in TEs. Moreover, the reported evidence emphasizes the importance of population genomics in wide-genome screening for non-variable DNA stretches, suggestive of selective sweeps. This methodology is prone to reveal adaptive mutations that, like those induced by TE transpositions, are otherwise difficult to detect.

The riddle might have been solved when genomic studies (Bailey *et al.* 2003) uncovered a burst of *Alu* transposition about 40 mya, during the primate radiation. At that time, researchers hypothesized that *Alu*s likely used, with high efficiency, the mobilizing machinery of LINEs, but, since then, this transposition ability has diminished along the primate lineage. Consequently, the current transposition rate of *L1*s, the most abundant LINE superfamily, and *Alu*s is likely 100 times lower than in the initial stages of the primate radiation. In spite of the huge numbers of *L1* copies (about 500,000) in mouse and human genomes, probably less than 100 are active in humans and only around 3,000 are active in mice. This explains the observation that only two in 1,000 new mutations are due to retroelement insertions in humans, while this rate rises to 10% in mice.

This low mobilization activity may explain the low mutagenic impact of retroelements in the human genome and has led many researchers to postulate that their role in evolution was also negligible. Yet, this assessment has been contradicted by many recent studies. First, retroelements need not be active to engage in recombination (unequal or ectopic) between copies occupying different sites. These recombination events promote large genomic rearrangements ranging from duplications to inversions (Fig. 1.10) as well as rearrangements involving shorter tracts of DNA. Recent whole genome studies revealed that the human genome contains about 5–6% of low copy repeats (LCR, also named segmental duplications SD) of DNA sequences up to 250 Kb long (see Chapter 1). These LCRs often contain genic sequences frequently associated with hereditary diseases. Interestingly, the ubiquity of *Alu*s, especially in genic regions, makes them likely candidates for LCR formation through **unequal recombination** between homologous copies. The prime role of *Alu*s in the evolution of LCRs is supported by their excess presence in these large genomic blocks and at their perimeters. In sum, *Alu*-mediated mechanisms of gene-rich expansions of segmental duplications are likely to be responsible for content differences in gene family members among humans, mice, and other mammals.

The second class of observations in favour of the evolutionary value of retroelements refers to their presence in coding and conserved non-coding regions of genes (see above). Here, the insertion of

Alus in **introns** deserves our attention. We know that a large fraction of human genes (40–60%) engage in different patterns of alternative splicing, generating different messenger RNAs that translate into several proteins. In this way complex organisms, like humans and mice, evolve new protein functions without the need for increasing the number of genes. Interestingly, it has been shown that *Alus* in human introns play a crucial role in splicing patterns. Whole-genome studies have shown that some *Alus* inserted in introns have evolved into **exons**, a process known as 'exonization' (Fig. 3.5). It is estimated that over 5% of all alternatively spliced exons in humans are the result of *Alu*-exonizations (Nekrutenko and Li 2001). Such alternative splicing represents an evolutionary mechanism to test new protein isoforms without losing the adaptive function of the normal protein.

Alu-exonization is likely a common process. An estimated number of about 80,000 *Alu* elements located in introns may become exons just by a single base-pair mutation in an intron splice site. While the staggering numbers of potential *Alu*-derived exons show a great source of genome variability for evolution, it must be pointed out that a substantial number of these mutations are likely deleterious. In fact several human genetic diseases are produced by *Alu*-derived exons that are present in all transcripts derived from the involved gene. This is the case of the Alport syndrome that results from a base pair mutation that activates a cryptic splice site in an *Alu*-containing intron of a collagen-type gene. The end result of this mutation is the presence of an *Alu* exon in all transcripts, which are translated into defective collagen.

In sum, the *Alu* (and *L1*) evolutionary history illustrates the capacity for TEs to generate genetic variability through such diverse mechanisms as unequal recombination and exonization. It cannot be denied that new genomic functions originate from the activity of TEs, but also that this same activity is responsible for anomalous phenotypes that are at the source of human diseases. In this sense, as it is in all mutagenic agents, TEs possess a double life, both positive and negative, and it is natural selection that determines whether a TE insertion provides a selective advantage to the individual for going ahead in evolution.

Exon shuffling by transposition

Reorganization processes underlie much of the evolution of the genome. They can be observed ranging from gross, i.e. chromosomal or whole genome

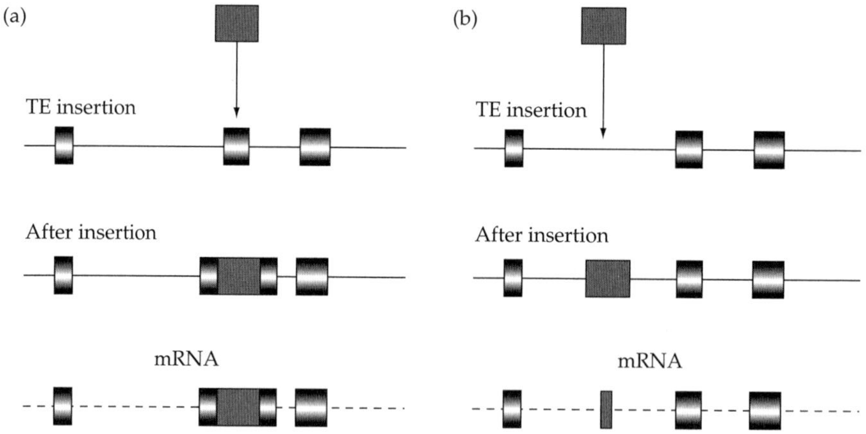

Fig. 3.5. Two possible mechanisms of transposable element (TE) insertion into protein-coding regions. (a) TE is inserted directly into a protein-coding exon. The effects of the direct insertion are likely to be deleterious because TEs often contain multiple stop codons and would destroy the target exon. (b) TE is inserted into an intron. Later a portion of the TE is recruited as a novel exon. This scenario is preferred for two reasons: first, in many instances the novel exon is alternatively spliced and may not destroy the function of the gene; and second, typically only a fragment of the TE insert is recruited, which is less likely to contain stop codons. From Nekrutenko and Li (2001) with permission from Elsevier.

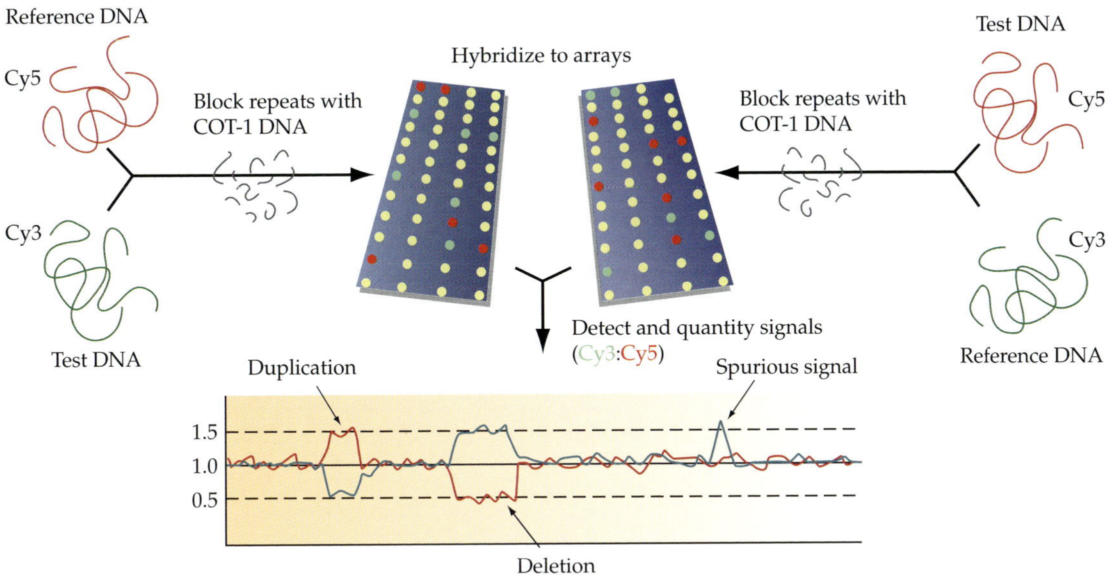

Plate 1. Schematic illustration of the array-based, genome-wide method for identification of CNV. Reference and test DNA samples are differentially labelled with fluorescent tags (Cy5 and Cy3, respectively), and are then hybridized to genomic arrays spotted with one of several DNA sources (BAC clones, PCR fragments, or any kind of DNA fragments) (left side). After hybridization, the fluorescence ratio (Cy3:Cy5) is determined. A lower ratio indicates an absence of DNA in the test DNA relative to reference DNA, i.e. a deletion; likewise a higher ratio reveals a duplication. To detect spurious signals a reversed labelling test is carried out, which must show a reciprocal signal (right side). (See Fig.1.13a) From Feuk *et al.* (2006) with permission from Nature Publishing Group.

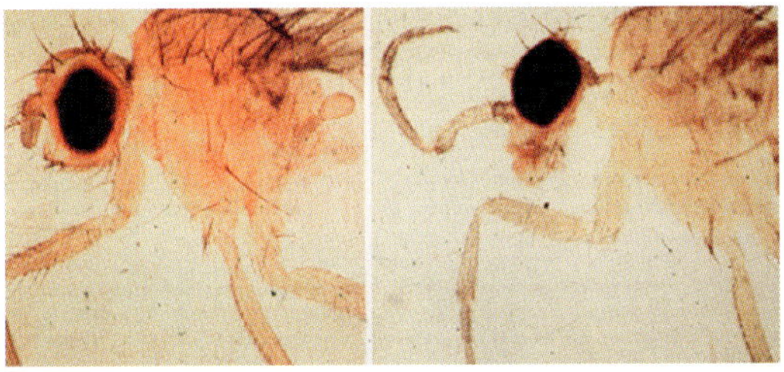

Plate 2. Antennapedia mutant (right) compared with a wild type fly (left). Notice how an antenna is transformed into a leg in the mutant. Courtesy of Ginés Morata.

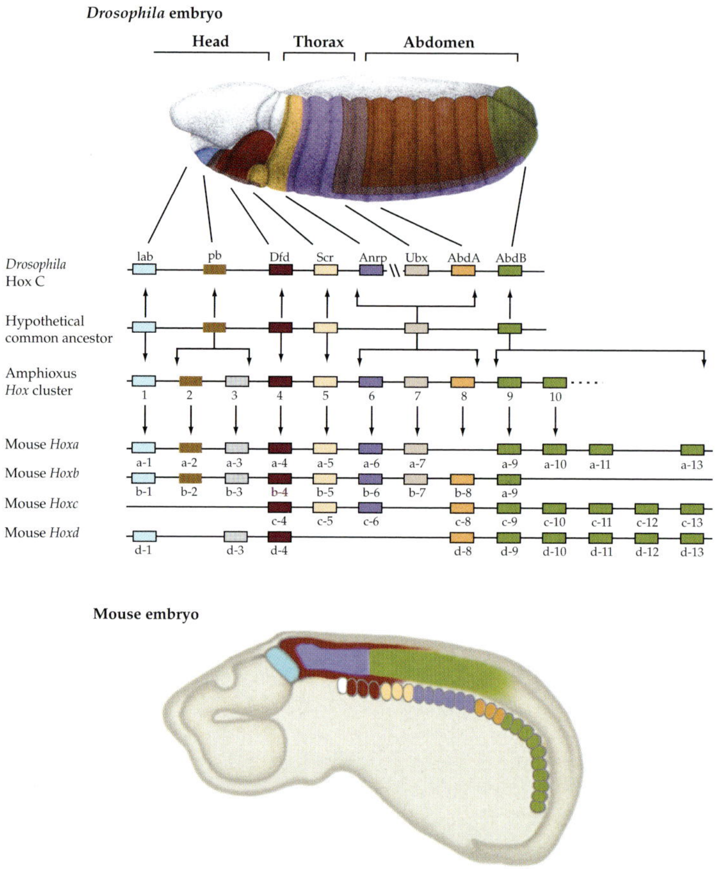

Plate 3. Homologous *Hox* gene organization and expression. The anterior–posterior (A–P) body domains of *Drosophila* (top) and mouse (bottom) of *Hox* gene expression correspond to gene order within the *Hox* complexes. The middle of the figure depicts the homologous gene relationships (arrows) between Drosophila, Amphioxus, and mouse *Hox* clusters, and also the deduced *Hox* complex in the common ancestor of arthropods and chordates. (See Fig. 2.3.) From Carroll (1995) with permission from Nature Publishing Group.

Plates 4. Polyphenism in *Byciclus anyana*. This species shows two seasonal morphs: the dry season morph (DSM) and the wet season morph (WSM). Adults of the DSM (a) possess uniformly brown wings with largely reduced ventral eyespots in hindwings. These butterflies are well camouflaged when they rest amongst dried leaf litter in the dry season. On the other hand, WSM adults (b) are coloured with conspicuous eyespots on their ventral wingside that are exposed when at rest, probably for deflecting predatory attacks from the body or serving as an intimidating function. Photos courtesy of André Coetzer (a) and Oskar Brattström (b).

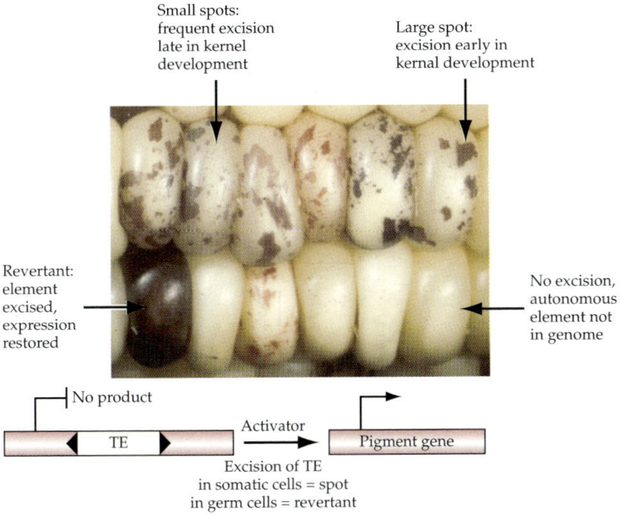

Plate 5. Using kernel phenotypes to study transposon behaviour. Kernels on a maize ear show unstable phenotypes due to the interplay between a transposable element insertion in an activator gene and an activated gene that encodes an enzyme in the anthocyanin (pigment) biosynthetic pathway. Sectors of revertant (pigmented) aleurone tissue result from the excision of the TE from the activator gene that restores the encoding expression of the pigment gene in a single cell. The size of the sector reflects the time in kernel development at which excision occurred, i.e. the larger the size the earlier the excision. An understanding of the genetic basis of this and similar mutant phenotypes led to the discovery of TEs. (See Fig. 3.2). From Feschotte *et al.* (2002) with permission from Nature Publishing Group.

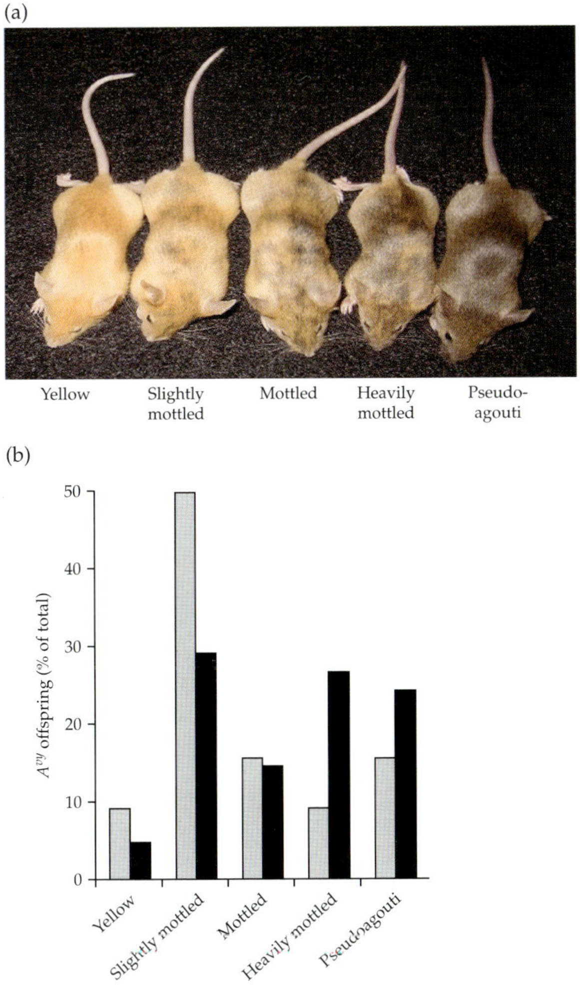

Plate 6. Maternal dietary methyl supplementation and coat colour phenotype of A^{vy}/a offspring. (a) Isogenic A^{vy}/a animals representing the five coat colour classes used to classify phenotype. The A^{vy} alleles of yellow mice are hypomethylated, allowing maximal ectopic *agouti* expression. A^{vy} hypermethylation silences ectopic *agouti* expression in pseudoagouti animals, recapitulating the agouti phenotype. (b) Coat colour distribution of all A^{vy}/a offspring born to nine unsupplemented dams (30 offspring; shaded bars) and 10 supplemented dams (39 offspring; black bars). The coat colour distribution of supplemented offspring is shifted toward the pseudoagouti phenotype compared to that of unsupplemented offspring. From Waterland and Jirtle (2003) with permission from the American Society for Microbiology.

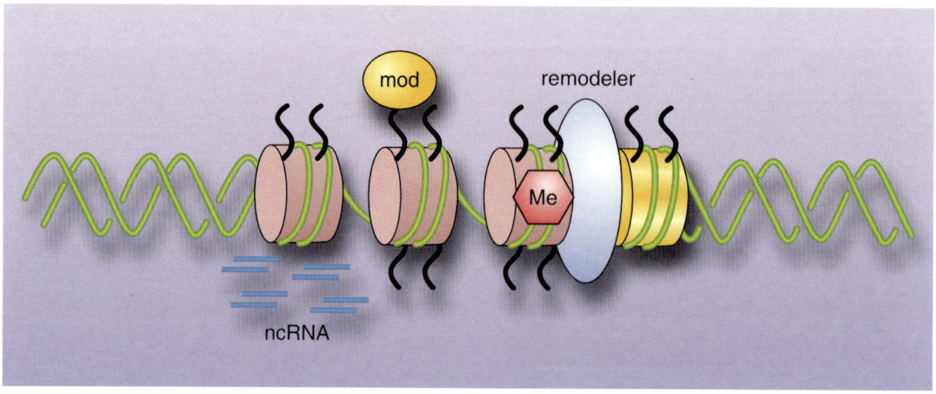

Plate 7. A summary depiction of epigenetic variations by (1) histone modifications (mod), (2) chromatin remodelling (remodeller), (3) hystone variant composition (yellow nucleosome), (4) DNA methylation (Me), and (5) non-coding RNAs. (See text for further details). From Allis *et al.* (2007) with permission from Cold Spring Harbor Laboratory Press.

Plate 8. A sample of Petunia phenotypes due to RNA interference. White areas in flowers are generated by silencing a pigment gene through RNAi mechanisms. (See Box 3.1, Fig. A.) From Grosshans and Filipovicz (2008) with permission from Nature Publishing Group.

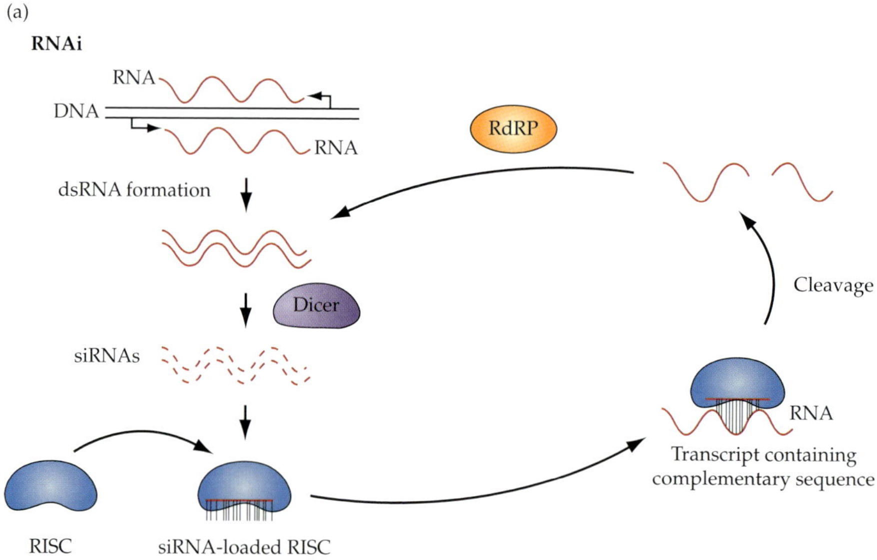

Plate 9. Post-transcriptional silencing by RNAi. (See Box 3.1, Fig. B). (See text for further details) From Slotkin and Martienssen 2007 with permission from Nature Publishing Group.

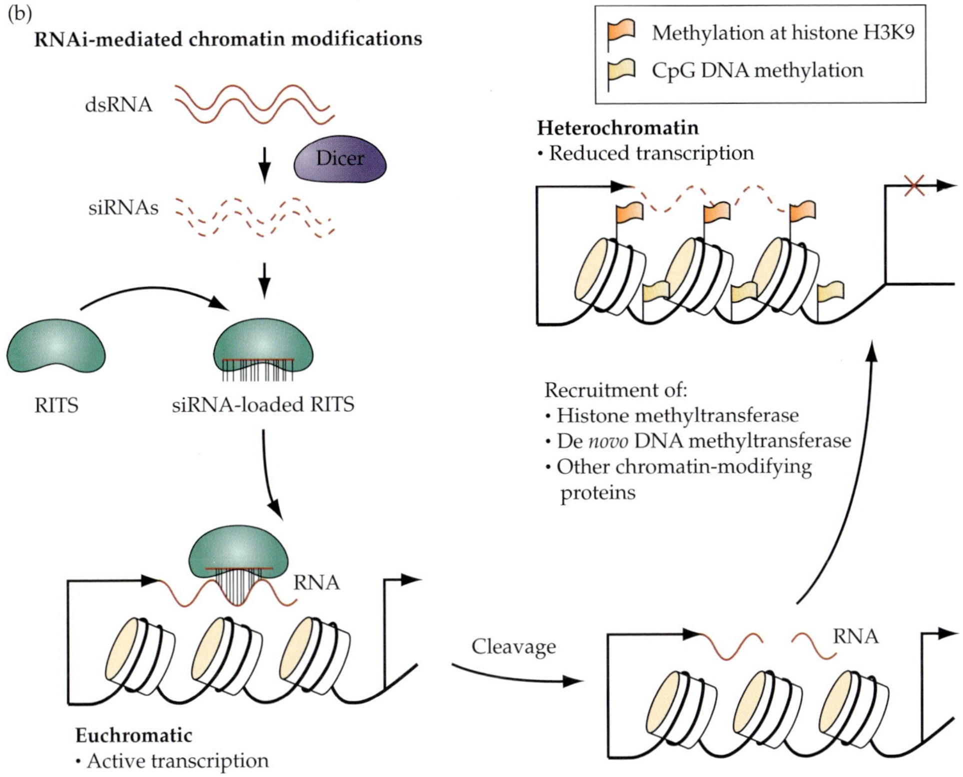

Plate 9. *Continued*

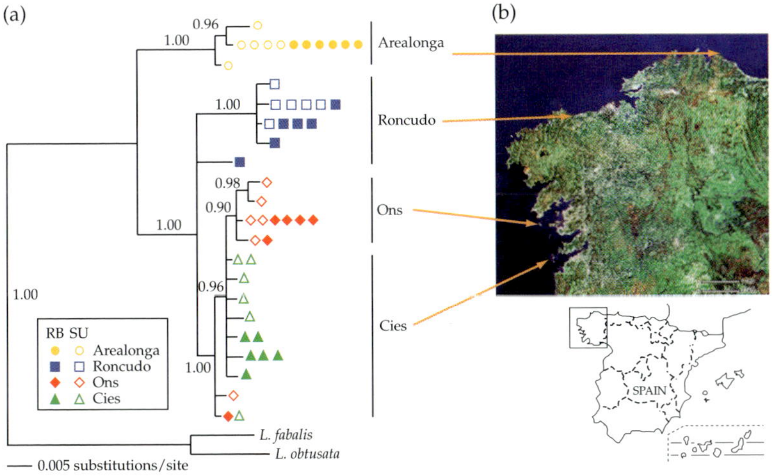

Plate 10. Phylogenetic tree using mtDNA nucleotide variants of the two kinds of *L. saxatilis* ecotypes (RB close and SU open symbols) in four different localities of north-west Spain. Note that both ecotypes cluster together in each locality, which indicates a common geographical origin, a result incompatible with the allopatric model. (See text for further details). (See Box 4.1, Fig. A). From Quesada *et al.* (2007) with permission from John Wiley and Sons.

Plate 11. (a) *Heliconius melpomene*: note the red forewing band controlled by the *HbM* gene locus. (b) *Heliconius heurippa*: this species originated by homoploid hybrid speciation between *H. melpomene* and *H. cydna*. Note the double band (red and white) that combines the bands from its parental species. The HbM-controlled red forewing band is signalled by an arrow. (c). *Heliconius cydna*: note the white forewing band. Courtesy of Mauricio Linares. Photography by Juan G. Montañes; produced by Mauricio Linares.

Plate 12. A parasitic dodder (*Cuscuta californica*) in flower, penetrating intracellularly a host tomato plant by means of haustoria. Courtesy of © Barry Rice, sarracenia.com.

Plate 13. Flowers of *M. lewisii* (a), an F_1 hybrid (b), *M. cardinalis* (c), and examples of variation in floral traits found in F_2 hybrids (d–l). From Schemske and Bradshaw (1999) with permission from the Proceedings of the National Academy of Sciences (USA).

Plate 14. (a) A threespine marine stickleback fish with pelvic spines (red arrow). (b and c) Two freshwater stickleback fishes that have lost their pelvic spines present in the marine (and probably ancestral) population. Courtesy of Job de Roij.

Plate 15. Some developmental abnormalities appearing in *Hsp83* mutant stocks: (a), deformed foreleg and transformed 2nd leg with an ectopic sex-comb (arrow); (b), deformed eye with an extra antennae (arrow); (c), smooth eyes with black facets; (d), eye margin transformed into scutellum. Abnormal F1 hybrids produced from crosses between *Hsp83* mutant stocks and marked laboratory strains: (e), left eye has black facets; (f), disorganized abdominal tergites; (g), small wings; (h), extraneous tissue growing out of tracheal pit (arrow); (i), eyes absent; (j), wing margin material growing into wing. Heteroallelic *Hsp83* combination: (k), severely deformed legs; (l), severe black-facet phenotype. Abnormal F1 hybrids produced with wild-type laboratory stocks and *Hsp83* mutants: (m), thickened wing veins; (n), transformed wing and extra scutellar bristle (arrow). Abnormalities in wild-type lines raised on geldanamycin: (o), notched wings; (p), deformed eye. From Rutherford and Lindquist (1998) with permission from Nature Publishing Group.

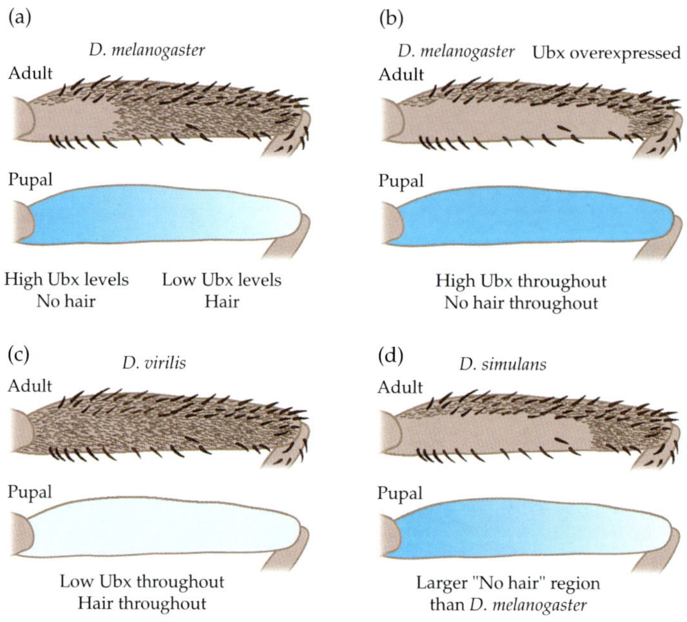

Plate 16. Depiction of the correlation of the *Ubx* expression level (blue intensity) with hairiness of femur in Drosophila species. From Barton *et al.* (2007) with permission from Cold Spring Harbor Laboratory Press.

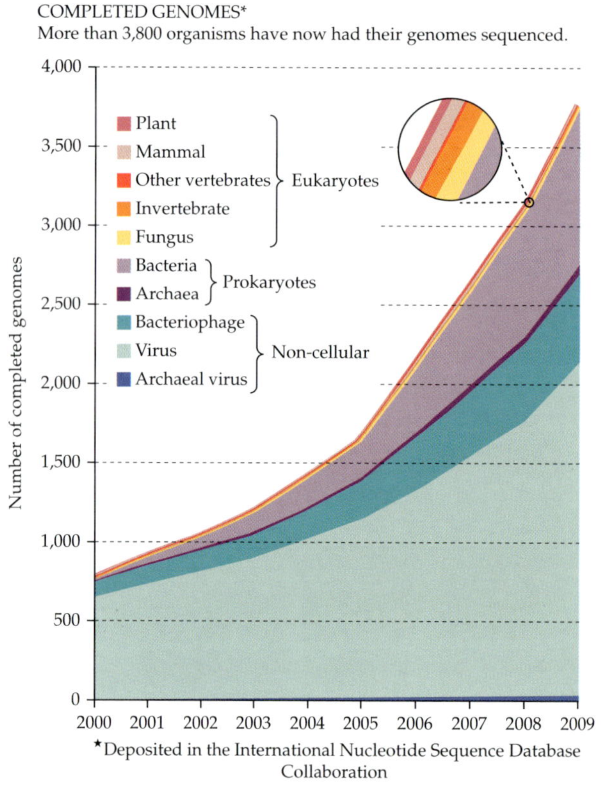

Plate 17. The figure depicts the ever-increasing number of deposited completed genomes in the last decade. (See Fig. 5.7.) From Venter (2010) with permission from Nature Publishing Group.

duplications, to small rearrangements, such as LCRs and TE transpositions. In the late 1970s, the discovery of the interrupted structure of the eukaryotic gene, in which stretches of coding DNA (exons) are separated by segments of non-coding DNA (introns), prompted many evolutionists to propose that exons could be the building units of genes. Assemblage between exons was advanced as one mechanism of gene evolution. Two main observations support this **exon shuffling** hypothesis. First, often each domain of a protein corresponds to an exon of its coding gene. Many complex proteins are built by assembling different domains, some of them repetitive and each one with a specific function, that are coded by a corresponding number of exons (Fig. 3.6). Second, introns are known to recombine because of their homology, in a manner analogous to *Alu*-mediated unequal recombination discussed above. The end result of intron recombination would be the shuffling of exons to produce a new gene.

The intriguing side of the evolution of the eukaryotic gene is also highlighted by the evolutionary origin of introns. Were they present ever since the first common ancestor of all lineages (the intron-early

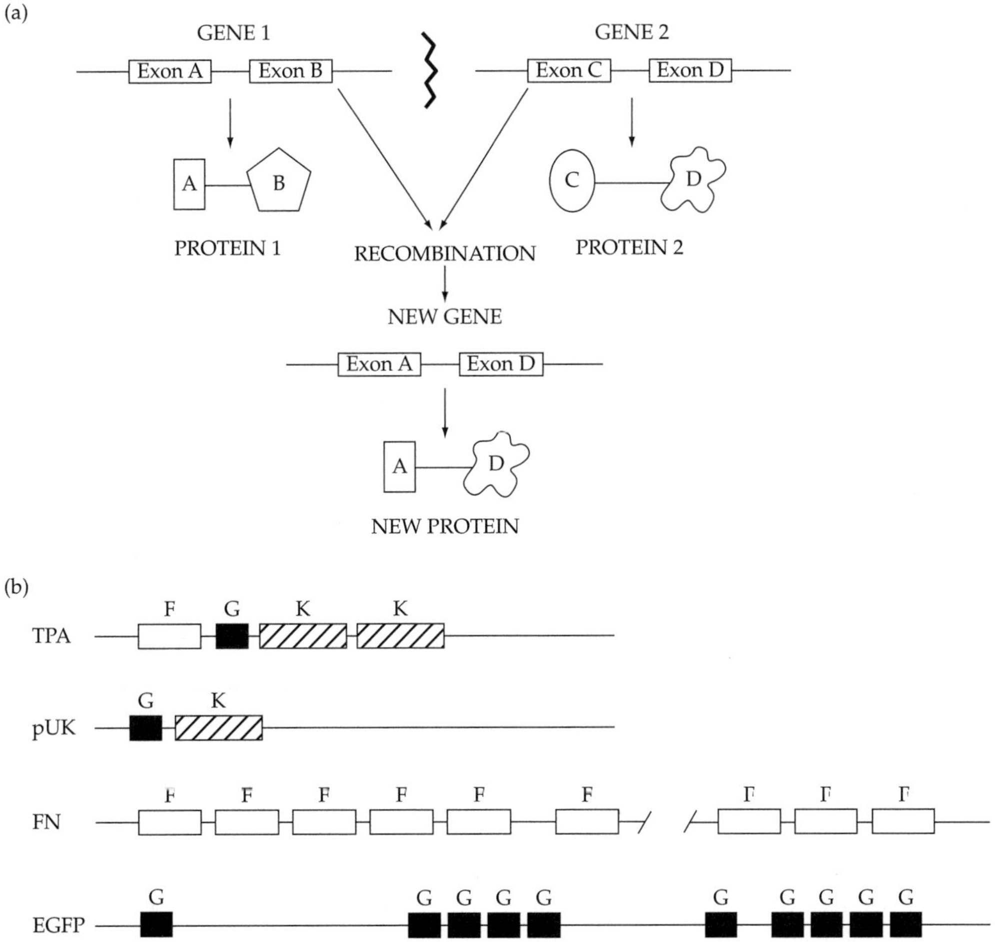

Fig. 3.6. (a) Sketch of protein evolution by assembling protein domains (A, B, C, D) as a result of exon shuffling through recombination. (b) Examples of domain assemblage by exon shuffling and duplication. Human fibronectin (FN) and epidermal growth factor precursor (EGFP) genes evolved by duplication of multiple copies of exons encoding protein domains (F and G, respectively). These domain-coding exons have been found in many other genes; for instance the tissue plasminogen activator (TPA) gene evolved by exon shuffling of three domain-coding exons: F (from FN), G (from EGFP), and K (Kringle domain in the ApoA protein-encoding gene). Also, the prourokinase (pUK) gene evolved by assembling two exons (G and K).

hypothesis), as Gilbert (1987) proposed? Or, are they the end result of late insertions (the intron-late hypothesis)? One can refer to recent reviews of this unresolved issue (see Lynch 2007, p. 242), but here it suffices to say that even the hard intron-early defenders now concede that a large number of introns are of recent origin. The lack of exon-domain exact correspondence in many proteins, the demonstration that introns can insert into genes, and the fact that exon-shuffling is almost exclusively shown in protein-genes of recent evolution in higher eukaryotes, are some of the observations in favour of the hypothesis for the late origin of introns. One potential source for the late origin of introns is TEs. In particular, although TE insertions are usually deleterious, some cases have been reported in which transposons are spliced from plant (Wessler 1989) and *C. elegans* pre-mRNAs (Li and Shaw 1993; Rushforth and Anderson 1996) prior to translation, in a manner analogous to intron splicing. These suggestive mechanisms for the origin of introns notwithstanding, this continues to be a controversial issue that must not obscure the creative role of exon shuffling through intron recombination and exonization, in which the involvement of TEs is likely to be real.

Exon shuffling is also mediated by insertion of coding and regulatory DNA through TE-transduction events. **Transduction** originally referred to the transfer of short pieces of DNA from donor to recipient bacteria via a phage. Currently, however, this term is used more widely to designate the TE-mediated transposition of stretches of DNA inside the genome. This mechanism is well documented since retrotransposon transcription often skips the termination polyadenylation signal of the element and continues downstream until a further 3′ genomic termination signal is found. The end transcript contains the retrotransposon together with a flanking genomic DNA stretch that may include coding, regulatory, or intronic sequences. After retrotransposition, this flanking sequence is transduced to a new position, resulting in shuffling of new duplicated genomic DNA, where it could potentially affect gene function.

In the late 1990s the first clear-cut experimental evidence of exon-shuffling by transduction was reported in highly modified *L1* elements to achieve high retrotransposition rates in human cell cultures (Moran *et al.* 1999). Since then, a few human whole-genome studies using young *L1*s, i.e. those of recent insertion, suggested that about 15% of all insertions originated via transduction, which by extrapolation led the authors to predict that the amount of DNA transduced by *L1* represents around 1% of the genome (Pickeral *et al.* 2000). This is not a low figure at all, but the impact of 3′ transduction in the human genome may be even higher. Recently, Mark Batzer and his collaborators (Xing *et al.* 2006) have discovered SVA, a new hominid non-LTR retrotransposon that represents the youngest retrotransposon family in primates which is currently active in humans. This activity is responsible for transducing downstream (3′) but also upstream (5′) sequences. Interestingly, whole-genome analyses indicate that SVA-mediated transduction is responsible not only for exon shuffling, but also provides a mechanism for gene duplication and the evolution of new gene families. One group of genes have been identified as deriving from the source locus AMAC, a condensing enzyme, by SVA-mediated transduction events that occurred after the divergence of African apes and orangutans but before the divergence of humans, chimpanzees, and gorillas (about 7–14 mya) (Fig. 3.7). Moreover, the AMAC gene was found to be subjected to strong purifying selection before the duplication events, but all AMAC copies experienced a relaxation of selection after the transduction-mediated duplications. This is in good agreement with the fate of new duplicated copies whose redundancy initially allows a less constrained evolution (see Chapter 1).

The transducing potential of SVA elements for exon shuffling and gene duplication seems to be stronger than that of *L1* elements, possibly because the downstream termination signals are less efficient within SVAs. It is estimated that about 10% of human SVA elements (out of a total of 3,000 copies) are involved in transduction episodes. This is a much higher proportion than that active in *L1*. However, other mechanisms of DNA capture and transduction have been described in other non-retrotransposon TEs. While in bacterial evolution the transfer of genes mediated by DNA transposons and bacteriophages has been long recognized, in eukaryotes it was only recently acknowledged that DNA transposons contain pieces of host genes.

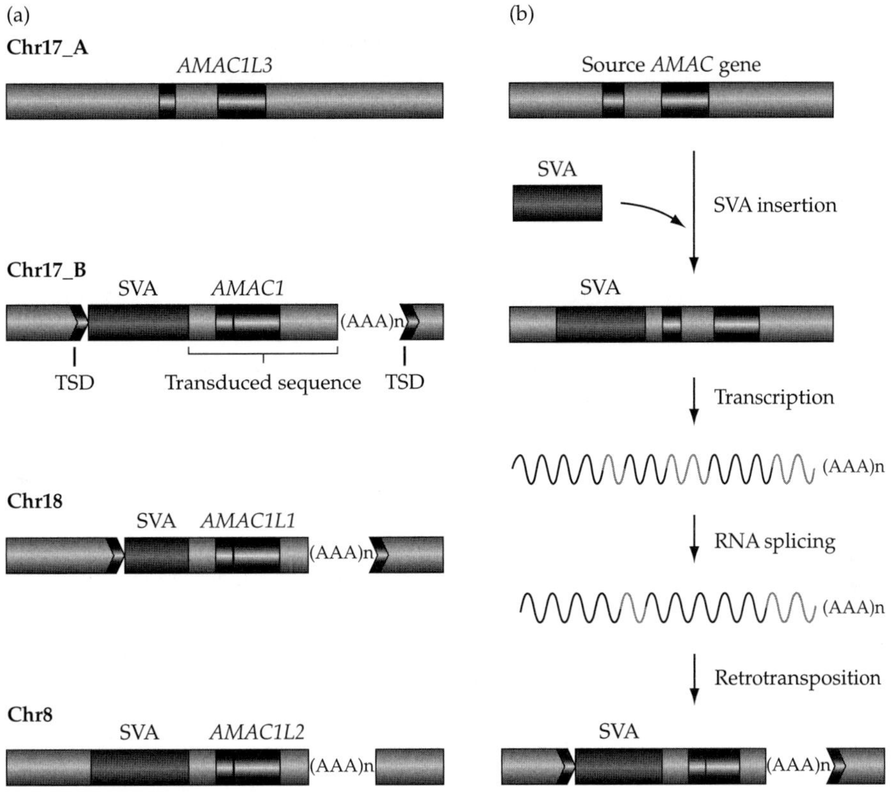

Fig. 3.7. Origin of a gene family by retrotransposon-mediated transduction. (a) Three transduced sequences located on chromosomes 8, 17, and 18 originated from the same source locus that is located elsewhere on chromosome 17. The ancestral AMAC1L3 gene copy at the source locus on chromosome 17 consisted of two exons (depicted as dark grey boxes) separated by an intron. By contrast, the three transduced copies of AMAC1L3 (AMAC1, AMAC1L1, and AMAC1L2) were intronless as a result of the intron splicing during the retrotransposition process. SVA elements are shown as dark grey boxes, and the SVA element oligo(dA)-rich tails are shown as '(AAA)n'. (b) The proposed evolutionary process is shown in this flow diagram. First, non-LTR retrotransposon SVA inserted in 5′ position of the source AMAC gene. Later the whole complex is transcribed and retrotransduced into different genome loci, giving rise to the AMAC family. Interestingly, analyses of RNA transcripts (depicted as waving lines) have shown that AMAC promoter sequences have also been duplicated (and retrotranscribed) along with coding regions. This explains that retrotransposon-mediated duplication can lead to conservation of gene function, as it has been confirmed by the expression of at least two of the AMAC1 copies. TSD means target side duplication. From Cordaux and Batzer (2009) with permission from Nature Publishing Group.

In plants, transposon-mediated exon shuffling and gene duplication has been documented, as exemplified by mutator-like and helitron-like elements. Some mutator elements (*Mu*) belonging to a family of DNA transposons discovered in maize mutants, that contained a fragment of a gene, were found to increase in numbers, indicating that this captured DNA does not impede transposition. Recently, studies of whole plant genomes, including Arabidopsis and rice, have led to the discovery of mutator-like TEs (named MULEs) that capture many gene fragments. These chimeric structures, known as Pack-MULEs, are abundant in many eukaryotic genomes, especially in plants. In rice, more than 3,000 copies containing fragments from over 1,000 genes have been documented, some of them identified in cDNA studies. Many of the structures found in Pack-MULEs are complex chimeric combinations of multiple gene fragments that have been rearranged and amplified over time, showing their potential to generate exon shuffling (Fig. 3.8).

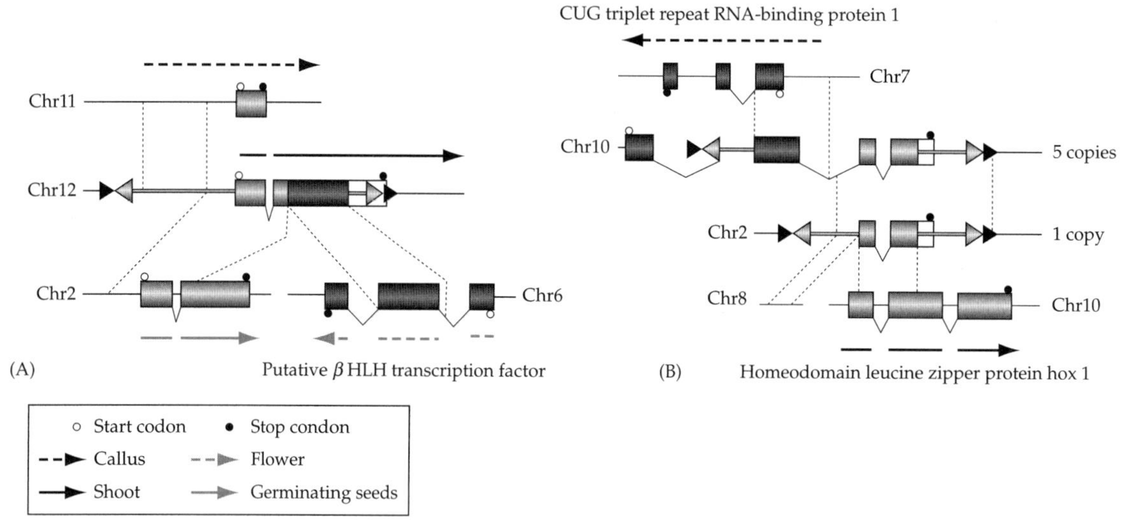

Fig. 3.8. Two examples of rice chimeric Pack-MULEs depicting their structure and genomic origin. (A) A Pack-MULE on chromosome 12 containing gene fragments from three genomic loci in chromosomes 11, 2, and 6. (B) A Pack-MULE on chromosome 10 showing its possible step-wise formation containing sequences from four loci in chromosomes 7, 8, 10, and 2. Chromosome 2 provides an intermediate fragment containing gene fragments from chromosomes 8 and 10. Terminal inverted repeats (TIRs) are shown as grey arrowheads and target site duplications are shown as black arrowheads. Homologous regions are associated with dashed lines. Long dotted arrows indicate sequences matching cDNAs from the designated tissues. Exons are depicted as grey boxes and the introns as the lines connecting exons. The gene name is given for putative genes and hypothetical proteins; all other genes encode unknown proteins. From Wessler (2006) with permission from Oxford University Press.

Mobilization in nature: the population approach

The involvement of TEs in long-term genome evolution has been documented in the preceding paragraphs. But what is the role of TEs in recent adaptive evolution? How much of local adaptation is the outcome of TE insertions? Perhaps one of the greatest paradoxes in modern evolutionary biology is why it took so long to discover TEs. Today we know that their low mobility in most extant populations may be the reason for this delay. Many site-occupancy profiles in chromosomes showed a general low-frequency distribution of TEs, which gave support to the tenet that TEs were of no adaptive value to the host. Otherwise, why weren't TEs driven to fixation if they provided increasing fitness to their host?

In *Drosophila melanogaster* several cases exist of TE insertions at high population frequencies, but in only a few of them has a close relationship with adaptation been shown (see Box 3.3). Although the paucity of examples of TE-related positive adaptation in populations is in agreement with the view that most of the TE insertions are deleterious and maintained at low frequencies, some researchers argue that a systematic genomic search for adaptive TE insertions has never been done. Recently, Dimitri Petrov and his team have screened the *D. melanogaster* genome for recent TE insertions, that is, after the expansion of this species out of Africa some 10,000–16,000 years ago. This team (Gonzalez *et al.* 2008) reported that out of 902 insertions a group of 13 putatively adaptive TEs was found. These insertions are mostly responsible for regulatory changes in genes involved in out-of-Africa adaptation to temperate climates. The authors estimate that the real number of TEs that have contributed to adaptation of *D. melanogaster* in its worldwide colonization may rise to 25–50. If true, this represents about one adaptive insertion every 200 to 1,250 years since the out-of-Africa migration: quite a remarkable rate.

Indeed, the rate of TE-induced adaptation is of the same order as the overall genomic rate of adaptive evolution (Macpherson *et al.* 2007). However, if all adaptive TEs are destined to attain fixation, we should see about 5,000 TEs in the Drosophila euchromatic genome instead of the 114 fixed TE-insertions observed. There may be several likely explanations for this paucity of fixed insertions. Some of them may rely on the evolutionary scenarios, namely non-constant transposition rates, environment-differential adaptation, and lack of detection due to faster sequence divergence caused by individual positive selection. However, other element-characteristic molecular strategies of genome invasion and perpetuation may be involved as well.

The natural scenario of TE invasions

One plausible scenario of DNA transposon invasion and persistence in populations includes the following phases. At the beginning functional elements are rare and crosses are mostly between a parent with an element and a parent without it. These crosses, though rare, will show higher rates of transposition that promote an increase of copy numbers in their progeny. The widely documented phenomenon of hybrid dysgenesis (HD) in natural populations (see Box 3.4), observed in the *I*, *P*, and *hobo* TEs, gives support to this initial phase. This initial transposition increase is also bolstered by numerous cases of experimental transgenic experiments. For example, it has been shown experimentally that the introduction of an *I* element (by crossing or injection) in a Drosophila strain increases the number of copies across several generations, but not indefinitely. The same increase in transposition has been observed when engineered copies of *Tc1/mariner* DNA transposons from frog and fish, and LINE1 human retrotransposons were inserted into the foreign mouse genome.

The second phase of the invasion is characterized by a levelling off of the rate of insertions that seems to be inversely related to the number of copies. The system reaches an equilibrium copy number that is 10–15 per haploid genome in the case of *I* invasion, but it may be larger (40–50 copies) in the *P* invasion.

Box 3.4 The history of recent TE invasions

In the 1970s, almost simultaneously two groups of researchers, one led by L'Héritier in Clermont-Ferrand (France) (Picard 1976) and the other in Providence (USA) led by Kidwell, (Kidwell et al. 1977), noticed that *Drosophila melanogaster* crosses between recently field-collected males and old laboratory-stock females, a routine procedure to study the genetics of natural populations, yielded an offspring that showed a high degree of sterility. Moreover, the progeny of those individuals that escaped sterility showed a set of genetic abnormalities, named hybrid dysgenesis, which included high frequencies of mutants, recombination in males (absent in normal males), and an increasing number of chromosomal rearrangements. Interestingly, crosses between old laboratory stock males and collected females did not show this syndrome.

Although both hybrid dysgenic events have similarities, they show striking differences. The French syndrome (named I-R) occurs only in hybrid females, while the American syndrome (named P-M) affects both sexes but only at 28 °C. Further in-depth research has shown that both systems are caused by two independent TEs, named I and P, respectively. Namely, P element is a DNA transposon (class 2) while I element is a non-LTR L1 retrotransposon (class 1). Notice that, regardless of the system, I-R or P-M, mobilization occurs only in germ line cells, resulting in first-generation hybrids that appear normal, but sterile.

The I and P elements are stable in males of natural populations, but become active when inserted (by fertilization or injection) into eggs devoid of these elements (in type R and M females, respectively). On the contrary, when an I- or P-containing egg is fertilized by a sperm devoid of the element, the element remains stable in the zygote. These findings explain the hybrid dysgenesis. Thus, males collected in the wild (I- and P-type) contained the I- or P-element, and when crossed with element-free old stock females (R- and M-types) the elements became activated in the progeny germ line, either producing gonadal atrophy (the P–M system) or egg abortion (the I–R system). Occasionally some gametes can be produced in the progeny, as in the low temperature of development in the P–M system, that produce developing zygotes in the next generation. Then these offspring show all the dysgenic traits produced by TE mobilization in the gametes (Fig. E).

continues

Box 3.4 (*continued*)

The hybrid dysgenesis phenomenon afforded researchers a way to probe into the mechanisms that repress the mobility of TEs in the genome, something that I will discuss later, but also made possible experiments to understand the dynamics of TE spreading in populations. Obviously, the first enigma to answer is why did recently collected populations contain the P and I elements, while old stocks (collected much earlier) did not? Examination of hundreds of worldwide population collections showed that prior to 1950 all *D. melanogaster* populations carried no P elements and that P elements likely started to invade this species by this date. The first invasion was located most probably in SE United States or in Central America, then spread to Europe, with the remainder of continents following until worldwide completion. Figure E shows a dynamic map of the *P* population invasion (white dots) in *P*-free geographic areas (black dots) during the 1980s. Although less well documented, the worldwide I invasion likely started a little earlier (maybe in the 1930s).

The second enigma refers to the origin of the P element. This enigma is not completely solved but the most reliable evidence points to a horizontal transfer between species.

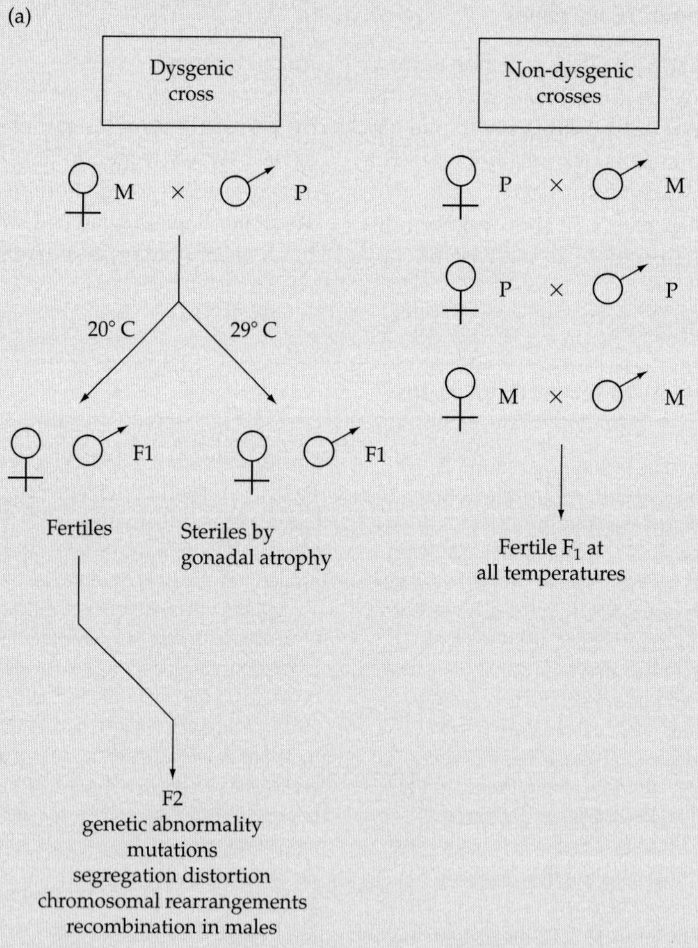

Fig. E. (a) Flow chart depicting the scheme of dysgenic and non-dysgenic crosses. (b) Geographic localities of P (empty circles) and M (full circles) strains in natural populations of *Drosophila melanogaster* collected between 1980 and 1990. Arrows depict the inferred invasive migration of flies with the P element. Redrawn from Anxolabéhère *et al.* (1988) with permission from Oxford University Press.

continues

Box 3.4 *(continued)*

(b)

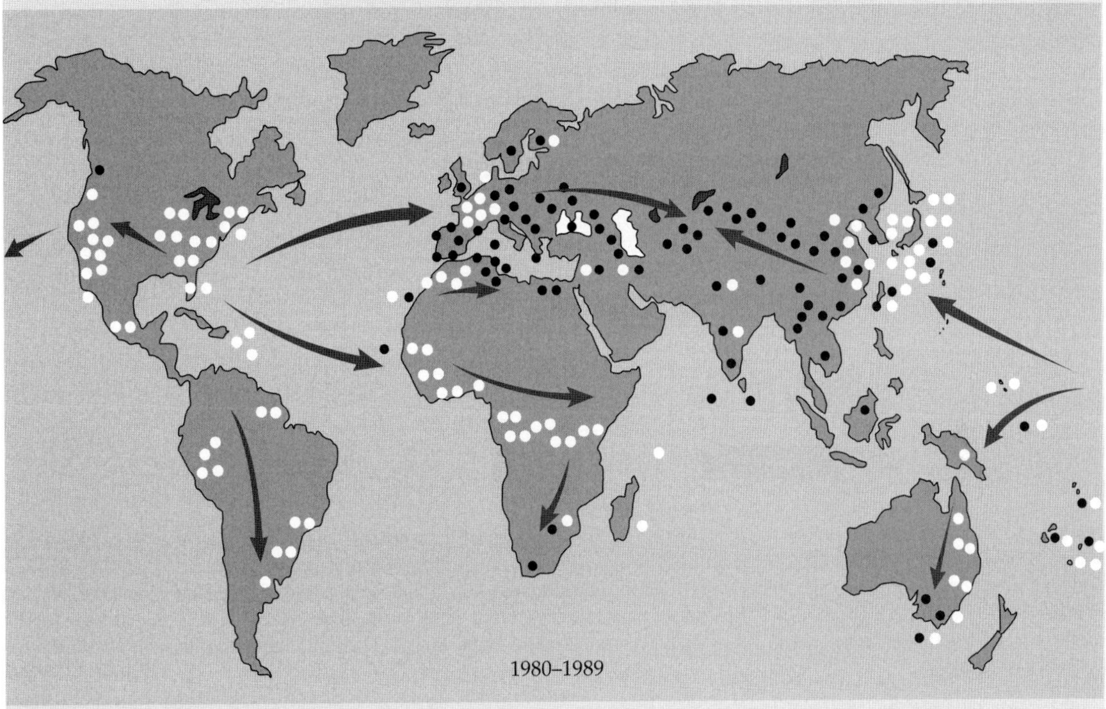

Fig. E. *Continued*

When the presence of P was analysed in species of the genus Drosophila, researchers found to their surprise that this element was carried by none of the species closely related to *D. melanogaster* that belong to the same group. Contrarily, the P element was present in more distantly related species that belong either to rather close groups, like *obscura*, *willistoni*, and *saltans*, that diverged about 55 mya, or even to some more distant diptera species. The discordance between host phylogeny and P distribution is important evidence in support of the hypothesis of horizontal transfer, but is not the only one. A second argument in favour of the *D. melanogaster* P invasion from *D. willistoni* arose when researchers from Tucson (USA) (Daniels *et al.* 1990) cloned a P element from this species and found that its sequence was almost identical to that of an intact P copy of *D. melanogaster*. Last, but not least, evidence for the recent invasion from *D. willistoni* is bolstered by the overlapping range between both species in the Americas after the recent colonization of the American continent by *D. melanogaster*, estimated not much earlier than the nineteenth century (David and Capy 1988), and also by the low molecular variability of *P* in *D. melanogaster*.

How is this regulation achieved? Is it due to self-regulation by the element itself or by factors from the host? There may be both kinds of regulation involved, but, at least in the case of DNA transposons like *P*, the increase of transposase, the element-encoded enzyme that promotes transposition, seems to induce a high number of breaks that generate defective elements which act as repressors (Fig. 3.9). In the case of DNA transposons, the excision of a copy is achieved by transposase-induced double-strand breaks that leave behind a gap that must be filled by copying the element from the one left at the sister chromatid. Since this gap repair is prone to error, copied transposons are usually inactivated by mutations. As non-autonomous and repressor elements accumulate, they drive the intact elements to extinction.

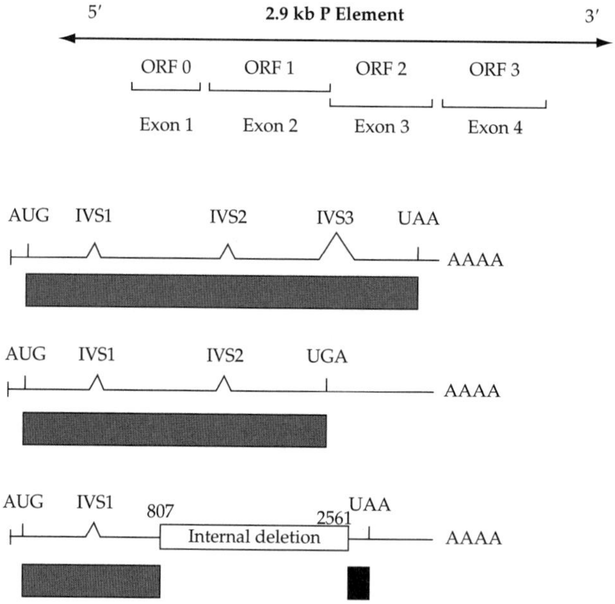

Fig. 3.9. A schematic picture of P element transcripts. The active, complete P element (top) contains four exons that encode a 751 aa transposase protein, the expression of which only occurs in the germ line. In somatic cells the third intron (IVS3) is not spliced out (in the middle of the figure), producing a short functional mRNA that encodes a short transcription-repressor protein (576 aa long). Other truncated proteins produced by internally deleted P elements, exemplified by KP type at the bottom of the figure, also act as repressors of transcription. AUG denotes the initiation codon, UAA and UGA the termination codons. Full rectangles denote the mRNA length. From Rio (2002), with permission from the American Society for Microbiology.

Recently, specialized genome defense mechanisms against TE spreading have been described. Remember that dysgenic crosses introduced *P* elements from males into *P*-free female genomes that were devoid of the epigenetic machinery for *P* silencing and a burst of transposition ensued, which was responsible for the HD syndrome (Box 3.4). The reciprocal cross, *P*-free males crossed to *P* females, does not produce HD, which shows that the ability to prevent *P* elements from transposing is maternally inherited. The acquisition of this silencing ability, designated as the *P* cytotype, takes several generations. Now we know that a large component of the *P* cytotype is due to a telomeric trans-silencing effect (TSE), which detects *P* sequences inserted in telomeres as markers for silencing other homologues elsewhere (Josse *et al.* 2007). Thus, this telomere heterochromatin seems to act as an attractor of fragmented transposon elements, protected epigenetically from elimination, that serve as a source of certain RNAs (named piRNAs) that provide the extrachromosomal component necessary for silencing. This component is produced in low concentrations in the female germ line of the first hybrid generation in a dysgenic cross, but acts as an inducer of more products in each subsequent generation until a repression concentration is reached after five to six generations. This type of silencing (see below), in which there is interaction between RNA silencing and heterochromatin formation, appears to be the basic mechanism for *P*-element repression.

These various, non-mutually exclusive mechanisms provide an efficient way to silence invasive TEs. How, then, can autonomous, active elements be maintained over long evolutionary times? The answer is likely to be the invasion of **gene pools** by horizontal transmission (see Chapter 1, p. 24). Thus, the autonomous, intact elements must jump to another species, escaping from inactivation by repressor elements. This would be the third phase in the long-persistence scenario of TEs. This scenario relies on a great deal of cross-species transfer. The horizontal TE transfer is well documented in the recent *P*-element invasion of *D. melanogaster*, and possibly, although less well documented, in the analogous *I*-element invasion, but the number of reported cases of horizontal transfer has increased ever since these pioneer observations. Not only have *P* element invasions from other organisms, such as the diptera *Scaptomyza pallida*, to Drosophila species been reported, but other DNA transposons, like *hobo*, have apparently invaded *D. melanogaster* recently. Perhaps the best case for horizontal transmission is provided by the *mariner* elements. Horizontal transmission of these elements has been well documented in insects and other invertebrates, vertebrates (between fish and frogs), fungi and plants (see for a review Burt and Trivers 2006, pp. 268–9). An extensive survey of hundreds of insect species (Robertson and Lampe 1995) revealed that distantly related species belonging to different families (*Drosophila ananassae*, a fruit fly, versus *Haemotobia irritans*, the hornfly), to different suborders (the last two species versus *Anopheles gambiae*, the mosquito), and even to different orders (all three species versus *Chrysoperla plorabunda*, a neuropter) shared a cluster of closely related *mariner* elements that differed by a few nucleotides. The divergence time between these taxa ranges from 100 to 265 myr, which clearly exceeds the nucleotide-estimated divergence time of the *mariner* elements found in them. Cases of LTR retroelement lateral exchange have also been reported. For instance, *D. melanogaster* and *D. willistoni* have horizontally exchanged the *copia* element, an LTR retrotransposon, during the roughly 200 years of recent range overlapping, but here the sense of transfer is from *D. melanogaster*, where *copia* is abundant, to *D. willistoni* that hosts few, and patchily distributed, copies (see also Box 3.4). Also, the *gypsy* LTR retroelement has likely experienced a complex history of invasions in Drosophila species in which vertical and horizontal episodes are present. Finally, although less frequently, non-LTR retroelements may also invade new species' genomes. Sequence divergence of LINE retroelements correlates nicely with the estimate divergence time of their hosts, but some outlier observations were detected for cows and snakes, snakes and silk moths, and fishes and plants, suggesting some horizontal transfer. These few observations notwithstanding, since the entire extrachromosomal phase of non-LTR retroelement replication involves only RNA forms, that are considerably less stable than DNA, it is not surprising that horizontal transmission has a reduced chance of success for these elements.

Bursts of transposition: genomic and environmental stresses

That TEs experience episodes of transposition bursts in populations is something well known to the researchers studying evolutionary processes. But the entrance of a new TE into a TE-free genome, as explained in the last section, is not the only trigger of these bursts. More than 25 years ago, McClintock interpreted her observations of transposition release in maize as a response to challenges of the genome. In her case the challenge was the introduction of broken chromosomes in the genome, but she expanded this genome response to other types of stress. In her own words, she anticipated 'unusual responses of a genome to various shocks it might receive, either produced by accidents occurring within the cell itself, or imposed from without, such as virus infections, species crosses, poisons of various sorts, or even altered surroundings such as those imposed by tissue cultures'. Since then, many external and internal challenges, including those cited above, have been shown to release epigenetic controls on TEs, some of which have already been discussed in this chapter.

Two general mechanisms of TE response to stress can be distinguished, depending on whether the challenge directly activates TEs or their activation is the indirect result of a general stress-mediated inhibition of gene-silencing in the genome. Perhaps the best example of direct activation is provided by the *Tnt1* retrotransposon in tobacco (Grandbastien 1998). In this example, when plants are subjected to abiotic (salycilic acid treatment or wounding) or biotic (viral, bacterial, or fungal attacks) treatments, *Tnt1* and other *Tnt*-family members are reactivated in a process mediated by their promoter regions. Interestingly, these promoters contain sequences similar to regulatory motifs also involved in the activation of stress-defence host genes. This similarity is not exclusive to *Tnt* retroelements. In Drosophila, several studies (Strand and McDonald 1985; McDonald *et al.* 1997) show that regulatory enhancer regions of *copia* and *mariner* TEs include motifs very similar to those of heat-shock protein gene (*hsp*) promoters. Moreover, these enhancers are variable in natural populations, induce differential TE expression, and are able to activate *hsp* promoters.

Since *hsp* genes code for proteins that are activated by high temperature (see Chapter 5 for a fuller account), and *copia* and *mariner* are also responsive to heat stress, these results support the hypothesis for a TE response to stress.

These early observations raise the question of the evolutionary origin of the TE stress-responsive promoter motifs. Do they originate from TEs themselves or have they been co-opted from host sequences? TE-mediated rearrangement processes are pervasive during genome evolution, as discussed in previous sections; thus, it is not implausible that TEs have provided stress-responsive sequences to host genes. On the other hand, TEs may have acquired regulatory sequences from host genes by mechanisms similar to retroviral transduction. Neither of these two possibilities can be discounted, and nor can a scenario involving a combination of both.

Stress can also inhibit genome-wide gene-silencing mechanisms. Some of these general mechanisms are temperature-dependent, as in **position effect** variegation in the fission yeast (*Schizosaccaromyces pombe*) and *D. melanogaster*. In fission yeast, abiotic stress conditions activate genes and TEs that are similar to those activated in histone deacetylation mutants. Moreover, stress-response factors are involved in RNAi-mediated heterochromatic silencing. Both observations strongly suggest that stress acts on primordial heterochromatic-mediated epigenetic mechanisms of gene silencing, such as position effects (see Slotkin and Martienssen 2007 for a review). Temperature may also be a factor influencing the rate of transposition in *Drosophila simulans* and latitudinal clines in TE copy numbers observed in populations of this species have been attributed to temperature gradients (Vieira *et al.* 1998). When explaining the epigenetic basis of TE regulation (see above: What is epigenetics?), the case of dietary-supplemented mice was described as an example of the role of a TE (IAP LTR) to contribute to the regulation of methylation levels at the *agouti* locus. This may also represent an epitome of the TE response to dietary-induced stress.

However, we do not yet know how much of the TE response to environmental stress is indirect, that is genome-wide, or directed to a specific element. What we do know is that this response exists and may have important consequences in the adaptive fate of populations in 'difficult' times. This does not mean that TEs evolved for the 'good' of host genomes. Knowing the opportunistic nature of natural selection, the story must be quite different. As Kidwell (2005, p. 205) states, 'a more accurate and enlightened approach is to consider them (TEs) and their hosts in the coevolutionary terms of host–parasite relationships'. Since TEs behave as parasites at the beginning of host invasions, their invasive, selfish behaviour must be controlled by the genome. As time goes by, a series of episodes, including those genome–TE exchanges already discussed, provide the opportunity for co-option of TE functions by the genome machinery and then some insertions become beneficial. Moreover, the host–parasite relationship may in turn be mutualistically beneficial 'in which the element and the host genome are interdependent on one another', in Kidwell's words. In the same vein, Burt and Trivers (2006, p. 296) 'do not see a contradiction in calling TEs "selfish" or "parasitic" and entertaining the possibility that . . . the great majority of extant insertions may be beneficial'. Their definition as selfish is based on the observation that the average effect of transposable element activity must be negative in the short term. But why then must TEs be activated at times of stress? 'Perhaps TEs have evolved to be active when host defences are low and a new insert is likely to be beneficial' as Burt and Trivers (2006, p. 295) point out. We do not know the answer to this question, but we do know that stress-induced TE mobilization is a highly pervasive phenomenon, as anticipated by the genius of McClintock (see Wessler 1996 for a review in plants).

Hybridization: a major challenge to the genome

When talking about the genome response to stress, Barbara McClintock (1984, p. 633) states that 'species crosses are another potent source of genomic modification'. This insight was bolstered by early observations in species hybrids, which showed large chromosomal rearrangements not present in their parental species. Species of the tobacco genus (Nicotiana), the wheat genus (Triticum), and the rye genus (Secale) were among those whose experimental hybrids were long known for their high level of instability. Yet, McClintock's real insight was to link hybrid-mediated genome instabilities with transposable elements.

Studies of hybrid instability have a long history. Pioneer investigators noticed that hybrid crosses could induce mutations. Alfred Sturtevant (1939), a Drosophila geneticist, crossed two then-called 'races' of *Drosophila pseudoobscura*, and observed, in backcrossed progenies, frequencies of lethal and morphological mutations much higher than those expected by normal spontaneous mutation rates. He clearly inferred that 'there is . . . a persistent feeling that perhaps interracial crossing also induces the production of new mutants'. These races were later recognized as two different sibling species: *D. pseudoobscura* and *D. persimilis*, so what Sturtevant actually observed was an increase of mutation rates in backcross interspecific hybrids. Apart from this early case of hybrid instability, other examples of increases in rates of chromosomal rearrangements in hybrids were described in plants, as indicated above, and also in animals, such as grasshoppers (Caledia) and midges (Chironomus) (see a review in Fontdevila 2004). Interestingly, a pioneer relationship between production of new rearrangements and TE transposition was already observed in hybrids between subspecies of *Chironomus thummi* (Schmidt 1984). Some circumstantial evidence of enhanced transposition in hybrids was also produced in Drosophila (see references in Labrador and Fontdevila 1994), but in all cases it was largely obtained by indirect methods, based on observations of morphological reverse mutations due to specific site-targeted TE excisions. These experiments lacked the potency to reveal a genome-wide effect of hybridization.

Perhaps the first well-documented case of increasing transposition rates in species hybrids by direct observation of new insertions was produced in my laboratory at the University Autònoma of Barcelona (Spain). *D. buzzatii* and *D. koepferae*, two sibling species coexisting in the arid zones of Bolivia and north-west Argentina, have long been used by my group as model organisms for evolutionary biology studies. They are able to hybridize, yielding sterile males and fertile females. Successive backcrosses of hybrid females to *D. buzzatii* males generate, after several generations, a collection of **introgressed** hybrids: that is, individuals that carry genomic portions of *D. koepferae* in a *D. buzzatii* background genome. In the mid-1980s, Horacio Naveira, then a graduate student in my lab who was studying the genetic basis of male hybrid sterility with these species, observed that progenies of some of these introgressed hybrids displayed a large number of new chromosomal rearrangements (Naveira and Fontdevila 1985). We immediately interpreted this observation as an episode of hybrid instability similar to those described above, suggestive of a hybrid-induced release of TE transposition (Fontdevila 1988).

The experimental test of this hypothesis was carried out by Mariano Labrador, also a graduate student in my lab, using *Osvaldo*, then a new retrotransposon isolated and characterized by my group (Pantazidis *et al.* 1999). *Osvaldo* seemed an ideal tool for this research because we had previously observed its transposition by direct cytological methods (namely chromosomal in situ hybridization), which allowed us to clone an active copy of the retroelement that was used as probe for cytological site detection. The prediction was confirmed. Table 3.1 shows that transposition

Table 3.1 Transposition rates (TR) of Osvaldo in *D. buzzatii* (Bu) and in hybrids between *D. buzzatii* and *D. koepferae* (Ko)

Line/Hybrid	N	LNI	NI	TO	TR = NI/TO
Bu line	301	15	36	4,224	8.5×10^{-3}
Hybrid with Ko-2.6 line	163	15	29	1,836	1.5×10^{-2}
Hybrid with Ko-SL line	174	39	87	2,230	3.9×10^{-2}

Note: N, LNI, and NI stand for number of analysed larvae, number of larvae with new insertions, and number of new insertions, respectively. Transposition rate (TR) is defined as the number of transpositions per element per generation. Transposition opportunities (TO) are the grand total of the number of times that each element has passed through a chromosomal generation. Results are reported for two experiments with two D. koepferae lines. (Adapted from Labrador *et al.* 1999.)

rates are one order of magnitude higher in hybrids ($1.5–3.9 \times 10^{-2}$) than in parental species (8.5×10^{-3}). To my knowledge, these results (Labrador *et al.* 1999) are the first quantitative evidence of transposition increase in species hybrids. Their strength is bolstered by two experimental approaches. First, insertions are detected by direct cytological methods, and new insertions are deduced by comparisons between progeny (introgressed lines) and parental individuals (nonhybrid), in which original positions were previously characterized. Second, the statistical treatment was based on paired tests of homogeneity that show a highly significant heterogeneity in all cases, which confirms that differences in insertion rates are real.

A similar direct observation linking increasing TE mobilization and 'de novo' chromosomal reorganizations in hybrids is the research work of Rachel O'Neill's team at the University of Connecticut (USA). Kangaroo species (Macropodidae family) easily produce hybrids, and O'Neill took advantage of this propensity. Initially she and her collaborators observed that kangaroo hybrid chromosomal reorganizations included extended centromeres that contained many highly repeated copies of a novel retroviral element (KERV). Interestingly, these hybrid element sequences were highly unmethylated compared to homologous sequences in the parental species (O'Neill *et al.* 1998). Recent work (Metcalfe *et al.* 2007) has confirmed that instability included amplification of KERV element and some satellite sequences in centromeres, changes in chromatin structure, and de novo whole-arm rearrangements. Whether TE mobilization in hybrid genomes is directly responsible for these chromosomal and chromatin changes or both phenomena are the result of a general relaxation of epigenetic controls, such as DNA methylation and RNA silencing, is not known. Since incompatibility of RNA-silencing mechanisms between maternal and paternal species within a hybrid genome (Lippman and Martienssen 2004; Brennecke *et al.* 2007) may lead to TE activation and/or disruption of heterochromatic structure and ultimately to hybrid dysgenesis, this is a likely cause of all the observed instabilities in kangaroo hybrids.

The far-reaching evolutionary effects of hybrid-mediated TE reactivation

These results with animals are paralleled in plant experiments. For example, inbred rice lines introgressed with DNA from wild rice showed a direct causal link between **introgression** and mobilization of MITE transposable elements (Shan *et al.* 2005). In fact it has been widely documented that plant hybridization is often followed by an extensive genome reorganization, including gross chromosomal rearrangements and DNA sequence repatterning, a topic amply documented in Chapter 4 (p. 131) of this volume. While it has not been easy to relate all this reorganization to TE reactivation, some examples provide strong evidence of the role of TE in hybrid instabilities. This applies not only to allopolyploid species, such as those belonging to Gossypium (cotton), Triticum (wheat), Nicotiana (tobacco), and Arabidopsis genera (see pp. 110–15 in Arnold 2006 for references), but also to **homoploid species**, as in Helianthus (sunflower).

In some cases, as in Spartina species—a set of plants the allopolyploid origin of which is well documented—methylation changes co-occurring with chromosome reorganizations have been observed, but no evidence of TE reactivation could be detected (Salmon *et al.* 2005). In other cases, as in wheat amphiploids, TE reactivation—in terms of increasing transcription—has been detected, but no evidence of new insertion sites were observed (Kashkush *et al.* 2002). These inconclusive cases of hybrid-induced transposition notwithstanding, other studies have reported strong evidence for transposition in allopolyploid and homoploid species. *Arabidopsis suecica* allopolyploids were recreated in the laboratory by crossing the two parental species (*A. thaliana* and *A. arenosa*). The neoallopolyploid was similar to its natural counterpart in genome constitution and in remodelling of methylation and gene silencing. A significant transcriptional and insertional activity of several TEs was detected in synthetic allopolyploids. This TE reactivation was coincident with demethylation. The authors (Madlung *et al.* 2005) also reported a remodelling of the transcriptome and of the epigenetic landscape coupled with changes in chromosomal structure. Thus, the whole set of instability events is perfectly consistent with the hypothesis of TE release in early stages of hybridization.

Recently, the Helianthus case has become the subject of renewed interest in the role of TEs in hybrid homoploid speciation. The whole process of the hybrid origin of Helianthus species is explained in detail in Chapter 4 of this volume. Here, suffice it to say that several hybrid species, independently originated from two parental species (*H. annuus* and *H. petiolaris*), have been found to contain at least 50% more genome DNA than their ancestral progenitors. This increase in genome size did not result in an increase of chromosome number, although hybrid species are characterized by specific novel chromosomal rearrangements. The researchers (Ungerer *et al.* 2006) reported that a hybrid increase in copies of sequences similar to *Ty3/gypsy-like* retrotransposons accounts for a large amount of genome size differences (62–79%) between hybrids and their parental species. In a later work the same research group (Kawakami *et al.* 2010) found that a similar transposition proliferation occurred for *Ty1/copy-like* retrotransposons in Helianthus hybrids. Interestingly, the fact that the majority (70%) of sequences participating in this transposition explosion derived from a single lineage suggests that this proliferation is recent. These results suggest that, either through hybridization or environmental stresses, the reactivation of silent TEs have likely played a role in their speciation (see Chapter 4).

The relationship between hybridization, stress, and speciation has been advanced and also periodically dismissed. It was first stated explicitly by McClintock (1980), that 'since the types of genome restructuring induced by such elements (namely TEs) know few limits, their extensive release, followed by stabilization, could give rise to new species or even new genera'. Yet, the ups and downs for the acceptance of hybrid-induced speciation theory mirror the contentious nature of this issue. The initial enthusiasm in the 1980s, fuelled in part by the Drosophila hybrid dysgenesis similarity to the inviability and sterility that accompanies interspecific hybrids, was followed by a period of confusion. Some experiments failed to replicate the hybrid dysgenesis syndrome in other species and, after all, the end result of the population dynamics of *P* and *I* elements was not their induction of incipient speciation but, to the contrary, their worldwide genome invasion. The view of TEs as pure 'selfish elements', with no other evolutionary role, was reinforced by these observations and created a wave of scepticism in the 1990s. Some voices were raised, however, defending the role for TEs in speciation (i.e. Fontdevila 1992) and also in larger evolutionary transitions (i.e. McDonald 1990). Since then, new studies, some of them referred to above, have accrued, which clearly show increasing TE activity in interspecific hybrids and in immediate generations following the origin of allopolyploids (reviewed in Liu and Wendel 2003). This ever increasing new evidence is changing our view of the role of TEs in long-term evolution, and in speciation (Rebollo *et al.* 2010).

Though denial of hybrid-induced TE activation is rapidly becoming untenable, the scepticism towards TE-mediated speciation is still maintained by some early critics. The demonstration of the direct contribution of TE mobilization on postzygotic isolation barriers, i.e. hybrid sterility or inviability, still remains elusive. This is the last stronghold of opposition to TE-mediated speciation. Yet, the number of experiments whose results establish a close direct association between epigenetic loss of silencing and fitness lowering in hybrids, is rapidly increasing. As an example, the long-lasting work of Comai, and his collaborators, with Arabidopsis hybrids has recently confirmed that the silence release degree of a heterochromatic element (*Athila*), as well as of other regulatory genes, is directly related to hybrid seed inviability. Interestingly, the degree of up-regulation of this element is dosage-dependent in a manner that resembles hybrid dysgenesis. Precisely, Josefsson *et al.* (2006) proposed the 'dosage-dependent-model' that assumes that the number of regulator and target sites differ between species and that the female gamete does not possess a sufficient amount of repressive factors, likely small RNAs, to bind all male target sites or their binding efficiency is not sufficient (Fig. 3.10). These species differences in regulatory elements cause derepression of TEs and other deregulation events such as chromatin demethylation in the hybrids that led to the hybrid dysgenesis syndrome.

Admittedly, showing a direct causal effect of TEs to speciation is not easy, but even indirect TE contributions should not be disregarded. Among them, the

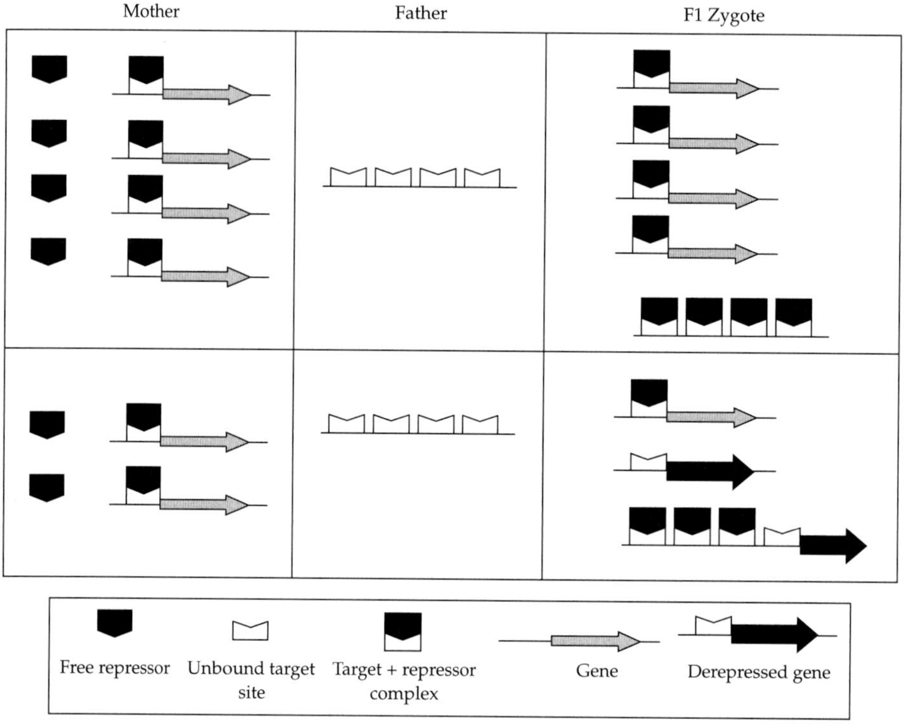

Fig. 3.10. The 'dosage-dependent induction' model propossed by Josefsson *et al.* (2006). Two crosses are depicted in which the maternal repressor activity differs in gene copy number, transcript abundance, or binding efficiency. In the top cross the female gamete provides an adequate quantity of repressor (for instance siRNA) to bind all the target sites in the male gamete. Then all target sites remain silenced. Yet, if the female gamete delivers an insufficient, or less binding-efficient, amount of repressor (bottom cross), redistribution of repressor factors will ensue and some maternal and paternal target sites will become unbound. Thus, their controlled elements (for instance TEs or gene chromatin demethylation) will be derepressed and escape silencing. This model would explain hybrid dysgenesis and also hybrid inviability and/or sterility by alteration of regulatory networks. From Michalak (2009) with permission from Nature Publishing Group.

documented TE facilitation in generating chromosomal rearrangements is of prime interest. New arrangements may in turn promote speciation lowering the hybrid fitness, a condition named underdominance. Fixation of **underdominant** rearrangements requires a high rate of chromosomal reorganization and a population scenario where selection and drift cooperate to fixation of new rearrangements. The author (Fontdevila 1992, 2005) has proposed that contact zones and/or inbred small demes may be the right arenas for hybrid speciation. Namely, contact zones would allow species to merge in hybrid swarms where 'a) hybrid bursts of TEs fuel genome reorganization; b) exogenous ecological selection allows the establishment of high hybrid fitness genotypes in novel ecotone habitats; c) endogenous selection favours reorganized genomes that show high levels of fertility and viability; and d) small effective population size in hybrid zones increases fixation by drift of those underdominant rearrangements that show high fitness in homozygous state' (Fig. 3.11). The TE-mediated increase of genome rearrangements can also be produced under non-hybrid stress conditions in small, isolated demes exposed to environmental stress, initiating a series of events similar to those explained above leading to speciation. Moreover, the anti-recombinational effect of inversions preserving the evolutionary divergence of adaptive gene blocks is gaining credibility as a mechanism promoting species isolation (Noor *et al.* 2001; Rieseberg 2001; Navarro and Barton 2003) (see p. 130). Thus, either by underdominance or anti-recombinational effects, or both, chromosomal rearrangements likely play an important role in speciation.

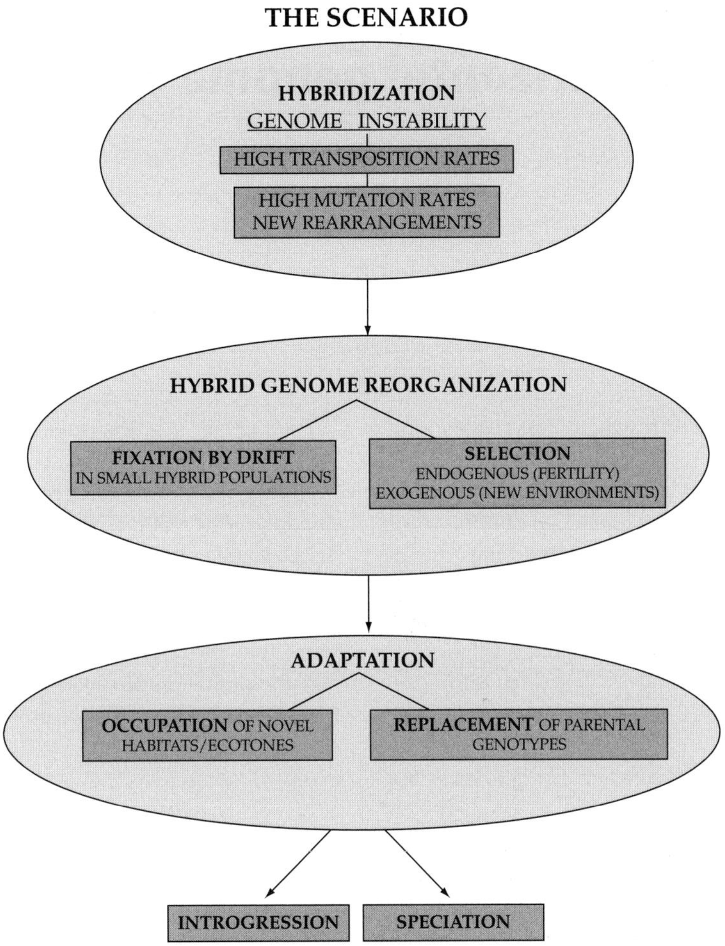

Fig. 3.11. Flow chart of the natural scenario of hybrid genome evolution by transposition. The initial (upper) part of the chart depicts how hybridization promotes genome instability through high transposition rates that induce new rearrangements and high mutation rates. The middle part shows how the ensuing hybrid genome reorganization is stabilized by drift fixation and selection (endogenous and exogenous). These processes lead to adaptation (the lower part) either by occupation of novel environments, new ecotones in many cases, or by replacement of parental species. The final step of this scenario is reticulate evolution by introgression or the evolution of a new species. From Fontdevila (2005) with permission from S. Karger AG Publishers.

Since TEs promote chromosomal reorganizations in the wild (Delprat *et al.* 2009), even if TE transpositions by themselves do not affect the hybrid fitness, 'they may predispose a genome to rearrangements that cause or facilitate speciation' as stated by Noor and Chang (2006).

CHAPTER 4

The horizontal genome

As buds give rise by growth to fresh buds, and these, if vigorous, branch out and overtop on all sides many a feebler branch, so by generation I believe it has been with the great Tree of Life, which fills with its dead and broken branches the crust of the earth, and covers the surface with its ever branching and beautiful ramifications.

—(Darwin, *The Origin of Species,* 1859, p. 130)

The Origin of Species received many criticisms, but perhaps one of the most derogative statements on Darwin's book is that '(it) is the law of higgledy-piggledy'. In a letter to Charles Lyell, Darwin attributes, 'by a roundabout channel', this criticism to John F.W. Herschel, a much admired philosopher and physicist of his time to whom he refers anonymously as 'one of our greatest philosophers' in the first paragraph of his book. In it, Darwin deals with the difficulties of understanding the origin of species, a process qualified as 'the mystery of mysteries' by this highly revered physicist. Herschel's criticisms and opinions impressed Darwin from their first meeting at the Cape of Good Hope, where Darwin stopped over on his return journey on the Beagle. While Herschel's disagreement with Darwinian natural selection does not justify his attack on the clarity of Darwin's argument, which is well accepted as a historical masterpiece of scientific reasoning, the contentious idea that speciation is a great mystery deserves a more profound consideration, as Darwin himself evidenced in his book's early statement. In this chapter I will try to shed some light on this issue, encompassing a historical perspective and some of the most up-to-date knowledge from genomic studies.

Precisely, it is in the origin of species that many evolutionists find the greatest difficulties. It has not been easy to reconcile the species concept as a natural, discontinuous reality with the continuous and gradual Darwinian population process that leads to species formation. Ernst Mayr has said on this contentious issue that has confronted biologists for more than 150 years, that 'the fallacy of assuming that the constancy and the definition of species are closely related has forced them to choose between evolutionary thinking (the species inconstancy), negating the species reality except as subjective fictions of imagination, and the adherence to the precise bounding of species (the species constancy), forcing the majority of classic naturalists to negate the evolutionary process' (Mayr 1957).

Darwin was perfectly aware of the indefinite concept of species imposed by the paradigm of descent with modification. Ironically, the title of his magnum opus, *The Origin of Species by means of Natural Selection*, summarizes his main purpose, probably not fully accomplished, of explaining the process that leads to a natural pattern of the discontinuous mosaic of lineages that we designate as species. Searching for an evolutionary basis to relate to these lineages, Darwin ends up by negating the species reality as a category when writing to his botanist friend Joseph Hooker:

> It is really laughable to see what different ideas are prominent in various naturalists' minds, when they speak of "species"; in some, resemblance is everything and descent of little weight—in some, resemblance seems to go for nothing, and Creation the reigning idea—in some, descent is the key,—in some, sterility an unfailing test, with others it is not worth a farthing. It all comes, I believe, from trying to define the undefinable. (Darwin, F. ed. (1887) *The life and letters of Charles Darwin,* vol. 2. p. 88. John Murray, London.)

The species un-definition is brought out by Darwin in many passages of *The Origin* when he discusses the difficulty in distinguishing between species and varieties; namely he states that 'the amount of difference considered necessary to give two forms the rank of species is quite indefinite'. However, this difficulty of establishing precise boundaries to species was overshadowed in Darwin's time by the interpretation of species variability under an evolutionary perspective. The typological or essentialist view of species, favoured by most naturalists of that time, which assigned this variability to mere ontogenetic interferences of the species 'kind', was challenged by the population thinking introduced by Darwin, which states that a species is not a 'kind' but a set of organisms that share a genetic pool by means of population relationships. This conceptual change prompted the death of species essentialism (Sober 1980); thus members of a species in a population were not simple deviations of an ideal type, like different members of a kind (e.g. a chair or a river); on the contrary, their variability is not a spurious deviation but the result of genetic shuffling of genotypes submitted to the laws of evolution, including natural selection, mutation, and drift. Precisely, it is this genetic variability which is the real stuff that evolutionary mechanisms act upon and by no means depicts 'ontogenetic interferences' devoid of biological significance.

The biological species: isolation as a product

One of the most relevant of Darwin's concepts is his view of biodiversity as a tree whose branches depict the continuous lineage-splitting promoted by the laws of evolution. After all, this conception of life is a natural outcome of population divergence. Yet how much divergence two populations must show to be considered different species is still a matter of controversy. However, intuitively, population thinking requires that, regardless of what mechanism is at work, two species lineages must keep their differential genetic identity by avoiding the conflating effect of introgression. Since the most ubiquitous mechanism of gene exchange, at least in sexual organisms, is sex, there can be no question that **reproductive isolation** occurred to the first evolutionists as a mechanism to keep species identity. Dobzhansky (1935) set off the whole isolationist thinking in speciation, proposing that 'a species is a group of individuals fully fertile inter se, but barred from interbreeding with other similar groups by its physiological properties (producing either incompatibility of parents, or sterility of hybrids, or both)'. This line of thinking was extended by him in his seminal work *Genetics and the Origin of Species* (1937) and was followed, and popularized, by Mayr (1942) who defined the biological species concept (BSC) thus: 'Species are groups of actually or potentially interbreeding natural populations, which are reproductively isolated from other such groups.' This definition represents the hard line position for the pre-eminence of isolation mechanisms in the making of species that led these authors to build an exhaustive list of these mechanisms (Table 4.1), currently present in any evolutionary text book.

Yet, several misconceptions pervade, in my opinion, the understanding of the evolutionary significance of isolating mechanisms. First, it is of paramount importance to distinguish between the process (speciation) and the product of the process (reproductive isolation) as a generator of a given pattern. This distinction is of utmost relevance since one may divine in some treatises a teleological attribution of a leading role in speciation to isolating mechanisms. After many decades of speciation research it has become clear that, apart from some putative processes of isolation **reinforcement** (see p. 149), reproductive isolation is not the direct target of a process of speciation but the by-product that accompanies a divergent process, either adaptive or not. This view has prompted some authors, Coyne and Orr (2004) among them, to change the term 'mechanism' to 'barrier'. Second, this conceptual distinction between process and pattern applied to isolation barriers allows us to question whether reproductive isolation is a necessary condition for the origin of species. Hard-line evolutionists, Mayr and Dobzhansky among them, posit firmly that it is impossible to maintain the integrity (cohesion) of a species in the absence of barriers to **gene flow**. However, this tenet is far from universal in view of the recent evidence of high rates of interspecies gene flow and introgression in nature, as I will explain below.

Table 4.1 Summary of reproductive isolating barriers

I. Prezygotic barriers impede the zygote formation either because sperm or pollen is not transferred (premating) or because they are prevented from fertilization after transference (postmating/prezygotic).

I.1 Premating isolation barriers

a) Reproductive individuals do not meet because:
- they occupy a different habitat (**habitat isolation**)
- they breed in a different reproductive period (**seasonal isolation or allochrony**).

b) Reproductive individuals do meet but:
- they lack a system of sexual recognition (**sexual or ethological isolation**)
- their sexual organs are incompatible in morphology or in recognition by pollinators, in plants (**mechanical isolation**)
- in organisms of external fertilization their gametes are not mutually attracted (**external gametic isolation**).

I.2 Postmating/prezygotic barriers

a) Inadequate behaviour during copulation to allow fertilization.

b) In organisms of internal fertilization (or in plant gametophytes) gametes of one species are unable to fertilize in the sexual canal of the other (**internal gametic isolation**).

II. Postzygotic barriers do not impede the zygote formation, but the zygote has a low fitness.

II.1 Hybrid inviability. Hybrids show a low viability, causing full or partial lethality, due to their inability to find a proper ecological niche and/or developmental impairment.

II.2 Hybrid sterility. Hybrids show a reduced or null fertility, due to defects in their morphological and/or behavioural sexual organs.

II.3 Hybrid breakdown. Back-crosses and/or second-generation hybrids show reduced fertility and/or viability.

The number of deserters from the leading role of reproductive isolation in speciation is increasing steadily as more empirical evidence of introgression is reported by the new genomic approaches to speciation genetics (see Noor and Feder 2006 for a review). Even the new 'hard-liners' already concede that a moderate amount of gene flow is acceptable and admit that 'distinct species are characterized by substantial but not necessarily complete reproductive isolation' (Coyne and Orr 2004, p. 30), thus departing from the 'hard-line' definition of the BSC. The role of reproductive isolation has been also thinned by many classic evolutionists in view of the increasing reports of gene exchange through introgressive hybridization among many related, sympatric species, with no effect in their morphological, ethological, or chromosomal identity. Carson (2000, p. 495), most probably influenced by the high rates of natural introgressive hybridization between two Hawaiian sympatric species, *Drosophila heteroneura* and *D. silvestris*, studied by him and his colleagues (Kaneshiro and Val 1977), writes: 'As stated by Mayr species are groups of actually or potentially interbreeding natural populations that are reproductively isolated from other such groups. I adopted this idea and for years did all my research with its assumptions in mind. I now advocate stopping the definition with a period after the word populations.'

After all these challenges to the role of reproductive isolation the question remains as to what preserves a species' integrity. The cohesive nature of species was already suggested by Mayr (1963, p. 460) when he defined a species as a set of individual organisms that share an integrated complex of epistatic and pleiotropic interacting genes. This concept does not show up in the BSC, where all the emphasis is on the reproductive isolation mechanisms. There seems to be a duality between concepts and definitions, the BSC being influenced more by the practicality of the role of isolation to set apart species in **sympatry**, which more easily allows the designing of a research programme to experimentally test species identity, rather than by the thorough, and often more difficult, analysis of the evolutionary mechanisms that promote species formation. But even the pragmatic value of the BSC may be useless if, as some evolutionists propose, a strong **diversifying selection** promotes, and maintains, the establishment of different sets of species-specific genes in front of free gene flow for other non-selective genes. Throughout this chapter I will discuss how much this and similar scenarios may be grounded in the current evidence, based mainly on recent genomic approaches, but for the moment the lesson to take away is that if the species cohesion depends more on the intensity of selection than of reproductive isolation, then the process of speciation is not uniquely dependent on the erection of isolating barriers.

Do we need geographic barriers?

Allopatric speciation, the most popular speciation model that proposes the necessity of a population geographical barrier to allow species differentiation, was envisaged by Moritz Wagner as early as 1868 and expanded in his posthumous book *Die Entstehung der Arten durch räumliche Sonderung* (The Origin of Species by Spatial Separation) published in 1889. His ideas were prompted by observations of sister species separated by geographical barriers over very short distances, such as different species of the wingless beetle Pimelia genus occupying both sides of many Algerian rivers. This biogeographical model, later generalized as the Riverine Barrier Hypothesis, was also observed by Wallace (1852), who reported that large rivers in the Amazonian basin act as natural barriers between different species of primates. Ironically, almost two decades earlier a similar observation was made by Darwin in the Galapagos Islands when he recalled:

> My attention was first called to this fact by the Vice-Governor, Mr Lawson, declaring that the tortoises differed from the different islands and that he could with certainty tell from which island any one was brought. I did not for some time pay sufficient attention to this statement, and I had already partially mingled together the collections from two of the islands. I never dreamed that islands, about fifty or sixty miles apart, and most of them in sight of each other, formed of precisely the same rocks, placed under a quite similar climate, rising to a nearly equal height, would have been differently tenanted; but we shall soon see that this is the case.

It turned out that what was true for tortoises was equally true for birds and plants. But Darwin never favoured the allopatric idea; he knew of Wagner's writings and his comments on them were highly derogative on occasions. In the last edition of *The Origin* one can read: 'Wagner . . . has shown that the service rendered by isolation in preventing crosses between newly formed varieties is probably greater even than I have supposed. But from reasons already assigned I can by no means agree with this naturalist, that migration and isolation are necessary for the formation of new species.'

Darwin's words look premonitory of the current discussion about the need of isolation and the role of natural selection in speciation, a theme that I want to develop in this chapter. Wagner's theory went largely unnoticed for more than 50 years and it took Ernst Mayr to unearth it, and to make it into the most popular speciation mechanism in most of the last half of the twentieth century. Mayr's early inspiration came from his three-year trip in 1928 to New Guinea, where he collected thousands of specimens of birds of paradise for Lord Walter Rothschild, the expedition sponsor, in order to clear up the confusion that pervaded the taxonomy of this group of birds. And sure enough he did it. More than that, Mayr studied the distribution and variability of these species, noticing immediately that sea and mountain ranges were often boundaries between sister species, as their fellow precursors did several decades earlier. His field work, first in New Guinea and later in the South Sea islands, was followed up by collections in other latitudes of other animals that reassured him that the geographical model of speciation, or allopatric speciation as he coined it, is the most frequent, if not the only, model to provide the diverging population with the necessary isolation to become a new species, as he magisterially explained (Mayr 1942).

Geographic isolation is a very plausible hypothesis of speciation that conforms to the observed geographic patterns in many natural instances, but this does not guarantee its reliability as the cause of speciation. The causative role of geographical isolation in speciation must be shown, by indirect evidence at least. Being a historical process we can't replay the scenario, but the current molecular theory of evolutionary genetics provides a means to measure the divergence time from the common ancestor in pairs of sister species, which must coincide with the time of the origin of the putative geographical barrier. This method, known as the molecular (or biological) clock (see Box 1.1), is grounded on the fact that a high proportion of DNA nucleotides are substituted along the divergence time at the same pace as the mutation rate, which may be considered constant. This clock is not ticking at the same pace in all evolutionary groups and genome fractions, but after correcting for lineage and nucleotide sequence, time of divergence may be estimated and compared to the geological time of the barrier. This is what researchers have been doing in cases such as in the Panama isthmus formation (Lessios 1998) and the glacier land forms that developed during the Pleistocene (Hewitt 2000).

Lessons from the past: plate tectonics and glaciations

Three million years ago, a massive crunch of tectonic plates culminated in the erection of a landbridge between South and North America. This isthmus, named Panamá, was to allow a faunal interchange between the two continents of dramatic consequence for the evolution of mammals in this area. Yet the evolutionary impact of the isthmus in the sea was equally great, but had opposite consequences. During millions of years the sea arm separating both continents had been hosting thousands of marine species, populations that now were sundered by the uplifting of the Panamá barrier, thus creating the ideal scenario for **allopatric speciation**. Evolutionists profited early from this natural set up, noticing that the two population isolates on the two sides of the isthmus, coined 'geminates' by Jordan in 1908, might be invaluable to speciation research. Since then dozens of studies with marine organisms, comprising mainly sea urchins, fishes, shrimps, and isopods, have shed light on morphological, isozyme, and DNA divergence between geminate species of Panamá. Altogether these studies conform to the allopatric model of speciation when the molecular clock gauge is applied to DNA divergence. For example, mtDNA nucleotide divergence between some geminate species of Alpheus, a kind of shrimp, is around 7%, which means a separation time between 3 and 4 my, since the rate of substitution of mtDNA is about 2% per million years. However, Knowlton and Weigt (1998) showed that not all pairs of trans-isthmian sister taxa became isolated simultaneously, their genetic divergences ranging over four-fold. Geminate species from mangrove environments were the least diverged, as expected if this habitat was the last to be divided 3 mya. But other species, namely those with deeper-dwelling larvae, may have begun their split much earlier along the whole time span of tectonic activity. Interestingly, these trans-isthmian studies provide a neat example of isolation without impermeable physical barriers, giving rise to non-simultaneous divergences of many organisms across barriers. As Knowlton and Weigt (1998) state 'where partial barriers do exist, biological responses to them are likely to be complex and drawn out in time, on land as well as sea'. This applies to divergence dates for North American birds (see below), to non-simultaneous divergences across barriers in rain forests (Patton and da Silva 1998), and to terrestrial taxa separated by the opening of the Strait of Gibraltar (Busack 1986).

Plate tectonics is not the only major cause of geographical barriers. Major climatic changes, documented mainly by the Pleistocene glaciations, have been also responsible for large episodes of isolation in natural populations (Hewitt 2004). Throughout the past 2.5 my about 40 glacial–interglacial cycles had occurred in the Northern Hemisphere, of which 8 cycles have been documented in the past 800,000 years. These last cycles were not only longer (about 100,000 years each) but also more intense. Interestingly, the four last ice ages (400,000 years) have been documented in detail, confirming that glacial phases lasted longer than interglacials and that during the last ice age (120,000 years) some 24 major warm periods and many shorter fluctuations were recorded. These alternate episodes of temperature changes greatly influenced the species distribution, as documented by the fossil record found in cores drilled in land and in the sea and lake bottoms. Most relevant was the assessment obtained from some pollen and beetle records, that in northern Europe the species turnover was often very rapid, spanning only hundreds of years (Birks and Amman 2000).

In view of this scenario one may wonder whether geographical barriers did not rise and fall at ease. In fact we know that during the last glacial maximum (LGM), 18,000–24,000 BP, large ice sheets occupied much of the north of the Northern Hemisphere and south of this was mostly tundra and permafrost. Northern species were packed in southern latitudes and present high-altitude species survived at lower elevations because mountain ranges were glaciated.

Traditionally, many biologists have tried to understand the present genetic diversity across Europe and North America by resorting to colonization of northern latitudes by species that occupied southern refuges when warming allowed ice sheets to disappear (Hewitt 2000). Moreover, there has been a great consensus that the current observation of bisected species pairs, one to the east and the other to the west of Northern continents, is a natural outcome of speciation promoted by glaciation barriers in the late Pleistocene. This is so because eastern and western species are the result of after-glaciation northern colonization from eastern and western populations isolated by ice sheets during glaciation times. But, again, this hypothesis must be subjected to test.

The clock test to geographical barriers

It is clear that the fossil record shows a long history of changes of species distribution in accordance with these major climatic shifts, but the real importance of ice sheet barriers in the origin of species is subject to debate. Beetle fossils are most illustrative in this respect. As an example, two sister species of water beetles, *Helophorus aspericollis* and *H. brevipalpis,* which currently occupy eastern Siberia and Europe, respectively, might be considered the product of a speciation event promoted by the glacier barrier that separated these two land masses in the LGM. Yet Coope (1979, 2004), a British palaeontologist, has shown that the fossil record tells a different story; the Siberian species occurred in Britain during the cold part of the LGM and the European species existed there during the previous temperate interludes. As Coope (1979; p. 254) says: 'the present day ranges of these two species thus reflect the different geographical locations of the environments that each finds acceptable; their modern distributions are not an indication of their recent evolutionary history'.

Unfortunately we do not have a fossil record for other groups of animals, birds among them, as complete as that of beetles, but modern DNA markers, such as mtDNA, allow us, using the molecular clock, to check if the glacial barriers were at work at the time of species origin. Birds in the Northern hemisphere often comprise sets of related species pairs, located to the eastern and the western side of each continent, that have been considered the result of allopatric speciation during late Pleistocene ice glaciations (less than 250,000 years ago). But Klicka and Zink (1997), using a mtDNA clock, challenged this paradigm, showing that only 11 of 35 assayed pairs of sister species of North American songbirds conformed to a Quaternary separation; the remaining species pairs diverged much earlier, some of them before the ice ages that started about 3 mya (Fig. 4.1). The authors concluded strongly that 'the entrenched paradigm proclaiming that many North American songbird species originated as a consequence of these glaciations is flawed'. This apparently powerful blow to the Late Pleistocene Origins (LPO) glaciation paradigm warns us to be more careful when trying to derive processes (the role of geographical barriers) from patterns (the current species distribution).

A reasonable interpretation of this result would be to protract the speciation time in American songbirds to the Pliocene. Avise (2000; p. 318), a pioneer evolutionist in the study of genetic patterns over space, a discipline called phylogeography, thinks that 'many species entering the Pleistocene would have been separated into distinctive intraspecific phylogroups, as are many extant bird species today. Such units', Avise's argument follows, 'would be likely candidates for subsequent evolutionary divergence during the Quaternary'. But even after correcting for intraspecific divergence, most of the songbird species (80%) initiated their phylogroup divergence at least 1 mya, which contradicts the almost exclusive importance of late Pleistocene glacial barriers as the LPO model proposes (Klicka and Zink 1999).

The sympatric challenge

The difficulties of assigning a strict and permanent geographical barrier to a speciation process, as the above case studies show, must not lead to the conclusion that allopatric speciation is improbable. Rather, the lesson to take away should be that even though geographic barriers are of great assistance in diminishing, or even impeding, gene flow among populations and facilitate species differentiation, by themselves they are likely not alone in the process

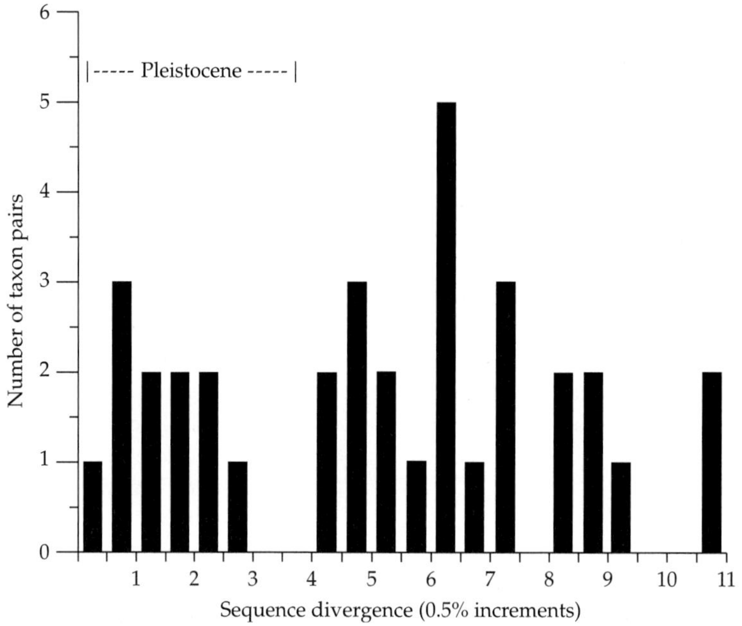

Fig. 4.1. Plot of estimated per cent of mtDNA sequence divergence (and inferred separation times) between 35 sister-species pairs of North American songbirds. From Klicka and Zink (1997) with permission from the American Association for the Advancement of Sciences.

of speciation. Recent phylogeographic approaches, as those discussed above, and, mainly, the use of experimental genome research, have changed our view of the allopatric model, a process that is not as simple as was early depicted by its proponents. Perusal of different biogeographic patterns of Pleistocene colonizations powerfully illustrate the complexity of the processes. Not only do glacial barriers form and disappear with great speed, changing the species range, but rapid climatic shifts also force the species to adapt to new environments. The combined effect of biogeographic and environmental changes allows the interplay of demographic and selective factors in the evolution of species.

How much gene flow exists in nature? In the strict allopatric model the answer is none, because the model is dependent on total geographic isolation. However, using recent genomic techniques and DNA markers, profuse gene flow has been detected in most of the natural scenarios from allopatric–parapatric to sympatric species, as will be illustrated below. Even though isolation may seem complete in extreme cases, such as in oceanic or habitat islands, some information is needed before asserting that gene flow is zero. Under a neutral **island model** the minimal gene flow of one migrant individual exchanged per generation ($Nm = 1$, being N the effective population size and m the migration rate) impedes the population differentiation. This parameter Nm can be estimated by computing the genetic differentiation across populations and has evidenced higher than expected values in scenarios where geographical barriers seem too insurmountable to allow migration of low-vagility organisms. Thus flightless beetles of the genus Stomion, endemic to the Galapagos islands, are distributed in spatially isolated island populations that exhibit significant gene flow (Nm greater than 3 in some cases) in spite of their low vagility (Finston and Peck 1995), showing that geographical isolation to the naked eye does not prove lack of gene flow. Analogously, Schilthuizen and his collaborators (Schilthuizen and Scott 2004) have studied a group of land snails, the Diplommatinidae, which require high calcium concentration and low acidity. In Borneo they occupy a karstic landscape, composed of about 300 hills of limestone separated by acidic and calcium-poor land. Until recently these hills were considered isolated islands for these snails, but new surveys show that many species of diplommatinids that were thought to live exclusively on limestone outcrops, have been found in the interspersed acidic territories. These and other studies suggest that geographical isolation, though an important force to induce species differentiation, is not the only one in many natural settings and cannot explain by itself the process of speciation. Other forces such as natural, and possibly sexual, selection may be at work.

The exclusivity allocated to geographical barriers to prevent gene flow in the initial process of speciation obliges us to disregard the possibility of sympatric speciation, the process of species formation by diversifying one population into two without geographical barriers. Yet, population genetics theory explains that certain ecological conditions may promote this population split in sympatry. In a single locus polymorphism this may occur when **habitat selection** favours the two extreme **homozygotes**, each in a different habitat, and the **heterozygote** shows an inferior fitness in either one. But this one-locus model is theoretically very inefficient in allowing population divergence if mating is at random, which makes a certain dose of **assortative mating**, associated to a second gene, a must for population sundering, as theoretical biologist Maynard Smith (1966) explains. This model of diversifying selection is dependent on several conditions to establish a stable polymorphism; habitat preference and homogamic selection among them. For decades many evolutionists have shown scepticism towards the accomplishment of these conditions, but new studies support them.

Mayr's multiheaded hydra and other sympatric histories

Early in 1917, Hopkins, an entomologist, described the fact that insects whose larvae fed on a certain type of plant tend to lay their eggs in this same plant species. This **habitat preference** allows the reinforcing of the habit of many insects to mate on the same host, usually between genetically related organisms that share the same host-adapted mutations (**homogamy**). Both behaviours, the **Hopkins effect** and the homogamic selection, probably constitute the two most important initial conditions to stabilize a genetic polymorphism that likely leads to an increased specialization to a new host and also to reproductive isolation without geographical barriers by incorporating, through pleiotropic effects, new adaptive characters such as **breeding allochrony** and sexual divergence. And this is likely what sympatric speciation is all about.

Many phylogenetic radiations of monophagous insects, such as those that produced each one of the more than 700 parasitic wasp species that breed specifically in only one fig species, may be explained by sympatric speciation. Actually, many examples of putative sympatric speciation by host shift have been reported in host races of phytophagous insects. The role of sympatric speciation in these insects was even admitted by Mayr (1963, p. 460) when he stated that 'host races are a challenging biological phenomenon, and constitute the only known case indicating the possible occurrence of incipient sympatric speciation'. Since a large proportion, up to 40% according to some authors (Berlocher and Feder 2002), of animal species may belong to this class of insects, the sympatric explanation cannot be disregarded as unusual. Well-studied examples include the hymenopteran *Pontania salicis*, whose races, monophagous on willow species, show experimentally a strong Hopkins effect; the nine closely related species (siblings) of treehoppers (genus Enchenopa) that show a specific synchrony of hatching with their host plants, triggered by specific flowering time, which suggests a sympatric evolutionary history of reproductive isolation by allochrony (Wood 1993); and the case of the ermine moths (genus Yponomeuta) that deserves a more extensive comment.

Briefly, the genus Yponomeuta comprises over 70 species of Palaearctic distribution (the ermine moths), most of them monophagous on the staff-tree family (Celastraceae), but several species of Western Europe are associated to rose (Rosaceae) and willow (Salicaceae) families. This change may be explained by a host shift in sympatry. The observation of several sympatric populations of *Y. padellus* associated to different host plants of the genera Prunus and Crataegus bolsters the sympatric argument (Menken and Roessingh 1998). But the strongest evidence comes when, starting in 1985, a massive spread of ***Y. padellus*** was witnessed all over the

Netherlands associated with rowans (*Sorbus aucuparia*), a plant never before occupied by ermine moths in this country. Initially Dutch researchers assumed that this was an invasion from Scandinavia, where it was known that ermines occur on rowan trees, but genetic markers identified the Dutch populations as autochtonous. The host shift was 'caught in the act' as Schilthuizen (2001) qualifies this usually unnoticed episode of sympatric speciation.

Perhaps the best documented case of incipient sympatric speciation refers to the dipteran *Rhagoletis pomonella* (the apple maggot fly). Its natural hosts in North America belong to the hawthorn genus (Crataegus), but around 1860 this species experienced a host shift in the north-east of the United States towards apple trees and since then has invaded other fruit trees of European origin. As early as 1864, Walsh, then an Illinois state entomologist and Darwin's former student mate at the University of Cambridge (England), was acquainted with the apple maggot invasion and wrote to Darwin about his impression that the fly might have evolved to a new incipient species due to its host shift. Darwin acknowledges this communication by stating:

> Mr B. D. Walsh, a distinguished entomologist of the United States, has lately described what he calls Phytophagic varieties and Phytophagic species . . . In several cases, however, insects found living on different plants have been observed by Mr Walsh to present, either exclusively in their larval or mature state, or in both states, slight, though constant differences in colour, size, or in the nature of their secretions . . . When the differences are rather more strongly marked, and when both sexes and all ages are affected, the forms would be ranked by all entomologists as species. But no observer can determine for others, even if he can do so for himself, which of these Phytophagic forms ought to be called species and which varieties. Mr Walsh ranks the forms which it may be supposed would freely intercross together, as varieties; and those which appear to have lost this power, as species. As the differences depend on the insects having long fed on distinct plants, it cannot be expected that intermediate links connecting the several forms should now be found. (Darwin, *The Origin of Species*, 1869, fifth edition pp. 57–8, chap. 2).

Darwin's reluctance to accept the species nature of these phytophagic forms notwithstanding, this is one of the first statements of sympatric speciation in the scientific literature.

Currently, these maggot fly populations associated to different hosts are true genetic host races that show female preferential oviposition for the host plant and genetic differences in developmental time, all these traits in line with the proposed scheme of sympatric speciation. Interestingly, habitat selection and the ensuing reproductive isolation have led to a significant genetic differentiation in less than 150 years after the host shift. Most, if not all, of the accomplishments in deciphering the Rhagoletis evolutionary history are the culmination of a personal endeavour led by Guy L. Bush, a former Harvard PhD student, who developed the courage to contest the ideas of Mayr, then a member of Bush's thesis committee. Mayr was totally opposed to speciation in sympatry but he enthusiastically endorsed the project of undertaking a thorough study of Rhagoletis for Bush's PhD thesis only 'to put that unresolved example of sympatric speciation to rest once for all', as Bush (1998) explains. To Mayr (1963, p. 451) it should no longer be necessary to devote much time to sympatric speciation, but since this issue 'is like the Lernean Hydra which grew two new heads whenever one of its old heads was cut off', he wanted to cut off all heads once and for all with his student's ultimate work. Ironically, during more than 35 years, Bush and his collaborators have unleashed a wave of thinking on speciation in sympatry, based on experimental data that ranges from identifying cues and traits responsible for host-plant recognition and mating behaviour (Bush 1993), to laying the genetic bases of the population structure of sympatric host races (Feder *et al.* 1988), to establishing the role of the diapause in host adaptation (Feder *et al.* 1997), and to finally building up a general sympatric speciation model in animals (Bush 1994). This body of work has challenged the uniqueness of **allopatry** in speciation, and has induced, rather than stopped, the recurrent growth of Mayr's hydra heads.

The allo-sympatric model

For decades this case study was taken as the paradigm of sympatric speciation. Recently, the Rhagoletis speciation story has been completed by investigating the phylogeography and the phylogenetics of the Rhagoletis species complex that comprises six or more sibling species ranging from the Altiplano highland in Mexico to the north-eastern United States. Using a set of 15 nuclear loci and several mtDNA sequences sampled from six populations in Mexico (two) and United States (four) a series of gene trees were constructed. These studies, summarized in Feder *et al.* (2005), revealed a dichotomous gene differentiation between Mexican and northern populations, depending on gene association with polymorphic chromosome rearrangements: namely, those genes linked to inversions show a higher differentiation than genes mapping in free inversion chromosomes. This result, coupled with the separate observation that these inversions are associated with diapause traits implicated in the Rhagoletis adaptation to variation in host fruiting time, has been interpreted as follows. First, an early isolation event, occurred in Mexico 1.57 mya, subdivided an initial population in two, north and south. Later, this isolation was followed by a long period of contact about 0.5–1 mya, in which southern inversions introgressed the northern population and formed adaptive geographic clines without allowing recombination among linked genes. These diapause latitudinal clines could have aided North American flies in host shifts to plants with different fruiting times, facilitating the origin of new host races in sympatry. On the other hand, genes not linked to inversions were introgressed and recombined freely. The story continues with a series of more episodes of isolation and secondary contacts, probably related to Pleistocene glaciations (Fig. 4.2).

The allo-sympatric speciation story of Rhagoletis is not the only one. The cichlids (Barlow 2002), a family of tropical fishes that includes some 1,500 species endemic to the great lakes of the Eastern African rift valley, also epitomize a speciation case in which allopatric and sympatric episodes have likely been intertwined. This story is documented as an explosive speciation. For example, Lake Victoria dried up to become flooded again only 15,000 years ago and since then some 500 species likely originated from a unique founding event, which raises the speciation rate to one species every 30 years!! Explosive cichlid speciation gave rise to large groups of closely related species, named species swarms or flocks. The causes of this species explosion are likely a combination of extreme founder events (often of an individual founder), followed by periods of lake contraction that favoured isolation in small basins, promoting divergence due to both founder events and also habitat selection. In this process, sexual selection would have played a prime role as well. Given that many male cichlids differ more in colour than in morphology, it has repeatedly been suggested that diversifying sexual selection would be responsible for prezygotic isolation in sympatry. Yet, habitat complexity in these great lakes and also their complex geological history makes it difficult to disregard putative allopatric events. Moreover, clear phylogenetic relationships that allow us to test whether recent sister species groups show distributions that are more overlapped than those of ancient sister species are not easy to establish, which makes the sympatric argument even more contentious. Fortunately, small volcanic crater lakes, due to their lower historical and biogeographical complexity, allow us to surmount some of these difficulties. Cichlid studies carried out in such lakes of Cameroon and Nicaragua (Barluenga *et al.* 2006) show that diversifying ecological selection related to food habitat, has likely been the primordial force of speciation in sympatry. These studies are not free from criticism, mainly because the possibility of more than a unique founding event followed by introgression has been advanced. However, they strongly suggest that once again speciation can occur in sympatry.

What can we conclude from all these speciation accounts? First, speciation in sympatry seems well established, at least when differentiation is driven by host adaptation, and may constitute a large fraction of speciation events in some taxonomic groups such as monophagous insects and possibly in others as well (see Box 4.1). Second, the use of DNA markers in phylogenetic and phylogeographic approaches has shown in many thoroughly studied cases that speciation is often a combined process in which episodes of sympatry and allopatry are intertwined. As stated by Mallet (2005a) 'theory and empirical evidence now argues for a more pluralistic view of the geographic mode of speciation, and indeed one which might readily occur without pure allopatry at all'. Although geographic barriers may play a role in speciation, even in those cases in which sympatry and/or **parapatry** are well

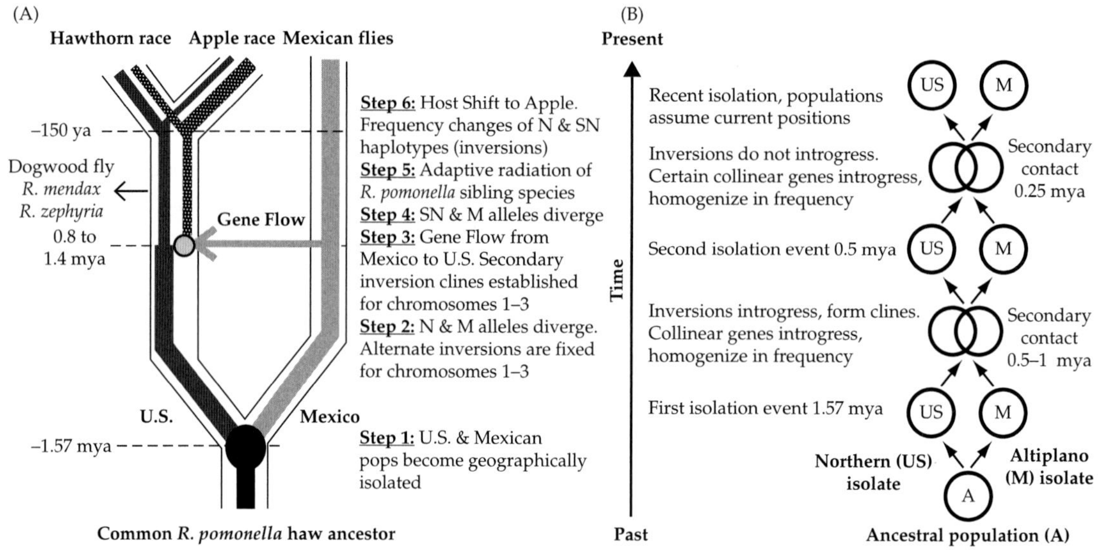

Fig. 4.2. (A) Introgression hypothesis accounting for *R. pomonella* in a geographic model with introgression (step 3b represents a series of gene-flow events between about 0.8 and 1.4 mya). Thin lines define population boundaries. Thick lines represent gene trees for alleles segregating within populations: grey line, Mexican haplotype; black line, Northern haplotype; dotted line, South/North haplotype (descended from introgressed genes from Mexico). The thicker width of the dotted line for the apple population denotes the generally higher frequency of SN haplotypes in the apple race at sympatric northern sites. Adapted from Feder *et al.* (2003) with permission from the National Academy of Sciences of the United States of America. (B) Biogeographic model depicting two cycles of isolation and differential introgression between Mexican Altiplano (M) and northern (United States, US) populations of *R. pomonella*. From Feder *et al.* (2005) with permission from the National Academy of Sciences of the United States of America.

Box 4.1 An ongoing sympatric ecological speciation event in intertidal populations of a marine snail

Most reconstructions of past speciation events are hampered by two main difficulties. First, the assumption that geographic distribution at the time of speciation can be inferred from current species ranges is often illusory. While phylogenetic comparative methods have shown their utility, they suffer from this predictive shortcoming. Other population approaches (i.e. coalescent methods) are still in an incipient stage, requiring some simplifying assumptions that cast some doubts on their utility. Second, ecological speciation studies require a thorough understanding of a) the ecological causes of divergent selection, b) the basis, ecological or not, of reproductive isolation, and c) how divergent selection and reproductive isolation are genetically related. Ideally both difficulties could be surmounted by studies of ongoing speciation events for which the geographic scenario is directly observed and the operation of selective and reproductive components can be experimentally approached. Such inclusive studies are rare, but the case of *Littorina saxatilis* may epitomize one of them. This case study, of which a summary follows, is the work of Rolan-Alvarez and his research group.

This ovoviviparous marine gastropod with low dispersal capabilities (1–2 m per month), displays a strong microhabitat-associated dimorphism in the intertidal shores of Galicia (north-west Spain). The form that occupies the upper shore has a large, ridged, and banded shell (RB), whereas the form inhabiting the lower shore presents a small, smooth, and unbanded shell (SU). Rolán-Alvarez and his team think that these ecotype morphologies, which have a strong genetic base, are highly adapted to the selective pressures associated with each intertidal zone. Namely, in the upper zone the robust RB snails are equipped to stand daily abrupt changes in salinity, high heat exposure, and predation by crabs; on the other hand, the main challenge in the lower zone is the strong surf against the rocks that the small-shelled and large-footed

continues

Box. 4.1 *(continued)*

SU snails endure without difficulty. Extensive evidence mainly based on mark–recapture and transplant experiments in nature, and also on laboratory experiments (summarized in Rolan-Alvarez 2007), bolsters the relationship between ecological adaptation and phenotypic distribution of ecotypes.

Two main additional pieces of evidence underlie the sympatric nature of Littorina differentiation; namely, the origin 'in situ' of reproductive prezygotic isolation and the local monophyletic grouping observed in molecular trees. Hybrid performance is intermediate in viability, sexual selection, and fertility compared with parental ecotypes. Thus, though inferior hybrid fitness in wild habitats cannot be excluded, postzygotic isolation is not likely the cause of reproductive isolation. On the contrary, there is ample evidence that premating mechanisms in the form of sexual isolation are at work. Microhabitat selection in upper and lower shores fosters an effective ecological isolation that is broken in the intermediate zone where both ecotypes meet and mate. However, the study of mating pairs in the wild has shown that matings are not random, providing an estimate of 70% sexual isolation (range 50–100%). This incomplete reproductive isolation that still generates a high gene flow between ecotypes (mating is random between pure ecotypes and sympatric intermediate forms), can also be inferred from molecular differentiation between ecotypes in sympatry, which does not exist between geographically distant populations of the same ecotype. In sum, divergent habitat selection probably has indirectly promoted size-based assortative mating.

The present geographical pattern of genetic variation cannot distinguish between past speciation scenarios. But, fortunately, the evidence from phylogenies built up using molecular markers (mainly mtDNA, but also AFLPs, microsatellites, and **allozymes**) sampled from four isolated populations is only compatible with a sympatric and parallel model of speciation. These trees show a systematic monophyletic origin of both haplotypes (RB and SU) at each locality, a result totally incompatible with an allopatric speciation model (Quesada *et al.* 2007) (see Fig. A and Plate 10). Moreover, further studies by the Rolan-Alvarez team have detected several genome regions that are, directly or not, targeted by natural selection at local scale. These regions amount to 3–4% of the genome, an important figure considering the high gene flow between ecotypes.

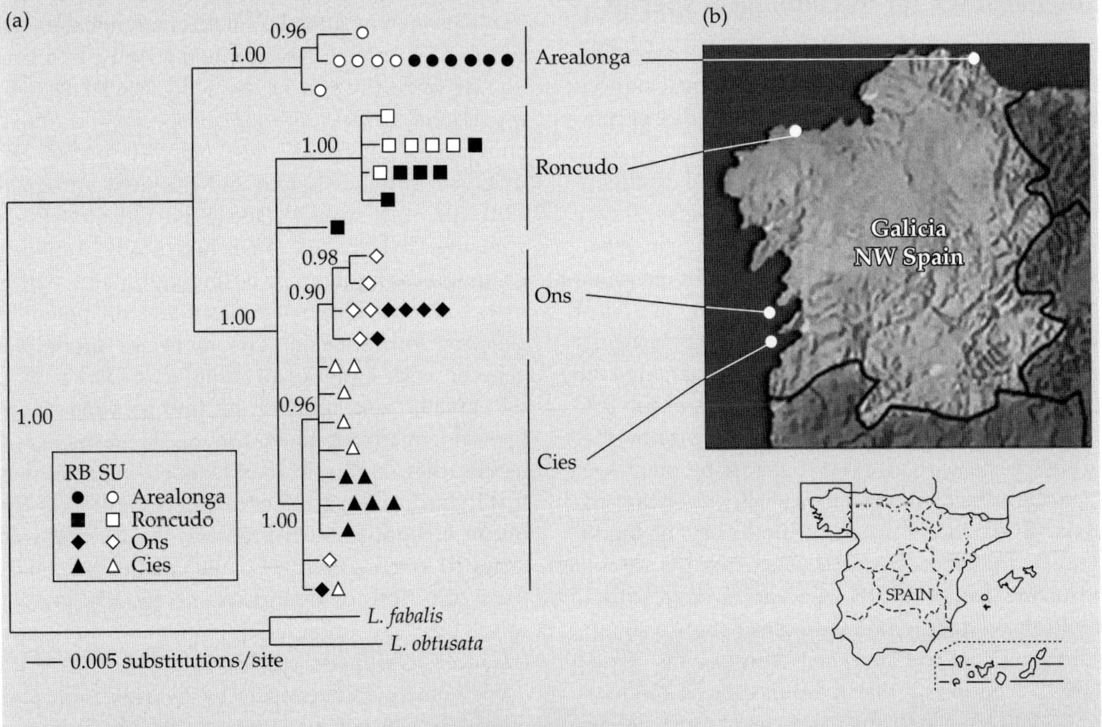

Fig. A. Phylogenetic tree using mtDNA nucleotide variants of the two kinds of *L. saxatilis* ecotypes (RB close and SU open symbols) in four different localities of north-west Spain. Note that both ecotypes cluster together in each locality, which indicates a common geographical origin, a result incompatible with the allopatric model. (See text for further details). From Quesada *et al.* (2007) with permission from John Wiley and Sons (see also Plate 10).

continues

Box. 4.1 *(continued)*
Combining all the data, the case of Littorina constitutes strong evidence that we have caught in the act an ongoing process of sympatric speciation. As Quesada and his collaborators (2007) state: 'The sympatric distribution of RB and SU ecotypes, the high gene flow estimated between ecotypes (Nm > 4), the evidence of reproductive isolation, the **monophyly** within regions, and the rejection of a putative allopatric scenario, satisfy all the criteria required to conclude a multiple sympatric origin of both ecotypes (Coyne and Orr 2004).' I fully agree with this statement.

established, the critical role of allopatry, strongly argued by 'allopatriots', is debatable. Only a thorough knowledge of the relative weight of gene flow and natural selection in the geographic scenario is going to enlighten our view of speciation. In sum, it seems to me that the sympatric argument has challenged the secondary role of natural selection in the initial steps of the origin of species sensu Mayr (1970), which brings us back to the more pluralistic, Darwinian view of speciation with a large appreciation for the role of natural selection.

Too much sex for the biological species

The finding that introgression through hybridization plays a role in speciation is not novel and has received a great support from recent studies of comparative genomics (Mallet 2005b). Historically, since the definition of the BSC, large groups of evolutionists have shown their scepticism towards the efficacy of isolation barriers as the unique mechanism to preserve species integrity. There is no doubt that plant biologists rallied easily to those evolutionists most reluctant to accept the BSC, their argument being the frequently observed capacity of hybridization in the plant world. Some botanists, defenders of the BSC such as Grant (1957), are not ashamed to admit that a high percentage of 'good' outcrossing plant species cannot be defined by reproductive isolation criteria. Ever since Darwin, evolutionary biologists have had to debate the fact that taxonomists are able to define species which co-occur in large natural hybridizing units. Often these units show a significant gene flow and are called '**syngameons**'. Grant (1981), who insists that hybridization is the most important issue in the plant species identity, defines the syngameon as: 'the most inclusive unit of interbreeding in a hybridising species group', and goes on 'solving' the problem by assigning to the syngameon the status of species under the isolation criterion, and to the taxa in the syngameon the category of semispecies. This elusive solution is unsatisfactory to most taxonomists, who view members of a syngameon as true species endowed with the required space-temporal stability of their morphologic, ecologic, and genetic characters. To make things more difficult, often such stability can be traced to long geological periods, as in the case of members of the genus Populus, whose hybridization and identity has been documented in the fossil record of the last 12 my (Eckenwalder 1984).

Admittedly, hybridization is reported more often in plants than in animals, but this does not allow us to consider animal hybridization a rarity. In a survey of hybridization in the wild, Mallet (2005b) reports that 'at least 25% of plant species and 10% of animal species, mostly the youngest species, are involved in hybridisation and potential introgression with other species'. Due to the difficulty of recognizing hybrids in morphologically uniform groups, these figures may be underestimates, which makes animals also a significant area for potential genomic introgression. The more we probe the genome with batteries of multilocus marker loci, the greater the number of introgression cases revealed, many of them coming as unsuspected results that contradict our instinctive perception that hybridization is either rare or a 'cul-de-sac' for future evolution. Most probably the leading role assigned to reproductive isolation in the BSC conditions our view of hybrids as necessarily low fit genotypes, an unfounded contention in many instances, as will be discussed below.

Fortunately, introgression by hybridization and other kinds of genetic exchange, discussed in this chapter, are amenable to testing (see Arnold 2006, Chapter 3 for a survey). Yet, conclusions from these methodologies are often vitiated by the researcher's underlying assumptions of the exchange process. Arnold (2006, p. 34) illustrates this bias stating that 'zoologists often assume that allele sharing between different taxa is due to incomplete **lineage sorting** while botanists are more likely to assume that the same pattern of allele sharing in plants is due to introgression' (Box 4.2), not a very unexpected behaviour in face of their divergent views on hybridization. In fact, discerning between both processes is a difficult task, which becomes even more complex when putative horizontal gene transfer could be at work, as many workers of microorganisms usually assume (see below). Regardless of mechanism, there is ample evidence of genetic exchange due to hybridization in many documented cases in the wild (see Mallet 2005b for a review) comprising not only plants (25% of vascular plant species) but also animals, from butterfly taxa (Rhopalocera: 12.4%; Papilionidae: 14–32%; Heliconiinae: 26.0%) to birds (9.3% average) ranging from ducks (Anatinae: 76.2%) to birds of paradise (Paradisaeidae: 42,9%), and to tits (Paridae: 28.6%). Even European mammals are reported to hybridize in significant amounts (6.0%).

A common objection to the evolutionary value of this pervasive hybridization is that the rate of hybridization per individual is low and never results in introgression. The rarity of hybrids notwithstanding, it is a well established fact, at least in some groups like insects and birds, that once a hybrid is produced, back-crossing to one parental individual occurs more easily providing that the hybrids are not completely sterile. Back-cross hybrids are much more difficult to detect and this

Box 4.2 The impact of lineage sorting vs introgression in gene genealogies

The true genealogy of three species S1, S2, and S3 is shown in Fig. B (top). Five alleles or haplotypes (A, B, C, D, E, F) are depicted and their true gene genealogy is shown bottom. Branching events in the gene tree (1, 2, 3, 4, 5) can either post-date or predate a species-level split. This asynchronic lineage sorting of alleles is responsible for assembling distantly related alleles (D, F) in the same species (S3) and closer alleles (D, C) in different species (S2, S3). The end result is that gene trees, as seen in Fig. B (bottom), and species trees (top of Fig. B) are not concordant.

Another distortion is caused by introgression (or horizontal transfer). In node 5 a new mutant allele B is produced that eventually introgresses from S2 to S1, so A and B, two distantly related alleles, coexist in species S1. This is at odds with the separation of C and D, two more closely related alleles. This may be interpreted as lineage sorting if insufficient molecular information is available.

In general gene flow is hard to set apart from ancestral polymorphism sorting, but in most cases where hybridization is suspected on non-molecular (mainly morphological) grounds, introgression has been bolstered by molecular data.

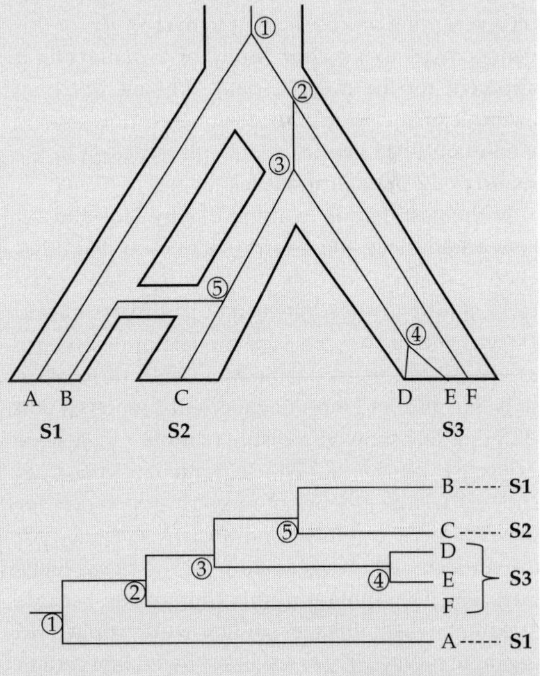

Fig. B. Diagram of the species tree (above) showing the details of the branching and the introgressive events in the gene tree (1, 2, 3, 4, 5). Note that the resulting gene genealogy (bottom) and the species tree (top) are not concordant. (See text for further details.) Drawn by Montserrat Peiró.

has traditionally been a handicap in assessing the importance of gene flow among species until DNA and molecular markers became available for genome analysis. The classic view, sponsored mainly by Mayr and Dobzhansky, that gene flow is negligible in speciation dynamics, is being challenged by present genomic analysis.

Divergence with gene flow: a Drosophila story

Ironically, the case of *Drosophila pseudoobscura* and allied species, one of the best examples worked on by Dobzhansky and their co-workers to defend the prevalence of isolation barriers in species integrity, has turned out to be an outstanding epitome of divergence in the presence of gene flow. Powell (1983), a former student of Dobzhansky, in a pioneering study using mtDNA, concluded that substantial gene flow exists between sympatric populations of *D. pseudoobscura* and its sibling *D. persimilis*, but several years later he seemed to change his view, suggesting that: 'Clearly, an alternative explanation, . . . is that the ancestral polymorphism/sorting process has led to **paraphyly** for this genetic marker. No clear choice of explanations is apparent for the mtDNA data' (Powell 1991); an example of how much the idea of gene flow among well-established species is difficult to accept in the world of the biological species.

In the past decade many works by Hey and his co-workers using a large array of nuclear and mitochondrial genes not only documented clear signatures of introgression between *D. pseudoobscura* and *D. persimilis*, mainly because models of divergence in allopatry were not compatible with population data, but also, and most unexpectedly, reported that different loci showed a distinct amount of genetic exchange (Wang *et al.* 1997). In particular, gene trees showed a large amount of recent introgression for the *Adh* locus, a limited and ancient exchange for the *per* locus, and no gene flow for the *Hsp82* locus (Fig. 4.3). The solution to this conundrum became apparent when linkage studies were applied to genomic surveys. These studies showed that certain genomic regions, generally associated with genes under divergent selection, contain strongly differentiated loci between species, whereas other regions devoid of divergent adaptations are highly introgressed. In the species pair *D. pseudoobscura*/*D. persimilis*, highly diverged genomic segments have been found to be associated with hybrid sterility loci (Machado and Hey 2003) and some divergent adaptations are also linked to inversions (Noor *et al.* 2001). Since inversions inhibit recombination, this association suggests that they contribute to maintaining those adaptive divergences characteristic of each species. The role of chromosomal rearrangements in speciation as inhibitors of recombination has been suggested also in the human–chimpanzee divergence, where the rate of amino acid substitution is significantly higher in chromosomes with rearrangements (Navarro and Barton 2003).

All in all, this evidence emphasizes that genomes are not impermeable to gene flow; rather, they can be described as semi-permeable, and that only those genomic segments critical for the species integrity are reproductively isolated. The model of Rhagoletis, an epitome of sympatric speciation, in which episodes of isolation in allopatry are interspersed with others of introgression in sympatry/parapatry is not much different from that of Drosophila and many species traditionally considered as paradigms of the allopatric model. Perusal of the genome reveals that hybridization is an ongoing process in species, generating gene flow mainly in those regions not involved in adaptive traits concerning reproductive, ecological, or behavioural performance. This view of reticulate speciation that makes gene flow and divergent selection compatible, challenges the exclusiveness of the reproductive isolation role in species integrity, and makes Mallet (2005b) state that: 'even if genetically isolated species play a role in diversification, today we know that evolutionary progress can continue while species undergo genomic invasions from other species'.

The creative power of genetic exchange

The role of hybridization in explaining biodiversity goes far beyond the effect of gene flow in the genome architecture and its relevance to species integrity because hybridization, as well as other processes of genetic exchange, is often a starting point for generating new species and possibly new evolutionary lineages. It is common knowledge that the great majority of ferns (pteridophytes), and

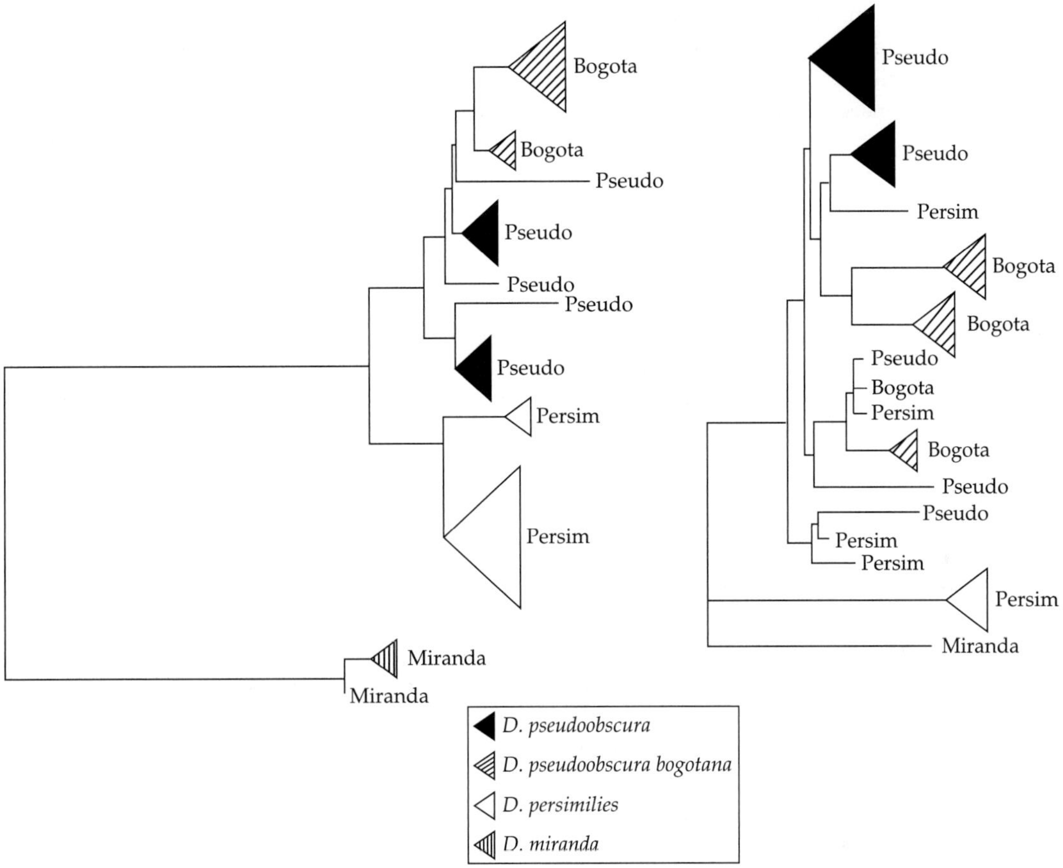

Fig. 4.3. Two phylogenies based on DNA sequences of the *Hsp82* gene (left) and portions of the *Adh* region (right) for *D. pseudoobscura* and close relatives *D. persimilis* and *D. pseudoobscura bogotana*. Each of the species samples form monophyletic groups in the *Hsp82* tree, with the exception of the clade spanning the *D. pseudoobscura* samples, which contains a subtree for the *D. p. bogotana* samples. On the other hand, the *Adh* tree reveals multiple instances where sequences do not cluster by taxon, showing reticulate placements. The minimum numbers of migration events suggested by this *Adh* reticulate tree are: three for *pseudo/p. bogotana*, four for *pseudo/persimilis*, and one for *persimilis/p. bogotana*, indicating large amounts of recent gene flow among the taxa, while little or no gene flow is apparent in the data from the *Hsp* locus. Triangles depict small clades of species lines. Adapted from Wang *et al.* (1997) with permission from the Genetics Society of America.

flowering plants (angiosperms) evolved by whole genome duplication (polyploidy) and that a substantial amount of their genomes is of hybrid origin (**allopolyploidy**). Usually hybrid sterility is a result of the uneven chromosome segregation in meiosis due to mispairing of non-homologous chromosomes, producing defective gametes. Chromosome doubling in hybrids by polyploidy provides a whole set of paired homologous chromosomes, which overcomes the hybrid sterility barrier. This is an extraordinary mechanism of one-step plant species formation, often referred to as '**instant speciation**'.

Hybridization is followed by genome reorganization

Although allopolyploidy is a mechanism that has been largely observed and reproduced in laboratory experiments since Winge first described it in 1917, it was not until recently that through using molecular markers we have understood that speciation by allopolyploidy is not so instantaneous, mainly because of the genome reorganizations that follow genome duplication. A striking characteristic of these reorganizations is their rapid occurrence after duplication, witnessed experimentally by the many genetic changes observed in the synthesis of artificial allopolyploids. This genomic dynamism has huge consequences for generating new recombinational variability useful to further ecological adaptation in new habitats and also to improve hybrid fertility.

Polyploid restructuring after hybridization involves both gross chromosomal rearrangements and DNA sequence reorganizations. Comparative genomic studies of many plants, including maize and cabbage, have provided compelling evidence of these changes. Thus, we know now that maize (*Zea mays*), a traditional diploid, is an ancient allotetraploid with more than 70% of its genome duplicated. However, the duplicated segments are fragmented and dispersed throughout the genome, unlike a conventional allopolyploid, suggesting that a large number of inversions and translocations have taken place after polyploidization (see Chapter 1, p. 20 and Figs 1.8 and 1.9). Similarly, in the genus Brassica, descended from an ancestral allohexaploid, a minimum of 24 chromosomal rearrangements must be assumed to have taken place since the original polyploidization event to explain differences in gene order observed among *B. nigra*, *B. rapa*, and *B. oleracea* (Lagercrantz and Lydiate 1996) (Fig. 4.4).

These and similar observations in other genera of the Brassicaceae family, though suggestively associated with allopolyploidy, do not tell us much of the underlying causes, namely whether they are induced by hybridization or other processes. McClintock (1984), inspired by her cytological

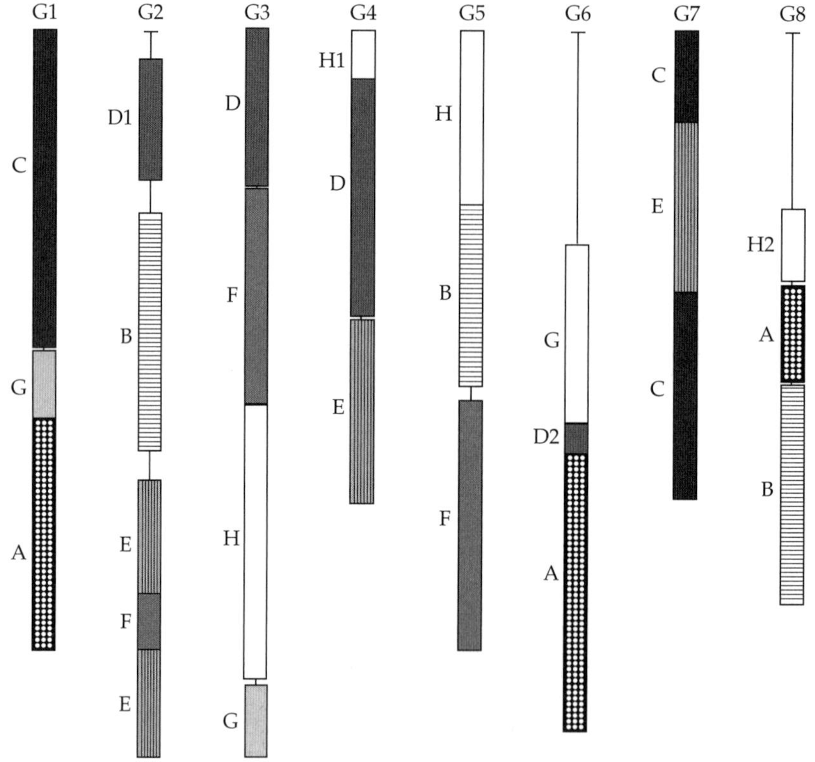

Fig. 4.4. Chromosomal map of *B. nigra* showing the distribution of the sets of triplicated chromosomal segments (**A-H**) in each of the eight chromosomes (linkage groups **G1–G8**). Triplicated chromosomal segments with the same shading share common sets of homologous loci. Virtually the whole *B. nigra* genome has a recognizably triplicated structure. In two cases, a segment is split in two different sites of the genome (D1/D2; H1/H2). This division and the different length shown by the homologous triplicated segments suggest that a large genomic reorganization was underway in allopolyploid speciation. From Lagercrantz and Lydiate (1996) with permission from the Genetics Society of America.

studies with maize, pioneered the idea that 'major restructuring of chromosome components may arise in a hybrid plant and continue to arise in its progeny'. Recently, several experimental studies, mainly with the genera Brassica, Arabidopsis, and Nicotiana, have bolstered the McClintock statement. Following self-fertilization of synthetic allotetraploids between pairs of the above Brassica species for several generations, extensive genomic changes, mostly involving loss and/or gain of parental and novel DNA fragments, were detected by comparing nuclear DNA probes between second- and fifth-generation progenies (Pires *et al.* 2004). This observation is similar to those obtained with the above genera and suggests that considerable genetic change accompanies allopolyploid speciation, but still provides no evidence for the mechanisms responsible for these changes. Yet, in experimental Arabidopsis allopolyploids (Madlung *et al.* 2005) and in recently formed allotetraploid cotton (*Gossypium barbadense*) (Zhao *et al.* 1998) dispersed repetitive DNA sequences, some of them identified as retroelements, have experienced increased rates of expression and/or transposition. In Chapter 3 I dealt with this contentious issue, but here I want to reintroduce the idea that there may be a clear connection between hybridization and transposition rate increase. Although not all cases of allopolyploidy give evidence of this increase, there is a growing acceptance among researchers that bringing together divergent genomes results in a 'genomic shock' that likely reactivates transposable elements (Arnold 2006; pp. 112–14).

Do we need chromosome doubling in hybrids for speciation?

Although allopolyploidy is a frequent mechanism of speciation, many species also originate by hybridization without genome duplication. These homoploid species are the result of overcoming the 'inferior' fertility of interspecies hybrids, a process vigorously opposed by the defenders of the BSC, who stigmatize hybrids as low fitness genotypes due to the supposedly highly effective, reproductive postzygotic isolation mechanisms inherent in true species. Consequently, hybrids have been viewed as lineages devoid of evolutionary significance. Incidentally, this view has not been shared by some early naturalists, including Linnaeus, nor by several neo-Darwinian evolutionists, such as Stebbins and Grant, who consider hybridization as a source of new species. In the last section I gave reasons, mostly thanks to the use of genome-wide markers, to support the tenet that hybridization is common in nature and hybrids survive, in many cases, to generate genomic mosaics able to further evolution. The potentiality of hybrids to survive is reinforced by the experimental measurements of fitness components, showing that the overall hybrid fitness is often not inferior to that of their parental species (Table 4.2).

Rapid genomic reorganization after hybridization has been also observed in homoploid species. A most thoroughly studied case of homoploid species is illustrated by the sunflower species *Helianthus anomalus*, originated by hybridization between *H. annuus* and *H. petiolaris*. A comparison of genetic maps across the three species has revealed an extensive repatterning in the hybrid genome, requiring at least three chromosomal breakages, three fusions, and one duplication to explain the differences in gene order between hybrid and parental species. These changes could be responsible for reproductive isolation, adaptation to a novel habitat, and/or increasing hybrid fertility. Rieseberg and his collaborators (1996) have investigated the dynamics of new genomic design at gene level by mimicking the natural hybridization of Helianthus species in the laboratory. They were able to obtain three fertile and viable hybrid lineages after five generations of hybrid selfing and/or back-crossing. Using 197 **RAPD (randomly amplified polymorphic DNA)** genome-wide markers they studied the genomic composition of these synthetic hybrid lineages and found that, regardless of their different crossing procedures, they showed a high concordance in genomic content among them. Most surprisingly this genomic composition was also statistically concordant with that of the natural hybrid species *H. anomalus* (Fig. 4.5). This similarity suggested to the authors that genomic reorganization in hybrids is not only rapid but also repeatable, most likely due to endogenous selection for specific gene blocks that enhance hybrid fertility. The rapid increase in fertility observed in the synthetic lines, from 4% to more than 90% in only five generations, shows how

Table 4.2. A sampler of homoploid hybrid species, their mode of ecological divergence from parental species, and their hybrid fitness relative to parental species.

Taxon	Description	Ecological divergence	Hybrid origin	Hybrid fitness
Arygyranthemum sundingii	annual herb	habitat	homoploid	I
Daphnia mendotae	freshwater invertebrate	habitat	homoploid	no data
Gila seminuda	freshwater fish	habitat	homoploid	no data
Helianthus anomalus	annual herb	habitat	homoploid	L-E-S
Helianthus deserticola	annual herb	habitat	homoploid	L-E-S
Helianthus paradoxus	annual herb	habitat	homoploid	L-E-S
Iris nelsonii	annual herb	habitat	homoploid	I-E-L-S
Macaca arctoides	primate	no data	homoploid	no data
Machaeranthera genus	herb	no data	homoploid	no data
Metriaclima genus	cichlid fish	mate choice	homoploid	S (indirect)
Paeonia genus	perennial herb	habitat	homoploid, allopoliploid	S (progenitor populations displaced by hybrids in the Pleistocene)
Penstemon clevelandii	annual herb	pollinator	homoploid	no data
Pinus densata	perennial tree	habitat	homoploid,	S (1)
Rana esculenta	vertebrate	habitat	hemiclonal hibridogenetic	L-S (1)
Rhagoletis	fly	host shift	homoploid	no data
Senecio eboracensis	annual herb	temporal and pollinator	homoploid	S (in some traits)
Warramaba virgo	grasshopper	no data	homoploid parthenogenetic	S-E

(1) habitat dependent; Hybrid fitness: S: superior; I: intermediate; L: low; E: equivalent. Data from Arnold, M. L. (2006). *Evolution Through Genetic Exchange.* Oxford University Press, and Gross, B.L. & Rieseberg, L.H. (2005). The ecological genetics of homoploid hybrid speciation. *Journal of Heredity* **96**, 241–252, where references can be found for each taxon.

fast natural selection can act to increase fitness when genome reorganization is taking place.

As mentioned above for allopolyploids, a likely cause of this reorganization could be the reactivation of transposition induced by the bringing together of two divergent genomes. Early instances of transposition activation followed by rapid repression, such as those observed in introgressed rice lines (Liu and Wendel 2000), bolster the transposition role in hybrid homoploid speciation. In Chapter 3 (p. 113) I discussed the role of TEs in the homoploid speciation of sunflowers (genus Helianthus). This genus shows a large documented number of homoploid speciation cases whose genetics and ecology have been studied in detail. Whether TE mobilization is the result of the molecular derepression mechanisms (i.e. methylation) induced by hybridization, or of stress environments, is something to be elucidated. Hybridization is known to alter DNA methylation in synthetic hybrids (see Chapter 3) and the number of documented cases in which hybridization is associated to transposition is increasing not only in plants but also in animals. Depending on the frequency of hybrid occurrence in the origin of new species (see below), the role of TE mobilization cannot be disregarded in speciation. In fact some authors (see for example Michalak 2010) are seriously asking whether TEs should be considered active drivers of speciation processes. However, stress environments found by invasive organisms could also induce transposition and contribute to species differentiation. Hybrids are often colonizers and therefore their ecology must be understood if we are to understand hybrid-related speciation processes.

The ecology of hybrid speciation

The above discussion might give the reader the impression that the genome's intrinsic reorganization of fertility traits is all that is involved in homoploid hybrid evolution. Stebbins (1957) and Grant (1958), almost simultaneously, proposed early on

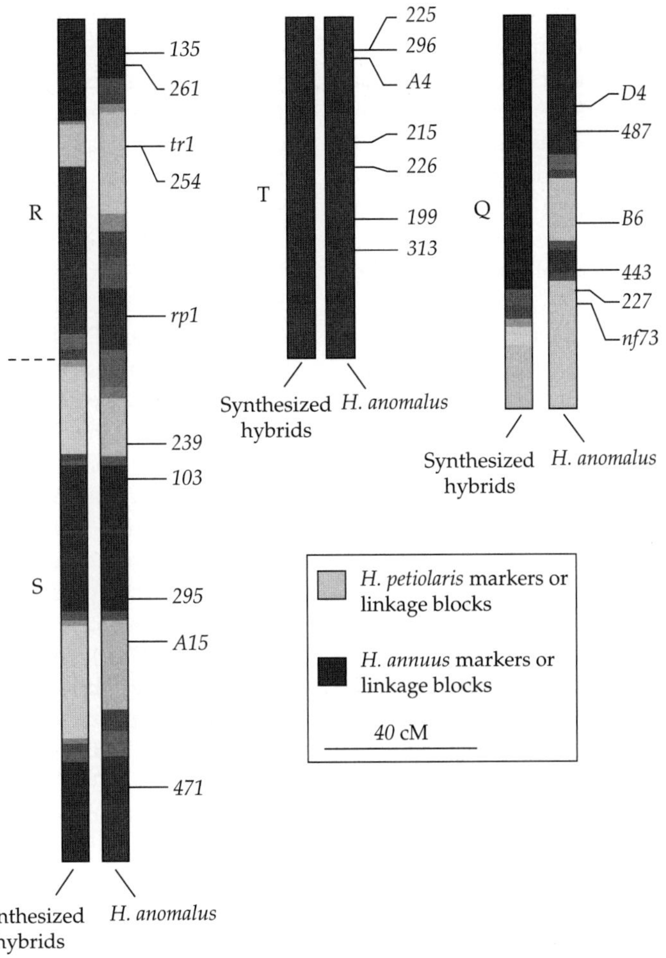

Fig. 4.5. Genomic composition of ancient and experimental hybrid lineages for three selected linkage groups. Letters at the left of each linkage group designate linkage blocks in *Helianthus anomalus* and indicate homology to linkages previously mapped in the parental species, *H. annuus* and *H. petiolaris* (the dashed line indicates the break-point between the R and S blocks of that linkage group). The distribution of parental chromosomal blocks in the synthesized hybrids (left linkage group) and *H. anomalus* (right linkage group) is indicated by shaded bars. Regions harbouring recombination points are indicated by a grey scale, with the intensity of shading indicating the likelihood that a particular region was derived from one parent or the other. From Rieseberg and Noyes (1998) with permission from Elsevier.

that only those hybrid recombining gametes carrying a whole, balanced genome will be able to produce homozygous fertile progeny reproductively isolated from the parental species. This early model, coined '**recombinational speciation**' by Grant (1981), was based on chromosomal recombination of microrearrangements that were responsible for parental reproductive isolation. During the following decades this model was considered as the only plausible one, although theoretical models showed it to be very unlikely (McCarthy *et al.* 1995). Homoploid speciation became clearer when evolutionists started to consider that external ecological barriers could also help to select some hybrid lineages originated by segregation, as Grant (1981) had already suggested. Unfortunately, the role of ecology in speciation was neglected in the second half of the twentieth century (but see Anderson 1949 for an early treatment of introgressive hybridization into an ecological context) producing a wide gap in the speciation theory progress, as Gross and Rieseberg (2005) explain, but in the past decade a reappraisal of ecological factors in speciation has been under way. Precisely, the current consideration that hybrids may show traits that allow them to adapt to new environments, different from those of their parent species, is fuelling the understanding of hybrid speciation.

There are already a number of instances in which ecological divergence between hybrids and their parental species has been documented by using molecular markers and ecological traits (Table 4.2). A long-studied case refers to the members of the iris family that inhabit the wet lands of Southern Louisiana (USA). Arnold (1993) confirmed the homoploid hybrid origin of *Iris nelsonii* from crosses between at least three parental species: *I. fulva*, inhabiting the shady, shallow waters of bayou banks, *I. hexagona*, found in sunny, deeper swamp marshlands, and *I. brevicaulis*, occurring in drier pastures and forests. Normally a large hybrid swarm, as a result of successive crosses and back-crosses, is observed in this scenario, which has promoted an extensive introgression of genetic material in areas of sympatry and into allopatric populations as well. Using a large variety of species-specific molecular markers, nuclear (ribosomal DNA, **RFLPs**, RAPDs, and allozymes), and cytoplasmic (**chloroplast** DNA) as well, Michael Arnold and his group were not only able to find genetic evidence of introgression but they also confirmed a hybrid origin for *Iris nelsonii*. Interestingly, this species was described almost 25 years earlier by Randolph (1966) as of hybrid origin, appealing to his naturalist's eye based on habitat, cytological, and morphological observations.

There are many experimental and natural studies with irises showing convincingly that hybrid fitness is related to some habitat components (see Arnold 2006, pp. 9–11 for a review), allowing hybrids in some cases to occupy new divergent habitats. Shade tolerance is likely one of these traits. Arnold and Bennett (1993) reported a significant association between the amount of genetic introgression from *I. fulva* in *I. hexagona*-like plants and the degree of natural light they experience, an observation according to the natural habitat of these plants. Moreover, *I. nelsonii*, the hybrid species, is found in ecotone habitats combining the shady and deep water of swamps that can be qualified as divergent. These novel, divergent ecological hybrid adaptations are likely to help the process of hybrid speciation and are quite widespread in the majority of homoploid hybrid species (see Gross and Rieseberg 2005 for a review).

Similar cases of ecological divergence have been documented in other homoploid hybrid plants, including species of the genera Stephanomeria, Paeonia, Argyranthemum, Penstemon, Senecio, Pinus, and Helianthus. Among them the case of the genus Helianthus deserves close scrutiny because it has been studied using both ecologic and genetic approaches, as I have introduced above for their genetic repatterning. At least three hybrid species (*H. anomalus*, *H. deserticola*, and *H. paradoxus*) are recognized, which occupy habitats quite divergent from those of their parental species that live in soils ranging from mesic, heavy clay (*H. annuus*) to dry-sandy (*H. petiolaris*). These two species are sympatric and produce hybrid swarms, with semi-sterile F_1 hybrids (<10% pollen viability and <1% seed viability) and F_2 individuals showing a wide range of pollen viability (13–97%). Repeated hybridization allows for stability in these swarms and the possibility of further evolutionary progress. In fact, *H. anomalus*, originated by hybridization between *annus* and *petiolaris*, is endemic to active sand dunes, providing a good example of hybrid invasiveness of novel restricted habitats. The hybrid origin of *anomalus* is bolstered through multiple studies using molecular markers, including the observation that *anomalus* gene linkage groups are interspersed with loci from the two parents in a 50:50 ratio, and that Helianthus phylogenies show reticulate evolution for ribosomal DNA (rDNA) and chloroplast DNA (cpDNA) sequences combining variants from *annuus* and *petiolaris* in *anomalus* (Rieseberg 1991) (Fig. 4.6). Similarly, *H. deserticola* is found on xeric habitats and *H. paradoxus* occupies desert salt marshes.

But the evidence of ecological divergence in hybrids is not a final proof that hybridization per se is responsible for the appearance of the traits under selection. It may be that these adaptive traits are the result of the gradual action of mutations accumulated after speciation. The possibility of producing

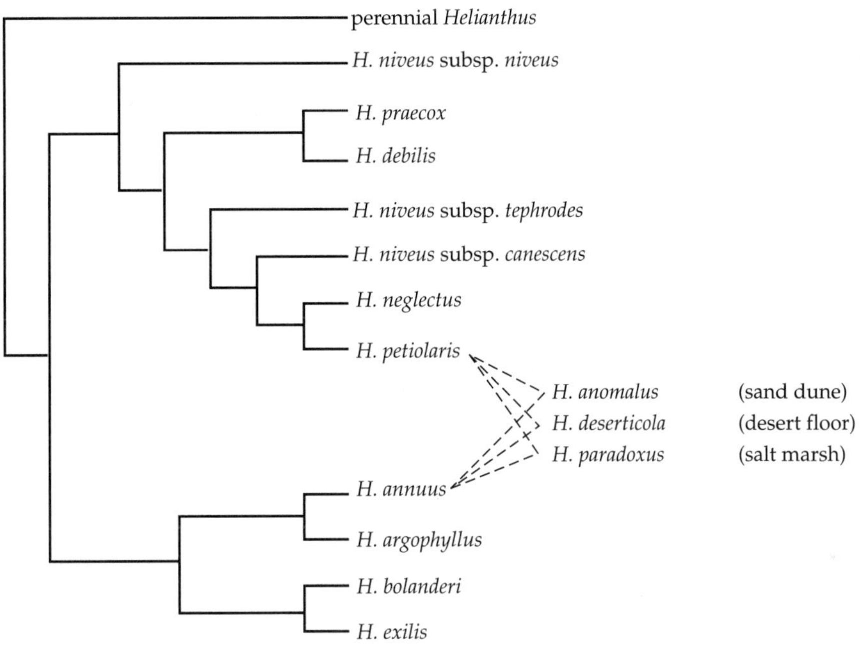

Fig. 4.6. Phylogenetic tree for sunflowers Helianthus (section Helianthus) based on combined chloroplast DNA and nuclear ribosomal DNA data (Rieseberg 1991), showing reticulation due to homoploid hybrid speciation. Dashed lines indicate parentage of homoploid hybrid species. From Gross and Rieseberg (2005) with permission from Oxford University Press.

synthetic hybrids renders experimental manipulation a feasible programme of research to detect whether hybridization generates traits prone to adaptive divergence. Experiments conducted in the greenhouse and in the field with Helianthus have recreated in the synthetic hybrids the same extreme traits that occur in natural ancient homoploid hybrid species. This means that when individuals of both parental and their natural hybrid species were planted in the hybrid species' native environment together with the synthetic hybrid, the synthetic hybrid overlapped the natural hybrid for adaptive traits. This has been tested for leaf succulence and leaf nitrogen content in *H. anomalus*, for leaf area, stem diameter, and flowering date in *H. deserticola*, and for sulphur, calcium, and boron content, leaf shape, and leaf succulence in *H. paradoxus*. Moreover, this work has been extended to prove by mapping of **quantitative trait loci (QTL)** that those extreme or transgressive traits can be generated by complementary gene action. Furthermore, Rieseberg *et al.* (2003) were able to predict more than 70% of the genomic composition of the synthetic species hybrids from the mapping data. All these results build up a formidable body of evidence that bolsters the main role of ecological genetics in the origin of homoploid hybrid species.

Is hybrid speciation restricted to plants?

While most thorough genetic and ecological studies of new evolutionary hybrid lineages have been carried out in plants, animals have also contributed with some well-documented examples. In animals, hybrid speciation is sometimes related to asexual processes, but in several cases new hybrid species are homoploid, bisexual species (see Arnold 1997 for references). Such is the case of *Gila seminuda*, a species belonging to the Cyprinidae (minnow species), a fish family that shows relatively high levels of natural hybridization (11–17%). It is well established that *G. seminuda* originated through introgressive hybridization between *G. elegans* and *G. robusta*. The whole genus Gila is now recognized as evolving via reticulate rather than divergent processes (Dowling and DeMarais 1993) and introgression seems to be continuing, as evidenced by the extreme mitochondrial (mt) DNA similarity of some Gila species. The role of ecological divergence in *G. seminuda* speciation is bolstered by the observation of its restricted occurrence in the Virgin River, a small tributary of the Colorado River in the south-west United States, where their parent species are never found in spite of their occurrence sympatrically along the Colorado and that no barrier seems to impede their migration. Recent reviews of evolution by homoploid hybrid lineage formation in animals report an ever increasing number of cases, some of them strongly verified, including organisms as diverse as water fleas (Daphnia), corals (Alcyonium), grasshoppers (Warramaba), frogs (Rana), flies (Rhagoletis), and monkeys (Macaca) (see Arnold 2006, pp. 144–47 for a review). This list is likely to increase rapidly as more molecular tools and new views on hybrid speciation will be adopted in future research projects.

An emergent case study that epitomizes the importance of homoploid hybrid speciation in animals is provided by the butterfly genus Heliconius. As early as the middle of the nineteenth century, naturalists were challenged by Heliconius forms whose wing patterns resembled a mosaic of those of other Heliconius species. For instance, *H. hermathena* shows forewings that resemble those of *H. erato*, and hindwings that mimic those of *H. charitonia*. Immediately, the hybrid picture that came to Hewitson's mind, the naturalist who named the species, was Hermathena, a mythological hybrid character from the gods Hermes and Athena. As Mávarez and Linares (2008) recount, this is probably the first scientific recognition of an animal hybrid species. Since then other Heliconius species have attracted the interest of evolutionists as an arena in which to study hybrid speciation in animals.

Heliconius butterflies have bright wing colouration that has a protective function from predators (aposematic), but also acts as a cue to mate recognition. The species hybridize in nature, generating mosaic colour patterns that can be recreated by introgression in experimental crosses. The best studied case is *Heliconius heurippa*, an inhabitant of the eastern slopes of Colombian Andes, where it coexists with the closely related species *H. melpomene* and *H. cydno* (see Plate 11). This two species are known to hybridize in nature and, occasionally, produce stable natural hybrid populations showing novel colour patterns, of which the heurippa pattern can be recreated through experimental crosses between the two sympatric races of melpomene and cydno. Interestingly, the heurippa-patterned artificial hybrids show a mating preference for their pattern versus that of either parental species. Thus, not only this mate choice evidence, but also the biogeography, the cross ability, and the extensive gene flow of these species, bolster the hybrid origin of *H. heurippa*. Yet the molecular evidence for the homoploid hybrid origin of *H. heurippa* remained inconclusive because the number and the kind of genetic markers used were insufficient to reject alternative explanations like retention of ancestral polymorphisms by lineage sorting (see Box 4.2) or simple introgression after speciation, among other complications.

Though molecular studies in plants (e.g. Helianthus) seem to be more convincing of the reality of the hybrid homoploid speciation, in animals most molecular approaches have traditionally faced great scepticism. Only recently are some studies combining ecological and molecular approaches gaining recognition in favour of this speciation process. Among them the *Heliconius heurippa* case stands in the front row. The key to its success relied on the direct study of those DNA stretches that directly control the adaptive traits for speciation. Then the Heliconius researchers (Salazar *et al.* 2010) focused on the locus (named *HmB*) controlling the wing pattern, namely its red forewing band. By using 29 additional control markers distributed in most of the chromosomes in *H. melpomene* they found that only inside the *HmB* locus was there a region introgressed from *H. melpomene* into the hybrid *H. heurippa*. This region comprises part of a gene whose expression is limited to the distal region of the forewing. On the other hand, all of the 29 control markers showed shared polymorphisms among the three species, a finding consistent with either extensive introgression after speciation or ancestral polymorphism (see Box 4.2) more than with hybrid speciation. This evidence of generalized introgression combined with the *Heliconius heurippa*-specific *HmB* introgression directly controlling wing pattern, a character implicated in mate-recognition isolation, largely supports the origin of *Heliconius heurippa* by homoploid hybridization. Note that this Heliconius hybrid speciation scenario involves introgression of one or just a few adaptive trait-loci into the genome of one parental species. This hybrid trait speciation, as named by the researchers, contrasts with what has been termed hybrid mosaic genome speciation in Helianthus sunflowers, where the hybrid genome is a rearrangement of DNA blocks originating from both parental species (see Fig. 4.5).

The Heliconius case could be the tip of the iceberg if similar additional studies are undertaken in the future. Although reported numbers of hybrid species in animals are approaching those in plants (see Table 4.2), many case studies are restricted to groups prone to hybridization, like fishes and butterflies, and there are still many groups waiting for a thorough approach. Whether animal hybrid speciation is a minority mechanism limited to certain groups or its difficulty of detection needs a multidisciplinary approach, combining morphological, geographical, ecological, and genetic data, is something that is difficult to predict. But if the hybrid trait speciation model is widespread among animals the use of batteries of neutral markers will make homoploid hybrid speciation more difficult to detect than in other organisms that show the hybrid mosaic speciation model. Be that as it may, the trend in recent speciation research suggests that hybrid speciation, once a neglected mechanism especially in animals, is a likely process in the origin of species.

The power of horizontal gene transfer

Sex is not the only mechanism to exchange genes among species. Negated for decades, **horizontal** (or lateral) **gene transfer (HGT)**, the transfer of genetic material from one organism to another that is not its offspring, is currently a widely accepted process in prokaryotes. Syvanen (2005), a pioneer of the idea, reminds us that 'it has been over 30 years since the suggestion that HGT may have been a factor in the evolution of life entered the literature'. While the first evidence that genes for resistance to antibiotics could move horizontally from one bacterium to another was announced in Japan in 1959, it took at least a decade to be recognized by the Western scientific community. This is not surprising, since the idea of HGT is at odds with the vertical transmission of phylogeny and the overall importance of reproductive isolation to preserve species integrity. Yet, outside the realm of prokaryotes, the importance of HGT did not capture much interest until the last decade, during which an avalanche of genomic data has been produced that shows beyond doubt its relevance in evolution, not only in bacteria and **archea** but also in eukaryotes.

The chimeric genome of prokaryotes

The world of gene exchange in microbes is by far better documented, and perhaps more pervasive, than in multicellular organisms. In no scenario is the power of recombination and lateral exchange more outstanding than in viruses. In particular, bacteriophages, or bacterial viruses, possibly the most abundant life forms in the biosphere, have captured the interest of many evolutionists because of their paramount role in bacterial evolution and adaptation, and also because they may encode virulence factors that integrate into the bacterial chromosome (Table 4.3). The potential of bacteriophages to act as intermediates for the bacteria to horizontally acquire new functions in a process named transduction is enhanced by their extensive genetic variability among lineages and extremely genome mosaicism, which must also be interpreted as the result of extensive phage genetic exchange (Kwan *et al.* 2005).

Transduction is not the only process of bacterial lateral exchange. At least two other well-documented mechanisms, transformation and **conjugation**, are available for bacteria to transfer genetic material horizontally, mediated by plasmid and other mobile genetic elements. Now we know that gene acquisition by HGT is the most potent source of evolutionary innovation in prokaryotes. Yet not all genes are transferred with equal frequency; while operational (housekeeping) genes, involved in general cell functions, such as genes encoding histones or ribosomal proteins, have experienced extensive horizontal transfer, informational genes, those participating in transcription, translation, and related processes, are horizontally transferred much less frequently. Lake and his co-workers propose that 'the complexity

Table 4.3. Examples of mobile genetic elements that encode virulence factors and are present in human pathogens

Type of mobile element	Pathogen	Virulence factor
Plasmid	*Bacillus anthracis*	Anthrax toxin
	Clostridium tetani	Tetanus toxin
	Enterotoxigenic *Escherichia coli*	Heat-stabile toxin, heat-labile toxin, and fimbriae
	Mycobacterium ulcerans	Polyketide toxin
	Salmonella enterica serovar Typhimurium	SpvR, SpvA, SpvB, SpvC, and SpvD proteins *
	Shigella spp.	Type III secretion system
	Staphylococcus aureus	Exfoliatin B
	Pathogenic *Yersinia* spp.	Type III secretion system
Prophage	*Corynebacterium diphtheriae*	Diphtheria toxin
	Enterohaemorrhagic *E. coli*	Shiga toxin and type III secretion effectors
	S. aureus	Staphylococcal enterotoxin A, exfoliatin A, and Panton–Valentine leukocidin
	Streptococcus pyogenes	Streptococcal pyrogenic exotoxins, DNases, and streptococcal phospholipase A_2 (Sla)
	Vibrio cholerae	Cholera toxin
Pathogenicity island	*Clostridium difficile*	Clostridial enterotoxin and clostridial cytotoxin
	Enteropathogenic and enterohaemorrhagic *E. coli*	Type III secretion system
	Uropathogenic *E. coli*	Fimbriae, iron-uptake systems, the capsular polysaccharide and -haemolysin
	Helicobacter pylori	Cag antigen
	S. enterica	Type III secretion systems
	S. aureus	Toxic-shock toxin, staphylococcal enterotoxin B, enterotoxin C, enterotoxin K and enterotoxin L

*Involved in intracellular survival. From Pallen and Wren (2007), with permission from Nature Publishing Group.

of informational gene interactions is a significant factor that restricts their successful horizontal transfer rates relative to high horizontal transfer rates observed for operational genes' (Jain *et al.* 1999).

Regardless of whether this 'complexity hypothesis' is true or not, the power and pervasiveness of HGT has changed our view of important concepts such as organism and species. Some evolutionists are regarding microbes as defined by wild communities, instead of individual organisms, that invade ecological niches—some of them new ones—due to their power to acquire and discard genes in response to their environment. To some authors like Carl Woese (Goldenfeld and Woese 2007), the pioneer proponent of the archea division, this continuum of genomic possibilities 'casts doubts on the validity of the concept of "species" when extended into the microbial realm'. The evolutionary implication of HGT in early life, when HGT must have been the primordial way of gene transmission, is very relevant to these authors. However, the question remains as to how much HGT has influenced evolution (and speciation) since the level of organismic complexity was attained, the 'Darwinian threshold' as Woese (2002) calls it, and forced it to proceed towards the current manner of predominantly vertical transmission.

In sum, nobody would currently deny that HGT is of paramount importance in prokaryote evolution. Its impact became apparent when appropriate methods revealed that a large proportion of bacterial genomes is of foreign lateral origin, amounting to up to 18% in the case of the familiar *Escherichia coli* (Lawrence and Ochman 1998). As stated above, the evolutionary potential of lateral gene acquisition confers on bacteria a large capability to invade new ecological niches, often inducing pathogenicity to new hosts due to infective factors encoded in horizontally transmitted genes (Table 4.3).

Gene trafficking in the eukaryotic world

In spite of its long-recognized importance in prokaryote evolution, the wide acceptance of HGT in the eukaryotic world had to wait until whole genomes were available for comparative studies. And even then the fact that bacteria seldom acquire eukaryotic genes from their hosts (but see Box 4.3) has contributed to the negation of the role of HGT in eukaryote evolution. However, evidence is accumulating that eukaryote genomes can likely integrate genes of prokaryote (and eukaryote) origin with ease, at least in unicellular organisms like **phagotrophic** eukaryotes (Andersson 2005) (Fig. 4.7). Among them, there are now many reports of HGTs that include Giardia, Trypanosoma, Entamoeba, Euglena, Cryptosporidium, and other **apicomplexans** (see references in Huang *et al*. 2004). In particular, many genes acquired by horizontal transfer are of mitochondrial origin or supply mitochondrial functions. Thus, many parasitic protozoans lack mitochondria and were once thought the ancestral forms of pre-mitochondrial eukaryotes. Yet the discovery of mitochondrion-degenerate organelles and mitochondrion-related genes in many amitochondrial protozoa casts serious doubts on their ancestral origin. The genome of *Entamoeba histolytica*, an amitochondrial protist pathogen responsible for most deaths from protist infections after malaria, contains at least 96 genes probably of prokaryotic origin, 56% of them encoding metabolic enzymes that not only replace functions normally assigned to mitochondria but also increase the protist's metabolic, and consequently adaptive, efficiency. This represents a preliminary insight into an amitochondrial protist genome that must await future comparisons with other amitochondrial genomes to provide strong conclusions on how HGT affects eukaryotic evolution, but 'it is already clear that unrelated anaerobic eukaryotes seem to use convergent metabolic strategies imposed by their environments', as the authors of this genome report state (Loftus *et al*. 2005). Of course, these adaptive strategies are greatly facilitated by HGT.

Box 4.3 Legionnaires disease made possible by eukaryote-to-bacteria gene transfer

In July 1976 an American Legion convention in Philadelphia stirred worldwide interest. The reason, not a military concern, was an outbreak of pneumonia that affected 182 attendants; of these 29 were fatal cases. Since then, more than 30 years of research have revealed many aspects of the way in which *Legionella pneumophila*, the Legionnaire's disease pathogen, takes control of the host to infect it. This is a piece of evolutionary interest in which many of these bacteria's unique adaptations have been acquired by HGT from eukaryotes.

When normal bacteria invade a cell, the macrophages engulf them, and transport them to the lysosomes in a phagocytic vacuole for destruction. But these intracellular pathogens manage to exploit secretion proteins mediating normal trafficking to prevent phagosome–lysosome fusion and vacuole acidification, and to recruit *L. pneumophila*-containing vacuoles. The bacteria proliferate within these compartments, destroy the host, and continue to infect new cells.

L. pneumophila shows an extraordinary ability to disrupt the organelle trafficking system of many host cells, ranging from protozoa to humans, and to survive in harsh environments. These abilities are likely to be related to its lateral acquisition of many gene sequences, some of eukaryotic origin. Among them many are involved in interfering in the various steps of the infectious cycle by mimicking functions of eukaryotic proteins. Recent analysis of the genome sequence of *L. pneumophila* identified a high number of these kinds of proteins. (See Fig. C.)

As for the Legionella ability to stand extremely harsh and toxic environments, certain secretion system genes and genes for processing toxic compounds and heavy metals are likely the result of genetic exchange. For example, there is a 100 Kb region in its genome including several such genes that is flanked by tRNA, phage related, and transposase genes, that suggests an alien origin.

In sum, the long evolutionary history of Legionella species in soil and water environments ranging from highly virulent to non-pathogenic (protozoan symbionts) lifestyles underlies a series of opportunities for lateral gene exchange, including gene transfer from their host eukaryotes.

continues

Box. 4.3 *(continued)*

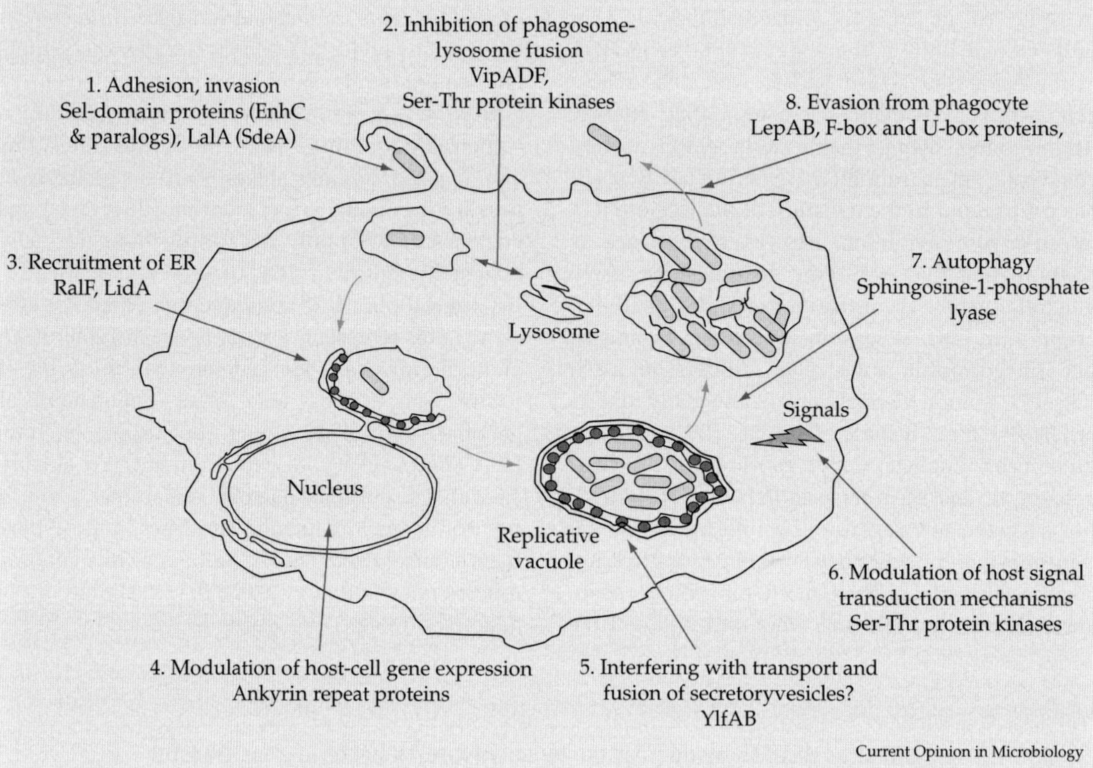

Fig. C. Schematic representation of the different steps of intracellular growth of *L. pneumophila* in macrophages and proteins relevant in Legionnaires disease. (See text for further details.) From Brüggemann *et al.* (2006) with permission from Elsevier.

Multicellular HGT: the tip of an iceberg

Among multicellular organisms the evidence of horizontally transferred genes is less abundant, possibly because the segregation of the germ line does not oblige the germ cells to acquire genes to survive. Nonetheless the evidence is mounting that HGT also occurs in complex, multicellular organisms (see Liu 2007). Palmer and co-workers reported in 2003 compelling evidence that mitochondrial phylogenetic discordances in angiosperms are likely due to frequent HGT between distantly related plants (Fig. 4.8). These transfers, reported for ribosomal and respiratory protein mitochondrial genes (Bergthorsson *et al.* 2003), created a wide array of genomic constructs comprising gene duplications, recapture of genes lost previously, and chimaeric genes. Apart from previously reported cases of HGT in eukaryotes mediated by transposable elements, this was the first unambiguous evidence that plants can transfer DNA to other plants. Moreover, as advanced by Palmer and his co-authors when they stated that the reported cases 'are merely the tip of a large iceberg', many instances of mtDNA horizontal transfer have been recorded since their pioneer study. Among them, those involving hosts and parasitic plant taxa are most illuminating because they provide evidence that HGT is occurring due to direct physical contact.

The cosmopolitan genus Plantago comprises numerous weeds that are parasitized by dodders, a group of leafless plants deficient in chlorophyll that belong to the genus Cuscuta. Using 43 species, Palmer and his team (Mower *et al.* 2004) constructed

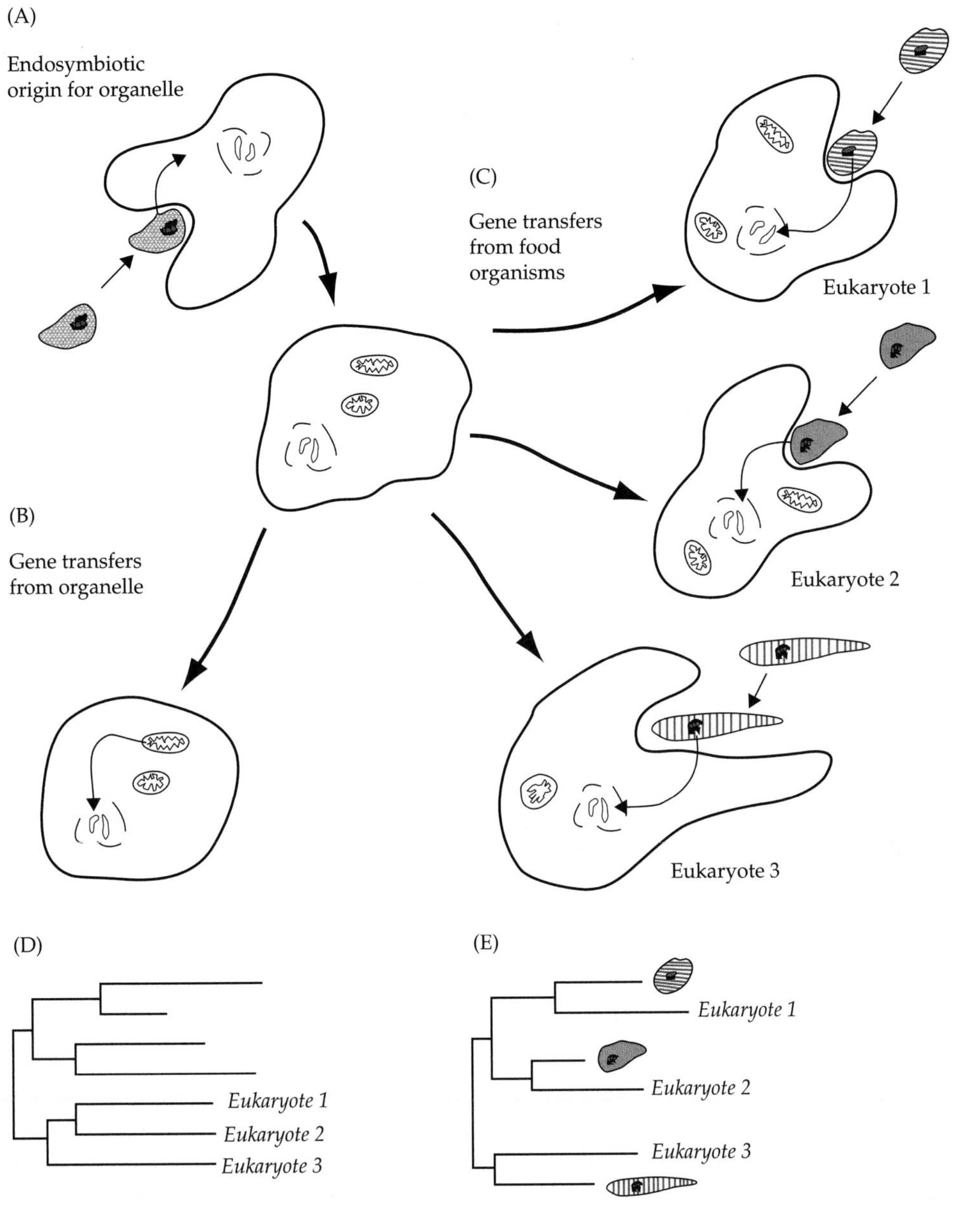

Fig. 4.7. Schematic drawings of processes that may introduce prokaryotic genes into the eukaryotic nucleus. The endosymbiotic origin of organelles (i.e. mitochondria and plastids) (A) may lead to transfer of genes from the organelle to the nucleus (B). Ingestion of prokaryotic cells may lead to transfer of genes from the food organism to the nucleus (C)—the 'You are what you eat' hypothesis (Doolittle 1998).The two pathways for the introduction of prokaryotic genes into eukaryotes have different predicted phylogenetic outcomes; transfer from an organelle is expected to give trees with monophyletic eukaryotes (D), while independent transfer from food organisms should give trees with polyphyletic eukaryotes (E), unless the transfers occurred before the extant eukaryotes diverged. From Andersson (2005) with permission from Springer Verlag.

a Plantago phylogeny based on the mitochondrial gene *atp*1 that agreed mostly with other independently derived phylogenies. Nonetheless, three related Plantago species contain an *atp*1 pseudogen that clustered with the distantly related Cuscuta genus, providing phylogenetic evidence of HGT from

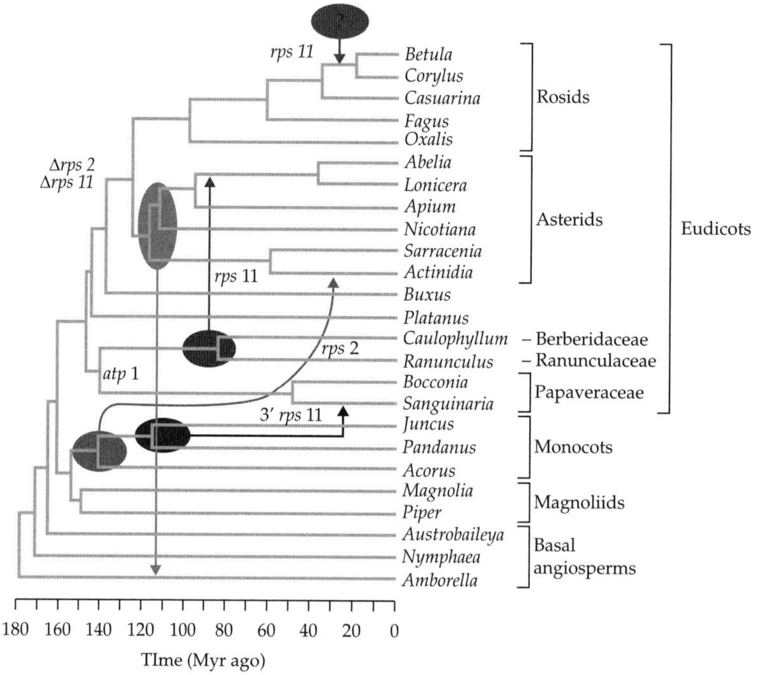

Fig. 4.8. Approximate timing and donor–recipient relationships of five HGT 'events' in angiosperm mitochondrial DNA. Shadowed ovals indicate rough identity of donor groups. The exact placement of arrowheads on recipient lineages is arbitrary. From Bergthorsson *et al.* (2003) with permission from Nature Publishing Group.

parasite to host. Interestingly, two other related Plantago species found only in the high altitude Andes also contain another *atp*1 pseudogen, that clusters with the family clade Orobanchaceae that also comprises parasitic plants. Some of them, belonging to the genus Bartsia and endemic also to the high Andes, are the likely donors by phylogenetic analysis. Since Plantago species that cluster with the dodders are native to Europe and North Africa—as are many dodders—and those clustering with Bartsia—an Andean genus—are restricted to the Andes, the phylogenetic evidence for HGT is strengthened by biogeography. The suggestion that both transfers occurred by direct plant-to-plant contact from parasite to host is bolstered by the common knowledge that parasitic plants penetrate their hosts intracellularly by means of haustoria (see Plate 12). Parasitic plants are counted by thousands, providing a large opportunity for HGT in both host–parasite directions. In fact, there is also evidence that parasitic plants from the genera Rafflesia and Sapria have acquired mtDNA sequences from their Tetrastigma host plants (Davis and Wurdack 2004; see Richardson and Palmer 2007 for a review on the massive plant-to-plant mitochondrial DNA transfer).

Although HGT events have been described above mostly within a single domain of life, and also between phagotrophic unicellular eukaryotes and bacteria, no case study between bacteria and multicellular eukaryotes has been reported. This kind of transfer was considered very rare until insect and nematode genomes were examined systematically for the presence of transfers from the **endosymbiont** bacteria *Wolbachia pipiens*. This study, performed jointly by seven teams of researchers (Dunning Hotopp *et al.* 2007), showed that insertions, whose sizes ranged from nearly the whole *Wolbachia* genome to short insertions of less than 1 Mb, were widely found in insect and nematode species. Interestingly, some of these inserted genes are transcribed within cells lacking endosymbionts. The fact that *W. pipiens* is one of the most frequent intracellular bacteria that occupies the most abundant animal phyla, prompts the authors to warn against 'the view that prokaryote-to-eukaryote transfers are uncommon and unimportant'. Moreover, 'recent bacterial LGT (i.e HGT) to eukaryotic genomes will continue to be difficult to detect if bacterial sequences are routinely excluded from assemblies without experimental verification', as the authors continue to warn. Since this pioneering work was reported the number of inter-domain and inter-kingdom gene transfer findings has steadily increased as more whole-genome sequences are becoming available. In some cases the functional acquisition from an alien distant species has been amply documented. For instance, although no animal had ever been reported to synthesize its own carotenoids, now we know that aphids (Insecta: Hemiptera) possess and express carotenoid biosynthetic genes that were acquired by an ancestral HGT from a fungus to an aphid ancestor (Moran and Jarvik 2010). In sum, I guess that many similar unanticipated events of horizontally acquired functions are waiting to be revealed.

TEs and viruses: the vehicles of HGT

Integration of viral sequences into eukaryotic genomes is also a well-characterized process. Most of this dynamism is mediated by retrotransposons, as I explained in chapter 3. This is a fascinating topic that allows us to speculate about the dynamic interplay between genomic DNA and the outside world (Leitch 2007). The interesting thing is that much stable genome-integrated viral DNA, long considered a relic of the past, can likely be released to form infectious particles (Lheureux *et al.* 2003) or, if they are retrotransposons, they can acquire *env*-like genes that make them infectious. In both cases a two-way traffic between species is facilitated by means of viral dynamics. Recently, well-described examples of horizontally transferred viral sequences into eukaryotes have been reported. In Nicotiana species geminivirus DNA is found integrated as tandem repeats that are likely due to two independent events. Possibly a mobile element, a *Helitron*, captured geminivirus DNA and helped the integration and amplification of a recombinant sequence (Murad *et al.* 2004). Similarly, integrated pararetroviruses sequences are also known in plant genomes comprising banana, tobacco, and petunias. The most recent integration described concerns the integration of a retrovirus into the Koala genome. The authors (Tarlinton *et al.* 2006) propose that sequence integration occurred only 100 years ago from a free-living retrovirus and state that ' this ongoing dynamic interaction with a wild species provides an exciting opportunity to study the process and consequences of retroviral endogenization in action'. In fact, many evolutionists believe that endogenization has been so pervasive that as much as one-third of our genome is from viruses.

The web of life

The 'long argument' of Darwin's *Origin* can be summarized by his tenet that evolution is a process of descent with modification. Two main conclusions related to this chapter stem from this statement. First, since modification in descendants implies the branching out of each lineage into two or several lineages, this ongoing process through time will inevitably lead to a view of life as a tree whose branches include all species, extant and extinct, that ever existed across the history of life. Darwin insinuated this metaphoric view very early on when, in his 1837 Notebook, he sketched a very primitive tree of life (Fig. 4.9) to explain the discontinuities between large groups (genera), but his overt depiction of the tree of life came with the only picture that he included in *The Origin*. After giving an extensive account on the diversifying process of branching depicted in that picture, Darwin concludes that

> 'the several subordinate groups in any class cannot be ranked in a single file, but seem rather to be clustered round points, and these round other points, and so on in almost endless cycles. On the view that each species has been independently created, I can see no explanation of this great fact in the classification of all organic beings; but, to the best of my judgment, it is explained through inheritance and the complex action of natural selection, entailing extinction and divergence of character, as we have seen illustrated in the diagram'.
>
> (Darwin, *The Origin of Species*, 1859, pp. 128–9).

This is, in my own perception, not only a magisterial illustration of how evolution proceeds by descent with modification, but also a last, but not least, definitive blow to the creationist view of life as a 'scala naturae' in which all organisms are delineated in a ladder, from less to more complex species, fixed from the time of their independent creation.

The second conclusion refers to Darwin's view of species. At the beginning of this chapter I discussed this contentious issue mainly framed in Darwin's reluctance to accept a precise definition of species. As a corollary of his continuous statements throughout his book against the notion of species, he asserts in his chapter XIV of recapitulation and conclusion, as a premonition, that:

'we shall have to treat species in the same manner as those naturalists treat genera, who admit that genera are merely artificial combinations made for convenience. This may not be a cheering prospect; but we shall at least be freed from the vain search for the undiscovered and undiscoverable essence of the term species.'

(Darwin, *The Origin of Species*, 1859, p. 485).

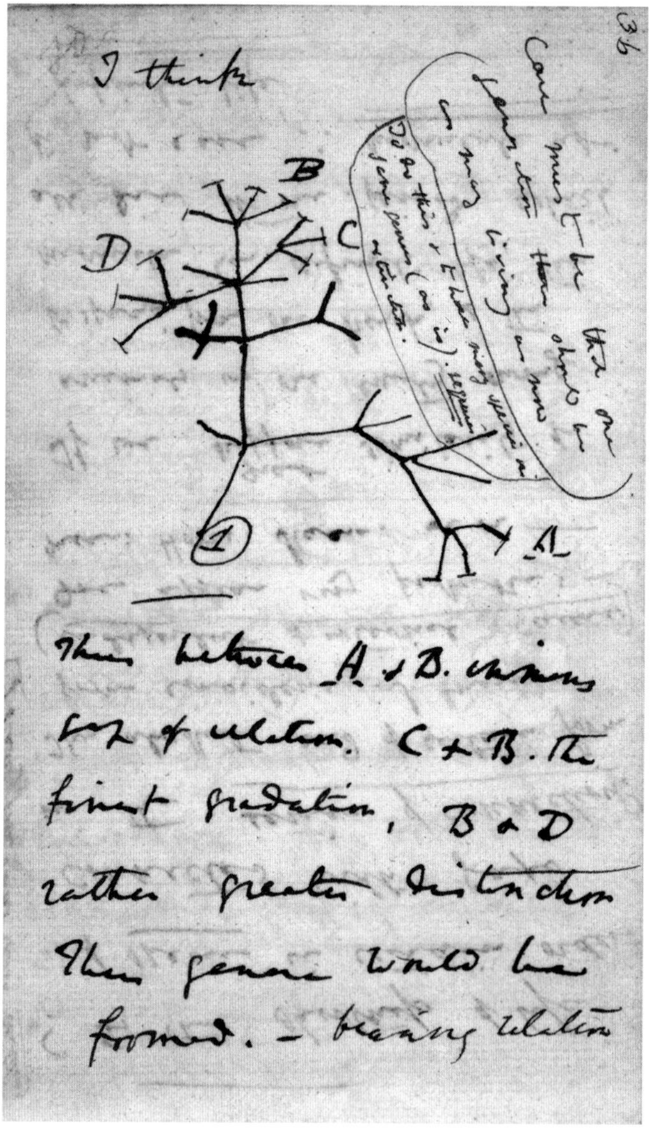

Fig. 4.9. Sketch of the tree of life in Darwin's notebook of 1837. From an ancestral species (1) several branches emerge that give rise to extinct and extant species. The latter are depicted by a segment across the branch tip. From Darwin Notebook B *Transmutation of Species* 1837–1838.

Isn't it ironic that 150 years later we are still discussing that matter? In view of all the recent contributions of new molecular and genomic data sketched in this chapter, Darwin's words may sound very contemporary, at least in what he refers to as the dubious value given to reproductive isolation as a criteria to species definition. Darwin was always a defender of treating reproductive isolation barriers as gradual characters as any others; even when sterility barriers are involved he states that:

> it can thus be shown that neither sterility nor fertility affords any clear distinction between species and varieties; but that the evidence from this source graduates away, and is doubtful in the same degree as is the evidence derived from other constitutional and structural differences.
>
> (Darwin, *The Origin of Species*, 1859, p. 248).

Currently, this idea underlies the explanation of so much genetic exchange revealed by present comparative genomics and molecular phylogenetics, of which a sample is given above. In sum, it seems to me that Darwin's vindication of the role of natural selection in speciation is bolstered by the pervasiveness of introgression through hybridization, and possibly horizontal transfer as well, something not alien to Darwin's early observations.

While the species definition and its applicability to all domains of life seems a formidable prospect, the effort is worth it because it gives unity to evolutionary biology and, most importantly, it provides a null hypothesis to orient a research programme. In fact, to some evolutionists the greatest advantage of the BSC is that it 'immediately suggests a research program to explain the existence of the entities it defines. Under the BSC, the nebulous problem of "the origin of species" is instantly reduced to the more tractable problem of the evolution of isolation barriers' as Coyne and Orr (2004, p. 39) state, to add immediately: 'we feel that the best species concepts produce the richest research program'. Of course some evolutionists argue against this criterion because they think that the choice of a species concept should not be guided by its pragmatic value. I have given reasons against the BSC that here can be summarized in three points. First, it cannot be applied to allopatric speciation, the most common speciation process according to Mayr (1942, but see pp. xxx–i in the introduction added in 1999) and other hard allopatric 'speciationists', because nobody can be sure that allopatric 'species' are reproductively isolated. Second, throughout this chapter I have explained that many species can exchange genes, which casts serious doubts on reproductive isolation as the exclusive criterion to species conceptualization. Third, when asexual reproduction is the predominant mechanism this concept is inapplicable. This applies not only to prokaryotes but also to the agamic complexes with reproductive mechanisms ranging from sexuality to **agamospermy**. No wonder that many evolutionists, in view of these difficulties, have put themselves to work on other reproductive-free species concepts.

The cohesive species

Some species concepts are based on the lineage criterion as the biological universal instead of the reproductive community. Thus the evolutionary concept defines a species as 'a single lineage of ancestral descendant populations or organisms which maintain its identity from other such lineages' (Simpson 1961; Wiley 1978). Cracraft (1989), using the phylogenetic criterion, thinks of species as 'an irreducible (basal) cluster of organisms that is diagnosably distinct from other such clusters, and within which there is a paternal pattern of ancestry and descent'. These insights are intended to apply to all, extant and extinct, biological groups, regardless of their reproductive system. While these concepts do not assign a primary role to interruption of gene flow in species cohesion, implying that other genetic, ecologic, or developmental processes are likely to be more important, they possibly contain a caveat that lies in their lack of criteria to pinpoint which traits define a species or to know what are the boundaries of the permissible diversity inside a given lineage. Besides, and no less important, they propose no evolutionary mechanisms responsible for species cohesiveness.

The duality of species definition by either splitting or integrating criteria may be responsible for the dissatisfaction inherent in the historical argument about species concepts. I have given reasons to justify the inapplicability of reproductive isolation (a splitting criterion) as a means to understand

and distinguish true species in a majority of cases. On the other hand, integrating criteria have captured the interest of many evolutionists in an attempt to overcome difficulties inherent in considering gene flow as the only mechanism defining the boundaries of an evolutionary lineage. In my opinion, the endeavour to consider cohesive evolutionary mechanisms to explain species integrity has been much more illuminating. Templeton (1989) states this purpose clearly when he defines 'a species as an evolutionary lineage through the mechanisms that limit the population boundaries for the action of such basic microevolutionary forces as gene flow, natural selection, and genetic drift'. The applicability of the 'cohesion concept of species', as coined by Templeton, is universal provided that the relative degrees of incidence of each force change according to the organism's life strategy. Besides its general applicability, the advantage of this concept is that gene flow holds no monopoly to underlie the evolutionary lineage.

As Templeton (1989) explains, the distribution of genetic variants in a lineage obviously depends on gene flow by means of processes of *genetic exchange*, mainly via sexual reproduction but not exclusively, because *demographic exchange*, the sorting of genotypes, driven by natural selection and genetic drift, is also a crucial mechanism in maintaining the lineage genetic identity. Demographic exchange is the main cohesive mechanism in those lineages where sexual reproduction is not prevalent, although its role is also important in any sexual organism (Fig. 4.10). In fact, many models in population ecology only take into account demographic exchange to define populations that occupy a well-defined niche, which means that the cohesive concept, at odds with the BSC, allows a fundamental role for ecology in species definition.

Templeton (1989) distinguishes two kinds of demographic exchange. In the first kind genetic drift randomly replaces alleles (or haplotypes) inside a lineage. The operation of genetic drift does not depend on genetic exchangeability between alleles, or individuals that carry them. This 'replaceability', as it was called by Templeton, is a cohesive mechanism that defines the population boundaries because its effect is that all alleles are descendent from an ancestral unique allele. The second type of demographic exchange is an outcome of natural selection and does not depend either on genetic

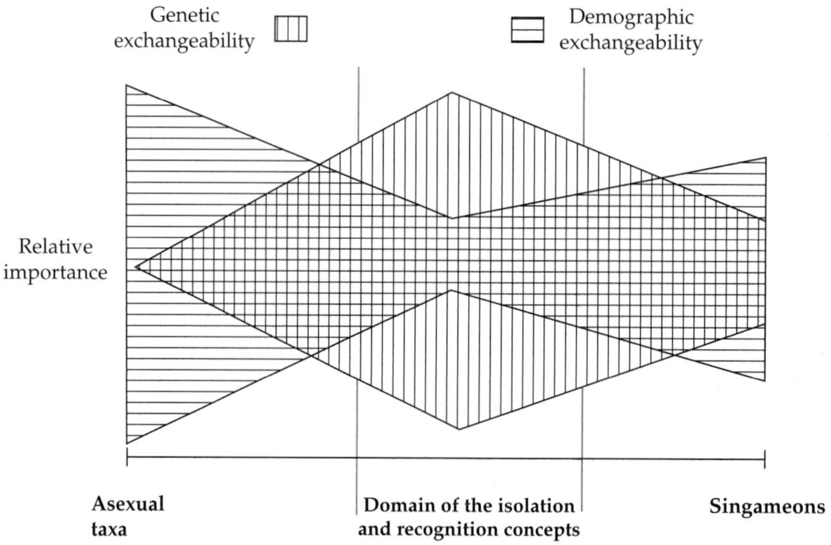

Fig. 4.10. The relative importance of demographic and genetic exchangeability over the reproductive continuum. The areas marked by vertical lines indicate the importance of genetic exchangeability (or exchange), with the width of that area at any particular point in the reproductive continuum indicating its importance in defining species. Similarly, the areas marked by horizontal lines are used to indicate the importance of demographic exchangeability. From Templeton (1989) with permission from Alan Templeton.

exchangeability because natural selection operates equally both in closed (asexual and agamic) and open (sexual mendelian) reproductive populations. This kind of demographic exchangeability is designated as 'displaceability' by Templeton because the offspring of the fittest individuals will displace those of other less fit individuals in the lineage, promoting genetic identity. The ecological niche requirements and the habitat availability for them are crucial for that displaceability, providing the selective forces that drive adaptive transitions. However, apart from this ecological constraint, other constraints may be in operation, namely developmental, historical, and populational.

Natural selection in speciation revisited

Two main advantages of the cohesive concept are relevant to the current discussion. One stems from the above primary role of the ecology in speciation, the other refers to the meaning of gene flow in species definition. Precisely, a fundamental difference between the cohesive and the biological concept of species focuses on the role of natural selection in speciation. Although Mayr (1970) favours the idea that each species occupies a differentiated niche, and that this is a key idea in evolution because it underlies adaptive radiation and evolutionary 'progress', he ever contends that natural selection plays no direct role in speciation, its meaning being limited to preserve species integrity through subsidiary isolation mechanisms. This view negates in principle speciation in sympatry, as discussed in previous paragraphs of this chapter. Dobzhansky (1937) assigns to natural selection a role in reinforcement of prezygotic isolation. He argues that upon secondary contact between two incipient species that diverged in allopatry, some inter-species crosses occur that produce unfit hybrids. If genetic variation underlies propensity of mating, as Dobzhansky's reasoning follows, the same genetic variants that favour homogamy will be favoured by natural selection because their carriers will yield fitter, non-hybrid progeny. The continuous operation of this selective process across generations would reinforce prezygotic isolation until completion. The reinforcement hypothesis has experienced a long history of fluctuation between acceptance and rejection, from total enthusiasm when formulated, to disbelief in the 1980s, to a wave of revival in the 1990s. These oscillations reflect, in my opinion, the difficulties in distinguishing reinforcement from other alternatives, such as ecological displacement, to explain prezygotic isolation. Perhaps the best pondered judgement on this issue was formulated by Coyne and Orr (2004, p.354), two actors in the revival of enthusiasm for reinforcement, when they stated: 'We believe the present data and theory show that reinforcement is possible—and must be taken seriously—but they do not show that reinforcement is common, much less ubiquitous.' Regardless of its importance, reinforcement does not assign a primary role in speciation to natural selection, nor it seems to protect ad hoc the integrity of species, as Dobzhansky posits in his 1937 opus.

Darwin's view on the role of natural selection in speciation is quite different. In a very extensive discussion on the circumstances favourable for the production of new forms through natural selection he states clearly his tenet on the relative value of ecology and isolation. Here I transcript some excerpts:

> That natural selection generally acts with extreme slowness I fully admit. It can act only when there are places in the natural polity of a district which can be better occupied by the modification of some of its existing inhabitants. The occurrence of such places will often depend on physical changes, which generally take place very slowly, and on the immigration of better adapted forms being prevented. As some few of the old inhabitants become modified, the mutual relations of others will often be disturbed; and this will create new places, ready to be filled up by better adapted forms; but all this will take place very slowly.

This vindication of the action of natural selection is followed by his belief that isolation, although important for divergence time, is not necessary to species divergence when he continues to state:

> 'The result would often be greatly retarded by free intercrossing. Many will exclaim that these several causes are amply sufficient to neutralise the power of natural selection. I do not believe so. But I do believe that natural selection will generally act very slowly, only at long intervals of time, and only on a few of the inhabitants of the same region.'

(Darwin, *The Origin of Species*, 1872, pp. 84–5).

'Intercrossing', a Darwinian synonym for gene flow, is greatly taken into account by Darwin as an opposing force to divergence, but he never assigns to its interruption a necessary condition for speciation, as he comments on Wagner's ideas (see above). On the contrary, natural selection is the leading force that originates and maintains species in front of other forces, gene flow mainly, that oppose population divergence.

The web in the genome space

I have presented many proofs that gene flow by means of different mechanisms of genetic exchange, ranging from hybridization to horizontal transfer, is a pervasive observation in nature, which casts serious doubts on the universal importance of isolation barriers in species origin and maintenance. There is no doubt that species cohesion is dependent on genetic exchange, especially in sexual organisms, but demographic exchange processes have also been advanced as important to species cohesiveness.

When considering the role of ecology and gene flow under the cohesive species, the reticulate depiction of species relationships appears less conflictive to the understanding of the tree of life. Under this model, species lineages may maintain their status in the face of genetic exchange provided that demographic exchangeability is at work. Hybridization is not a problem any more; rather, it is a new source of genetic variability which natural selection is acting upon. So it is in the pervasive horizontal transfer of genes in prokaryotes and also in other domains of life. This is something that Darwin could not foresee; otherwise I am pretty sure that he would have substituted his famous unique picture of a tree of life by a representation of biological diversity as a web, encompassing, rather than excluding, his early visionary idea of lineage branching with the present view of genetic exchange among lineages. But one thing has been made clear throughout this chapter's argument, namely that comparative genomics and phylogenetic studies allow us to vindicate the primary role of natural selection in species origin and maintenance, regardless of the species nature, an important, and highly disregarded, tenet of Darwin's long argument. Now we, evolutionists of the twenty-first century, are in the privileged position to ponder the several mechanisms taking part in the process of speciation, especially the role of isolation barriers versus other adaptive (i.e. natural selection) and non-adaptive (i.e drift) forces. And most importantly to distinguish between processes (speciation by itself) and their products, that generate a given pattern (i.e reproductive isolation). In sum, the title I have given to this chapter, the horizontal genome, may seem hyperbolic to many who still wholly adhere to the only classic view of vertical, static transmission of the genome. This view started to weaken when McClintock posited that the genome was mobile and subject to alien invasions, as explained in the previous chapter, and has ever since weakened further as we have deepened our understanding of the genome. Perhaps, the realization that species can exchange parts of their genomes horizontally, via hybridization and/or physical contact, will convince the sceptics.

CHAPTER 5

Reconstructing Darwinism: from Darwin to the genome via the Modern Synthesis

It is natural selection that gives direction to changes, orients chance, and slowly, progressively produces more complex structures, new organs, and new species. Novelties come from previously unseen association of old material. To create is to recombine.

The action of natural selection has often been compared to that of an engineer. This, however, does not seem to be a suitable comparison. (. . .) It works like a tinkerer—a tinkerer who does not know exactly what he is going to produce but uses whatever he finds around him whether it be pieces of string, fragments of wood, or old cardboards; in short it works like a tinkerer who uses everything at his disposal to produce some kind of workable object.

—(Jacob 1977, Evolution and tinkering, *Science* **196**, 1161–6.)

Among Darwin's many concerns about the public acceptance of his theory, the misunderstanding of the concept of natural selection comes first. He often bitterly complains that the term could be mistakenly taken as an act of volition by some kind of personal 'selector' instead of a mechanistic process devoid of purpose. He admits that his choice of 'the term of Natural Selection, in order to mark its relation to man's power of selection' (*The Origin of Species*, 1859, chapter III, p. 61) was unfortunate when he wrote to Lyell in 1860: 'if I had to commence de novo, I would have used "natural preservation"'. However, in the third edition (1861), and all subsequent editions of *The Origin*, Darwin was already defending his choice when he stated:

'In the literal sense of the work, no doubt, natural selection is a misnomer; . . . It has been said that I speak of natural selection as an active power or Deity; but who objects to an author speaking of the attraction of gravity as ruling the movements of the planets? Every one knows what is meant and is implied by such metaphorical expressions; and they are almost necessary for brevity. So again it is difficult to avoid personifying the word Nature; but I mean by Nature, only the aggregate action and product of many natural laws, and by laws the sequence of events as ascertained by us.' (*Darwin, The Origin Of Species*, 1861, p. 93).

Misunderstandings about the action of natural selection still persist in our time. Natural selection is taken as an all-powerful mechanism by some who criticize it, but this erring argument, as explained by Jacob in the opening citation of this chapter, forgets that natural selection acts more like a tinkerer than an engineer. Darwin himself often acknowledges this way of operation, particularly in the building of complex organs. In chapter VI of *The Origin* entitled 'Difficulties on the theory' he asserts that 'we should be extremely cautious in concluding that an organ could not have been formed by transitional gradations of some kind'. One way to evolve a complex, specialized organ is, following Darwin's reasoning, to start from less complex organs in ancestral organisms where they perform 'at the same time wholly distinct functions'. Darwin continues his argument by delving into the tinkerer strategy, 'In such cases natural selection might easily specialise, if any advantage were thus gained, a part or organ, which had performed two functions, for one function alone, and thus wholly change its nature by insensible steps.' This paragraph introduces the modern concept of co-option, amply discussed in Chapter 2, in which a structure that evolved for one function is then changed to perform another function. Moreover, Darwin deploys a large number of examples ranging from dragonflies and hydras to fishes and barnacles, a group of organisms that he knew very well, in which some ancestral simple organs have changed their functions to more complex organs. Perhaps, the most extreme case of co-option (often referred as 'radical' by internalists such as Gould) is that of vertebrate skull sutures, because no adaptive value can be assigned to the ancestral structure. This is Darwin's contention:

> The sutures in the skulls of young mammals have been advanced as a beautiful adaptation for aiding parturition, and no doubt they facilitate, or may be indispensable for this act; but as sutures occur in the skulls of young birds and reptiles, which have only to escape from a broken egg, we may infer that this structure has arisen from the laws of growth, and has been taken advantage of in the parturition of the higher animals. (Darwin, *The Origin of Species*, 1859, p. 197).

This is a beautiful example of the radical co-option of an ancestral vertebrate character that appears as a result of development (the laws of growth, in Darwin's parlance) and not by natural selection, but that in one descendent lineage (the mammals) has become highly adaptive. So Darwin was highly aware of the tinkering nature of natural selection, that never works from scratch, taking currently available materials, whatever adaptive or not, to build new adaptive structures.

Although the opportunistic manner in which natural selection becomes creative, as a tinkerer, is likely to be the least understood evolutionary mechanism by the public, academic, and layman alike, other operational subtleties of natural selection are also misunderstood or even disregarded. Phrases that try to define natural selection, like 'struggle for life' or 'survival of the fittest' (the latter coined by Herbert Spencer, a philosopher) have, in my opinion, sacrificed understanding for brevity. Admittedly, both definitions have truth in them, but a limited truth. Those who have tried to measure natural selection in nature know that this is not only a difficult task, but often an unanticipated source of intricate adaptive relationships that transcend those definitions (see Fontdevila 1995, and references therein). Contrary to the widespread view, those organisms that are more successful in some life history components directly related to the 'struggle' for life (for instance mating propensity) are often not those that transmit their genes to their offspring with the highest probability, because other more cryptic inferior life components in them, like lower fertility or fecundity, or even diminished cooperative behaviour in favour of relatives (dubbed **kin selection**), may decrease their fitness. Moreover, in the last half of the twentieth century, evolutionists have detected a wealth of adaptive characters previously considered without selective value (see Carroll 2006 for a popular account). Incidentally, without diminishing the important contribution of the neutral theory to the understanding of molecular and genome evolution, discussed in previous chapters (see for instance Box 1.1), the past decade has also witnessed an ever-increasing, and unanticipated, number of new adaptive molecular characters, ranging from coding (i.e. transposable sequences: Chapter 3) to non-coding (i.e. RNA transcripts: Chapter 1) DNA sequences in the genome (see Wright and Andolfatto 2008 for a review and relevant papers). All this evidence shows the subtleties of natural selection that makes its comprehension, and often its detection, a formidable enterprise.

Taking for granted that we are familiar with the operation of natural selection, then we could ask whether the insights in current genome investigation support the Darwinian paradigm. But first, the historical development of evolutionary theory since Darwin should probably be recounted to provide a reliable answer. We must understand that original Darwinism lacked a lot of basic knowledge on one of its pillars—the laws of descent, later coined 'genetics' by William Bateson. Also, although maybe to a lesser extent, the economy of nature, another Darwinian pillar, later named ecology, was still in its youth. The wide acceptance of natural selection as the channelling force of evolution had to wait until the middle of the last century when several investigators, theorists, and experimentalists alike, agreed that Mendelian genetics bolsters the bases of Darwinism. The incorporation of genetics coalesced with the naturalist (ecological) views into a theory that was later known as the Modern Synthesis, which I introduced in Chapter 2, and the history of its genesis is developed below. The great timely success of this evolutionary synthesis notwithstanding, a lot of criticisms have been raised since to do with its adduced insufficiency in explaining old basic evolutionary problems. Moreover, since then, each new scientific advance, first in molecular biology and later in the recent fields of developmental biology and genomics, has thrust a bitter controversy upon the validity of the Modern Synthesis postulates. But, what are these postulates? Why did this synthesis have to wait almost a century from *The Origin*'s publication to appear? What are these post-synthesis controversies all about? And, finally, why are some 'genomicists' still debating the validity of the Modern Synthesis? To answer all these questions with due precision would probably take me another book. Some partial answers have been discussed in the previous chapters. Now, I would like to discuss here, in some detail, the genesis and the current state of some criticisms on the orthodox evolutionary theory. Lastly, I will dare to delve into the main purpose of this text, namely to decide whether or not the genome shows dynamics that oblige us to change the Darwinian paradigm.

From Darwinism to the Modern Synthesis

Although Darwin was able to convince most of his contemporaries that species were not separately created, he failed to attract more than a few adherents to his theory of natural selection. Ironically, the advent of Mendelian genetics, following its rediscovery at the beginning of the twentieth century, did not provide the epiphany for the consolidation of Darwinism that might have been anticipated. In fact, early geneticists working on discontinuous characters of large effects, as Mendel did, adhered to a 'saltationist' view of hereditary characters that discarded the heritable nature of Darwinian 'gradualist' variation. This opposition lasted until the 1920s and the 1930s, as Mayr (1980) recollects when he states that 'during that period nearly all the major books on evolution . . . were more or less antiselectionists'.

Saltationism, which posits the origin of novelties only by discontinuous variation, was the realm of Hugo de Vries (one of the rediscoverers of Mendelism). Influenced by the large-effect mutations he found in *Oenothera*, the evening primrose, that later were identified as the genetic expression of an odd chromosomal system of linked translocations, de Vries was actually following a classical line of thought that was dominant until the nineteenth century. Even illustrious geneticists, such as the Nobel Prize winner T.H. Morgan, also had strong saltationist tendencies that influenced the rebuttal of **gradualism**. The writings of the first geneticists, such as Johannsen and Bateson, besides deVries and Morgan, who did not understand evolution in terms of population theory, influenced a whole generation of biologists. Their line of thinking is epitomized by Bateson's suggestion as early as 1894 that 'the discontinuity of species results from discontinuity of variation'. That is saltationary evolution after all.

Contrarily, naturalists (mainly including zoologists, botanists, and some palaeontologists) have always adhered to a more gradualist thinking of evolution. The classic fields of naturalists diverged into other experimental fields such as embryology, genetics, behaviour, and ecology, creating a wide gap between naturalists and experimentalists. This specialization, which had firmly started by the time of publication of *The Origin*, contributed, according to Mayr (1980), to the split of interests between the two fields. Mayr, based on the classic distinction between proximal and ultimate causes, has repeatedly divided biological studies into two wide subdisciplines: functional and evolutionary biology. 'The functional biologist, he asserts, is interested in the phenotype and its development resulting from the translation of the genetic program within the framework of the environment of the respective individual.' Thus, the genetic expression in a particular environment is a 'proximal cause', sensu Mayr, which is 'how' biological things operate. Yet, Mayr continues, 'the evolutionist is interested in the origin of the genotype, in the historical reasons of antecedent adaptation and speciation responsible for the particular genetic program that now exists'. In this case, it is 'why' things came to be (the 'ultimate cause') that interests the evolutionary biologist. Ideally, both approaches are necessary, and mutually enriching, to the understanding of the whole conundrum of life. However, it appears that during much of the first half of the twentieth century naturalists and functionalists talked past each other. Naturalists, perhaps not familiar with the recent advances in genetics, but certainly unconvinced by the presumptuous saltationism of the geneticists at odds with their experience in natural populations, adhered to gradualism and largely became the most fervent Darwinists.

On the other hand, geneticists were not much aware of the detailed accounts on geographic variation and speciation by early naturalists. Even more advanced geneticists such as Fisher and Haldane, and Morgan in his last writings, did not make much effort to explain the origin and divergence of species, not to mention the origin of higher taxa. It may be argued that these geneticists had a much more critical, and urgent, endeavour, to reconcile genetics with gradualism. During the last half of the nineteenth century until well into the twentieth century most evolutionists, even those supporting Darwinism, opposed natural selection, at least as the preeminent force in evolution. Contrarily, they were in favour of the direct effect of acquired characters (Lamarckism), of the inherent progressive tendency of some lineages to evolve in a specific direction (the 'orthogenesis' sponsored by Haeckel and Spencer), or of the variations of large effects (the 'sports', whose role was emphasized by Huxley and Galton). Perhaps August Weissmann, professor at the University of Freiburg in Germany, was the only renowned exception to place natural selection at the highest level. He argued strongly against the inheritance of acquired characters based on his tenet that, at least in animals, the germ line is separated from the soma from the early stages of development. This separation bolstered gradualism through natural selection on small random mutations in the germ cell line as the decisive mechanism in evolution.

All this landscape of confusion notwithstanding, some hopeful sparks, thrown out by new experiments, were enlightening the field of genetic variation. These crucial experiments, greatly supported by formal developments, paved the way towards the complete reconciliation of Mendelian genetics with Darwinian gradualism.

The pillars of the Modern Synthesis

In Britain a group of theoreticians, led by Weldon, a zoologist, and followed by statisticians like Pearson, believed in continuous variation as the stuff of natural selection. They set up a committee of the Royal Society to sponsor their research, coined 'Biometry' by Weldon. Naturally, they were the best allies of the naturalists and the worst enemies of saltationists like Bateson. The biometric school unleashed a thriving dispute with the Mendelians that was only settled by more genetic experiments using complex characters determined by many genes. But, in the meantime, the biometricians founded a new discipline, later named quantitative genetics, which uses highly developed statistical techniques to show how continuous variation changes through generations. This discipline is presently widely used to detect the intensity of selection and the number of quantitative trait loci of complex characters, among other important population characteristics. But at that time the biometric school was another field to be reconciled with Mendelian supporters.

Reconciliation came when researchers demonstrated that if a character is determined by many Mendelian factors, each one contributing a small amount to the character, the multiple combinations of factors coupled with random environmental effects gives the appearance of continuous variation (Fig. 5.1). Many critical experiments, among which those by Nilsson-Ehle on oats and wheat and by Castle on rats, eventually convinced most sceptical Mendelians that inheritance in continuous characters obeys the same Mendelian rules as large, discontinuous mutations do. Thus, there were not two kinds of variation: the variation which biometricians dealt with was the same with which Mendel did his experiments. Yet, how could the statistical results of biometricians with continuous characters, like the response to selection and the correlations between relatives, relate to the Mendelian segregations in multifactorial (polygenic) characters?

Meanwhile, theoreticians were building a solid mathematical edifice which bridged the gap between both fields. The first landmark was the work of Ronald A. Fisher (1918) who set the foundations of modern theory in population and quantitative genetics, not to say of many statistical techniques including the analysis of variance, that are widely used in current data analysis. Fisher's magnum opus was his 1930 book: *The Genetical Theory of Natural Selection*, a highly formalized treatise, where he showed how the continuous selection of small genetic variants in a large population can gradually change the population endowment. Almost simultaneously with Fisher's seminal work other theoretical researchers were

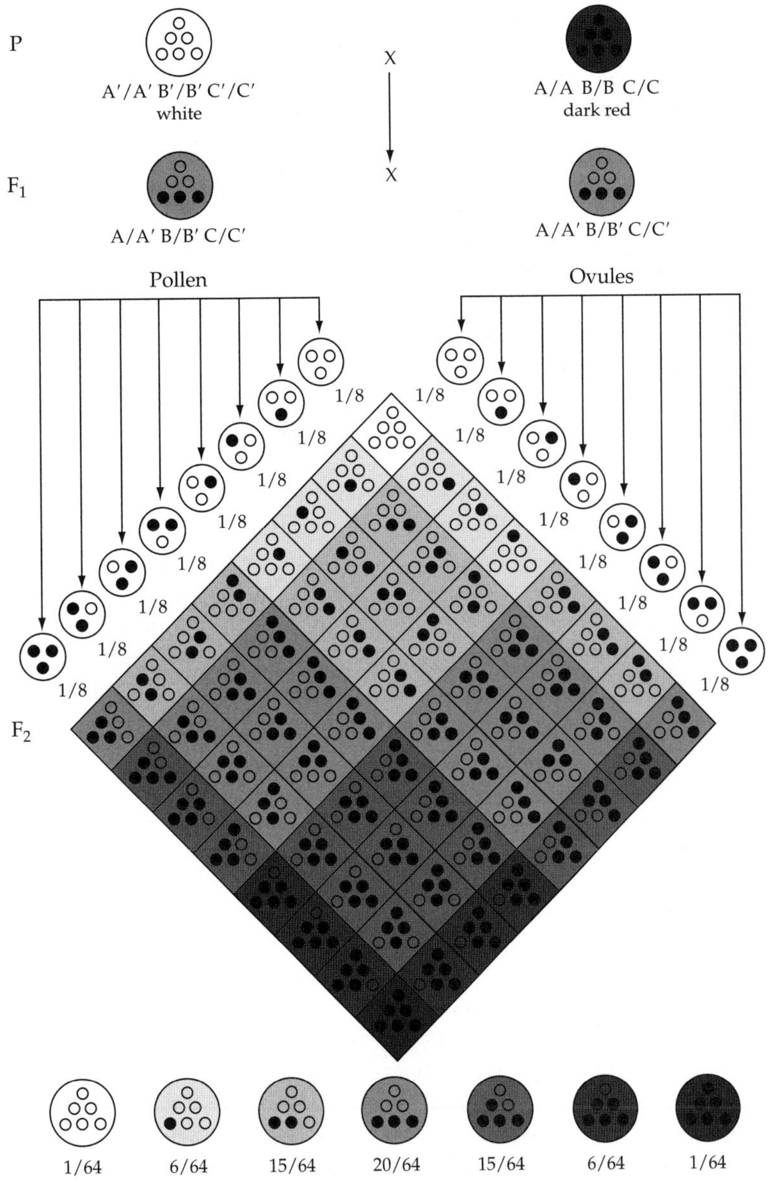

Fig. 5.1. Scheme of Mendelian crosses between two homozygous lines for alternative alleles at three genes that determine the colour (white or black). The F_1 hybrid is heterozygous A/A', B/B', C/C' (grey). A cross between two grey hybrids yields a progeny (F_2) that segregates because of allele assortment. If the colour depends on the number of alleles of each kind (black or white) in an additive way, all tones of grey will appear showing a virtually continuous distribution. If the number of genes increases and the trait is affected by the environment, the progeny distribution appears wholly continuous. From Barton *et al.* (2007).

completing the underlying foundations of modern evolution. Among them Sewall Wright, originally an experimenter with rats and guinea pigs and later a theoretician, became interested in the gamete-sampling effect in small populations, that he termed 'genetic drift' (see Chapter 1 and Box 1.1), and also on how matings between relatives (inbreeding) influences the future genetic composition of the population. This shift from Fisherian large populations to the Wrightian model with small populations introduced a new dimension in theoretical population genetics: the random fluctuations in gene frequency due to genetic drift, which, together with selection, mutation, and migration, complete four of the most fundamental 'processes' in any population genetics model. The third most famed theory builder was J.B.S. Haldane, a trained biochemist with a great diversity of interests ranging from the origin of life to how natural selection operates. He wrote a comprehensive series of papers on the Mathematical Theory of Natural and Artificial Selection about the way differences in survival and reproduction due to Mendelian genes would change a population. These three men showed differences in their specific interests, but they established the theoretical bases of population genetics and through it the reconciliation of the apparently conflicting fields of discontinuous (Mendelian) and continuous (Darwinian) variation. So, natural selection could not only act on continuous variation as biometricians claimed, but also a selected variation could be transmitted to next generations precisely because it was determined by Mendelian particulate genes.

If the systematists do not come, we can go to them

Realization that both 'kinds' of variation (discontinuous and continuous) were the same, even though this strongly supported Darwinism, did not accelerate the acceptance of natural selection. At the end of the 1920s many taxonomists, palaeontologists, and even geneticists still upheld evolutionary views that did not include natural selection as the supreme mechanism in the origin of species. In fact, species origin remained a Darwinian mystery, as Bateson recalls in his 1922 address to the American Association for the Advancement of Science, where he firmly stated that 'the claims of natural selection as the chief factor in the determination of species have consequently been discredited', to conclude that 'Our doubts are not as to the reality or truth of evolution, but as to the origin of species, a technical, almost domestic, problem. Any day that mystery may be solved.' Interestingly, he then insists in his address that the split between geneticists and naturalists is no good to the future of evolutionary science, as if their association were to solve the species mystery. He pointed out the mutual ignorance of recent findings from both fields by saying that 'systematic literature grows precisely as if the genetical discoveries had never been made and the geneticists more and more withdraw each into his special "claim"—a most lamentable result. Both are to blame', and advised as a remedy that 'if we cannot persuade the systematists to come to us, at least we can go to them'. Ironically, this conciliatory prospect occurred about a decade later although with a result that could only have half satisfied him (he died in 1926): genetics emerged as 'the common ground for the systematists (the field workers) and the laboratory worker (the geneticist)', Bateson's cherished wish, and natural selection was totally vindicated as the creative force in evolution, something absolutely opposed by him. This synthesis also tried to solve the big 'mystery', but in a way unanticipated by many mutationists, including Bateson himself (see Chapter 4 for a long treatment).

Although the experimental approach was badly needed to complete the evolutionary theory, the experimentalists continued to work on how variation was produced and transmitted, and rarely provided a hard explanation of how organisms become adapted, their speculations ranging from acquired characters to large spontaneous mutations. On the other hand, naturalists were more interested in the dynamics of natural variability, how it is subjected to natural selection, how it changes in populations, and how populations adapt and become species. Yet, they lacked the proper skills (or maybe the interest) to formulate hypotheses through which they could test their observations in a controlled environment, as experimentalists could.

The first tenet of Darwinism is the abundance of heritable variation in nature; the second is selection operating in nature. However, despite the intuitive claims by naturalists, was this variation present in nature in sufficient amounts to be the subject of natural selection? Or, was it just a laboratory curiosity? Finally, how could it be demonstrated that this variation is the stuff of the origin of species? These questions had to be answered by special people who could combine a 'lust' (and a deep insight) for nature with high skill to carry out experiments in the lab. Theodosius Dobzhansky (1980) was one of them. A Russian-born young naturalist interested in natural polymorphisms of elytra patterns in ladybirds (*Coccinellidae*), he became a Drosophila geneticist in the Philipchenko's Leningrad laboratory. In 1927 Dobzhansky left Russia to join Morgan's group at the California Institute of Technology, taking with him the naturalist approach that prevailed in the Darwinian tradition in Russia. Sergei Chetverikov, the epitome of that tradition, could be considered one of the founding fathers of the Modern Synthesis had he lived in the same environment as them. The interesting point is that Chetverikov and his students were, unlike most other groups like Morgan's, Darwinian naturalists who had to learn genetics to re-approach their natural populations with updated insight. Chetverikov was the first to show that Drosophila natural populations host a wealth of genetic variation in the form of recessive alleles that could be acted upon by natural selection. This evidence set off a whole series of similar surveys in natural populations of diverse organisms.

Although Dobzhansky never worked with Chetverikov, he was aware of his evolutionary interests and approaches, as well as of some of his students', like Timoféeff-Ressovsky. While in California Dobzhansky started a long-term field research project with Drosophila populations in Sierra Nevada and he immediately found an abundance of genetic variability. This project has extended, since then, to many natural populations all over the Americas. His interest centred on inversion polymorphisms in nature, for which he was able to show that seasonal and geographical frequency changes were caused by natural selection in nature, a postulate which he corroborated by observations in laboratory populations. Another central issue of his research was the basis underlying hybrid sterility between sibling Drosophila species. His work grew in importance in the middle 1930s because he was answering the two most important questions on Darwinism of that time: the presence of natural variability and its selective value. In 1937 Dobzhansky wrote one of the most influential books on evolution ever published: *Genetics and the Origin of Species*. Whoever has read this masterpiece will agree that it sealed the profound sundering that had for decades prevented a common understanding between naturalists and experimentalists, or between the 'why' and the 'how' seekers. Now the path to the new synthesis was finally paved.

What was integrated and what was left out in the Modern Synthesis

Mayr (1980) asks himself whether the Modern Synthesis was a scientific revolution, and his immediate answer was 'no'. Why? Because the synthesis did not discover new facts; rather, it removed misunderstandings by integrating all new discoveries, mainly Mendelian genetics, with Darwinian principles. Following the contributions from theoretical and experimentalist pioneers and those that followed, among whom Theodosius Dobzhansky (1937), Ernst Mayr (1942), Julian Huxley (1942), George Gaylord Simpson (1944), Ledyard Stebbins (1950), and Bernhard Rensch (1959) are most prominent, an international conference was convened at Princeton (USA) in 1947 to discuss evolution from the most diverse fields, including taxonomy, palaeontology, systematics, and genetics. The meeting turned out to be one of agreement 'on the gradual mode of evolution, with natural selection as the basic mechanism and the only direction-giving force', as Mayr (1980) recalls. The Modern Synthesis was born.

This agreement was the birth of a long pregnancy already announced by Julian Huxley's *Evolution: The Modern Synthesis* (1942). Huxley's epiphany contains most of the ingredients agreed upon in the Princeton conference, but it was necessary to wait five years more to reach a general consensus. If I may summarize, perhaps the two key issues of this consensus refer to: (a) the adequacy of Mendelian genetics to the Darwinian tenets that heritable variation is not induced directly by the environment, that its origin (the 'mutation') is random (isotropic) with respect to adaptation, and that, being particulate, it is maintained through generations; and (b) the supremacy of natural selection in adaptive evolution, the other Darwinian tenet, in front of other processes that, like drift, migration, and mutation, also influence evolution. Moreover, the power of natural selection is such that by its continuous action on small differences it can gradually build the most effective adaptations. Note that these tenets belie several then accepted concepts of inheritance (dubbed as 'soft' inheritance by Mayr), including the gradual change of the genetic material by the direct effect of the environment (Lamarckism), by use and disuse, by internal progressive tendencies (orthogenesis), or even by the unknown laws of inheritance (i.e. blending inheritance). Although Darwin always raised natural selection to the highest rank, he himself believed in some soft inheritance, compatible with natural selection. On the other hand, Mayr posits that 'it was perhaps the greatest contribution of the young science of genetics to show that soft inheritance does not exist'.

Despite the integrative effort of the Modern Synthesis, many fundamental issues were left out (or rather misrepresented). In Chapter 2 I discussed the absence of developmental studies from the Modern Synthesis and the ensuing divorce between embryology and evolution. Ironically, Darwin was very fond of embryology, which was considered by him a crucial pillar of the evolutionary theory and perhaps the best proof of it (see Chapter 2, p. 78). Fortunately, current knowledge in developmental genetics has been incorporated into the evolutionary theory through the new science of Evolutionary Developmental Biology (dubbed Evo Devo) and a new consensus, albeit not yet agreed by everybody, between Evo Devo and gradual evolution is gaining support (see Chapter 2 for a fuller account). Ecology was another field practically absent from the Modern Synthesis. Called 'the economy of nature' by Darwin and as 'the physiology of organisms' by Wallace, the term 'ecology' was introduced by Haeckel. This absence is quite ironic considering that Darwinism arose out of biogeographical and ecological premises. Darwin and Wallace were both expert naturalists, and Wallace can be considered the founder of biogeography. Natural history, the ancestor of ecology, has an ancient tradition in the study of the relationships between organisms and their environment, although the scientific method was often not employed in ancient studies, a fact that invalidated much of its findings. However, the interest in ecology ('the economy of nature') decayed with the emergence of the mechanistic-reductionist ideas that fuelled the scientific revolution in the sixteenth and seventeenth centuries. With some important exceptions, like Linnaeus and Buffon, natural history was demoted to second row by most biologists, who approached organisms mechanistically or just taxonomically, until Darwin's natural selection restored ecology to a preeminent position among the biological sciences. However, during most of the second half of the twentieth century many ecological studies (Margalef 1968) were trying to describe ecosystem dynamics, by applying methodologies from physical sciences and information theory. This endeavour, albeit praiseworthy, ignored the fact that evolutionary (historical) dynamics is a fundamental factor in our effort to understand ecosystems. In the 1960s and 1970s a large number of ecologists realized, fortunately, that Darwin's original ideas on adaptation by natural selection could be applied to ecosystems. Then the science of evolutionary ecology was officially born and has continued to grow ever since, claiming a place of privilege in the evolutionary synthesis.

Other questions remained unsolved at the time of the synthesis. Its Mendelian basis notwithstanding, the true nature of genetic variation was still a contentious issue. Two distinct views of the genotype structure were in competition. Some geneticists thought that in populations a wild type allele with highest fitness prevailed at each gene, while the gene variability consisted of low frequency deleterious alleles; this is the classical hypothesis sponsored by Morgan and his student Muller. On the other hand, Dobzhansky and his school favoured the view that there is no such a thing as a wild type: each gene hosts multiple alleles maintained by a kind of **balancing selection** that provide the variation for adaption. Although the detection of large amounts of natural variability suggested that the 'balanced hypothesis' was likely to be more plausible, the almost complete ignorance, at the time of the synthesis, of the nature of the gene and, consequently, the impossibility of observing directly genetic variability, left this question wholly open. The advent of molecular biology and its application to population genetics partially resolved this contention, although in an unanticipated way that did not satisfy many defenders of either side (see Chapter 1). However, only the recent advances in genomics are revealing the ultimate nature of genetic variation and we are beginning to throw light on how genes combine to influence complex traits and how variability is maintained in coding and non-coding parts of the genome.

Darwin posited that the level at which natural selection operates is the organism, and the Modern Synthesis adhered to this. Other levels, such as the group or even the species, were considered unimportant as the units upon which natural selection acts. Although Darwin in *The Descent of Man* gave a hint of the possible advantages of human groups whose members cooperate versus other groups that are composed by selfish individuals, he never proposed a mechanism of selection in favour of altruistic genes. Ever since Darwin many evolutionists have advanced mechanisms of **group selection**. These models range from Hamilton's proposal that altruistic genes could be favoured in groups formed by individuals sharing genes (kin selection), as in the beehive groups and other social organisms, to other more sophisticated models, such as reciprocal altruism, in which no kin is needed among individuals in the group. Other evolutionists, like Williams and Dawkins, do not think that group selection is of any importance and that only individual selection is operative. Although this topic falls out of the scope of this book, mainly because the genome studies did not contribute much to its resolution, it exemplifies how current evolutionary theory is extending Darwinism.

The multiple origins of genetic variability

Throughout the preceding chapters a survey on up-to-date advances in the molecular nature of genetic variability has been presented. Since the consensus of the Modern Synthesis, a long chain of significant landmark discoveries have provided insights to many questions that remained unanswered at the inception of the Synthesis. We now have a better understanding of key genetic concepts such as homology, mutation, regulation, development, and, in general, the structure and expression of genomes, which are no longer a black box. A key milestone in this chain of discoveries, albeit not the only one, was the DNA structure discovered by Watson and Crick (1953) that immediately buttressed all previous genetic knowledge and promoted the spectacular growth of molecular biology. This DNA discovery course, an ongoing endeavour, led to whole-genome sequencing (see Chapter 1) at the turn of this century, that unveiled many unanticipated facets of DNA dynamics. Evolutionary theory has profited enormously from molecular genetics and genomics to explore many contentious, and fundamental, issues that remained unsolved at the time of the Modern Synthesis. Among them the understanding of the origin and the expression of genetic variation emerges at a primary rank, impinging upon such fundamental mechanisms as development, the chain that links the zygote to the whole organism, dubbed the 'black box' in the Modern Synthesis. In Chapter 2 I discussed the thrust of Evo Devo into the evolution of the form and the degree to which this knowledge conforms to Darwinism. Here, I want to focus upon the Darwinian concept of gradualism, and how genomics and molecular genetics enlighten the means by which natural selection acts upon genetic variation.

Prodigal variation, niggard innovation

Darwin promoted gradual change because 'as natural selection acts solely by accumulating slight, successive, favourable variations, it can produce no great or sudden modification', citing his own words. Perusal of Darwin's *Origin* finds this adherence to gradualism in many sections. This has been a sticky issue, exploited by the critics of Darwinism until the modern genomic era. But, before getting into the analysis of what genomics tells us about gradualism, I would like to make a clear distinction between the origin and the evolution of natural variation. There is no doubt that the origin of variation is the traditional chance element in evolution (sensu Monod). In modern parlance it comprises any new random, non-adaptively directed change in the genome, be it nucleotide changes, any kind of duplication ranging from single nucleotides to whole genomes, insertions due to transposition or transduction, gross genome interactions like **endosymbiosis** that often incorporate bacterial genomes in cellular organisms, or horizontal gene transfers, a frequent event (discussed in Chapter 4). Despite the difference in occurrence along evolutionary time, more frequent in nucleotide changes and rarer in endosymbiosis, all these events contribute to the genetic variability, the rough stuff of evolution.

These variation-producing mechanisms, well documented by genomic studies (see Chapter 1), were totally unknown by Darwin and, for the most part, during the consolidation of the Modern Synthesis. Since many of these 'mutational' changes may produce a new phenotype that differs grossly from the old one, some evolutionists argued that evolution is not gradual and proceeds by large steps.

But, is the origin of variation equivalent to the fixation of variation? When Darwin in his most extreme statement, cited triumphantly by saltationists, posits that 'Natural selection can act only by the preservation and accumulation of infinitesimally small inherited modifications, each profitable to the preserved being', he was not aware of the many non-infinitesimal inherited mutations available in the genome. In other words, he was ignorant of the origin of variation as it is known today. At most, Darwin was aware of large mutations, dubbed 'sports', that proved to be non-viable for the most part. However, is his ignorance, natural at that time, a valid argument against the gradualness of natural selection as the channelling evolutionary force? Certainly not, if we are reminded that many mutations of large effect (i.e. *Pitx1* gene in sticklebacks), or large duplications that promote novelties (i.e. gene families) need a long series of small changes, often in regulatory positions, until they become fixed for new adaptive functions.

Gradualism has been attacked on different fronts. For instance, one relates to the 'alleged' impossibility of building complexity by gradual evolution; another stems from the idea that evolution is channelled by integrating big 'modules' that produce large steps. The former feeds the 'intelligent design' dogma; the latter is the province of internalists. Complexity is something that troubled Darwin to the point that he is very strict when he says that 'If it could be demonstrated that any complex organ existed, which could not possibly have been formed by numerous, successive, slight modifications, my theory would absolutely break down.' Thomas Huxley, a famous comparative anatomist and Darwin's vigorous defender (nicknamed Darwin's bulldog), in a famous laudatory letter to Darwin (23 November 1859), after saying: 'As for your doctrine, I am prepared to go to the stake', follows by opposing him: 'You have loaded yourself with an unnecessary difficulty in adopting *Natura non facit saltum* so unreservedly.' This Latin phrase (translated 'Nature makes no leaps'), used, ever since Greek and Roman times, by philosophers and naturalists like Linnaeus and referred to seven times in *The Origin*, might epitomize Darwin's gradualist view. However, is it necessary to adhere strictly to gradualism to explain complexity by the channelling drive of natural selection? I believe not.

Gould (2002, p. 155) distinguishes three strands in Darwin's gradualism: the rationalist argument against creationism, the validation of natural selection by insensible mediacy, and the slow pace of phenotypic change at geological scales. The argument of the unity of type, buttressed by current advances in Evo Devo largely discussed in Chapter 2, bolsters the evolutionary theory against independent creation. The recent delusive upheaval promoted by the apostles of 'intelligent design' (see Forrest and Gross 2005 for a documented story) collapses in front of current knowledge in developmental biology (and in the fossil record as well). However, Darwin's defence of gradualism in complex organs, exemplified in the eye evolution, is well known as a nineteenth-century pioneering battle against the adherents of the argument from design, who contended that some traits are too complex to have evolved gradually. In 1802 the Reverend William Paley, a Cambridge teacher, wrote a very influential book entitled *Natural Theology*, arguing the 'need' for an omnipotent designer of nature by using the 'watchmaker' argument. This book had a strong influence on the young Darwin and all his contemporary believers, but Darwin, in *The Origin*, convincingly dismantled the design argument. Contrary to rationality, the 'watchmaker' story haunted, albeit shyly, some educated circles during most of the twentieth century. Yet, when biblical creationism became untenable, the design argument was resurrected by the 'intelligent design' movement (see Box 5.1 for details).

The whole story of creationism (intelligent design included) is meaningless and easily dismantled by Darwinian arguments. This is why Darwin asserts that 'we can plainly see why nature is prodigal in variety, though niggard in innovation', following from his 'descent with modification' rule. 'But why

Box 5.1 Intelligent design: old wine in new bottles

In the early 1990s Phillip Johnson, a law professor at Berkeley (University of California) initiated the movement of 'intelligent design', which, using an updated 'watchmaker' argument, aims at debunking the evolutionary theory, the methodological naturalism, in Johnson's words, to be substituted by a 'deist realism' that assumes that 'the universe and all its creatures were brought into existence for a purpose by God', as Johnson states. Most leaders of the 'Discovery Institute', a pompous name for the administrative centre of the 'intelligent design' movement, located in Seattle, Washington (USA), are neither biologists, nor experimental scientists, and none work on evolution. The relevance for the inclusion of 'intelligent design' ideas in a place like this would be null were it not for the presence of some adherent scientists who may confound the public opinion with their tenets. Perhaps the most iconic of them is Michael Behe, a biochemistry professor at Lehigh University, Pennsylvania (USA), who maintains that metabolic pathways and some complex structures, the eukaryotic cilium and the bacterium flagellum among them, and also organic systems like blood coagulation and the immunologic system, are irreducibly complex. By this expression, Behe means that if the loss of just one part of a system implies its complete collapse then the system origin cannot be accounted for by gradual evolution, because intermediate steps would have been useless and then the whole incomplete system would be wiped out by natural selection. Behe's reasoning mirrors the ancient argument of design; it is the same old wine in a different new bottle.

The concept of 'irreducible complexity' may be appealing but is false. In the case of the flagellum we know that its various protein substructures have different functions in bacteria that can be independently selected (Pallen and Matzke 2006), which strongly suggests that, as in the eye evolution, each one of these substructures were likely incorporated stepwise by natural selection until the most complex flagellum was formed. Moreover, less complex cilia that lack some of their substructures (central microtubules) are perfectly functional in some unicellular algae (diatoms) and these substructures are parts of other less complex mobile structures, like the protozoan axostils. All these facts cannot be explained if the complexity is irreducible. Similar evidence has been pointed out for the blood coagulation factors by Doolittle and his group (Jiang and Doolittle 2003), who showed that the Fugu fish lacks at least three coagulation factors in its clotting system, otherwise present in the most complex clotting systems (like ours), without losing its functionality. The number of cases that contradict the irreducibility of complex systems is large and steadily increasing with the advance of evolutionary science, and can be found in a critical review by Cavalier-Smith (1997). Some of these cases, known for a while, have been cunningly omitted by Behe and other defenders of the intelligent design movement. For instance, the immunological system, a very well studied system in its evolutionary origin and development (see Box 3.2), has been used by Behe as proof of the irreducibility to underrate the power of the evolutionary science.

Though the intelligent design movement has no reputation in the scientific community, the average citizen, who often has a low, or absent, background in current evolutionary science, may be easily convinced by the false arguments deployed by the apostles of this 'creationism's Trojan horse', as denoted by Forrest and Gross (2005). The major strength of creationists does not come, however, from their fallacious arguments; rather, I think, it is the untenable acceptance of the finitude that predisposes the rejection of a materialistic explanation of life by most humans. Darwin experienced this fear early in propagating the 'dangerous idea' of evolution among his fellow citizens. This fear likely made him postpone for more than 20 years the publication of *The Origin* (see Chapter 2 for a longer discussion). To self-perpetuate adherence to creationism, the new creationists, disguised as pseudo scientists, try to introduce in the school curricula the 'intelligent design' doctrine as a science at the same rank as evolutionary science. Fortunately, all these attempts in the USA have been turned down by the American judges. Yet the offensive continues. I believe that evolutionists owe to their fellow citizens a clear explanation of the fundamentals of Darwinian evolution, denouncing the creationist fallacies, to let them decide what side they want to be. This is what I have tried to do in this book, emphasizing how the new genomics supports evolution more than ever, mainly because I feel very sorry when I see that the belief in the supernatural needs to be underpinned by fallacies like those proposed by creationists.

this should be a law of nature if each species has been independently created', he adds. Descent (inheritance) is the clue to basic body plans shared by all members of the same taxon; the phylum as the largest group. Evolution proceeds by adopting body plans that prove fit-to-life conditions and selectively modifying them by reusing ancient mechanisms and structures towards new adaptive needs. Yet, biodiversity hides, under its prodigality in variety, a niggard innovation, that is, a resistance to innovate due to the tinkering strategy of natural selection. Each member of a taxon, for instance any vertebrate animal or any flowering plant, has inherited the same basic body plan from its ancestral organism (an early vertebrate or angiosperm, respectively). Thus, early naturalists, like Linnaeus, were able to group all living beings into a finite number of taxa that were eventually reduced to four independent archetypes by Cuvier. But Darwin went further to postulate a unity of type (see Chapter 2), which demolished the independent creation, and he linked all types through a gradual passage. Isn't this a most convincing anti-creationist argument?

Gradualism in adaptive evolution today

In Chapter 2 I argued that the same gradual evolutionary mechanisms that we witness in natural populations, involved in the formation of current adaptive structures—like insect wing eye-spots or pelvic development in aquatic animals—are found in ancient organisms, though often used for other purposes. Remember, for instance, how modern eye structures have their ancient photoreceptor homologues in the ancestors of the three-layer organisms (the urbilaterians) that were incorporated in vertebrates to perform different functions (p. 59). Most of these changes are located in cis-regulatory regions, like the high prevalence of deletions at *Pitx*1 regulatory enhancer (Chan *et al.* 2010). The interesting points of this case study are two-fold: first it illustrates how major morphological changes can be accounted for by single mutations in regulatory regions, and second how natural selection co-opts a deficient structure (for instance, the reduced pelvis) under environmental conditions where it turns out to be fit. Although the number of similar worked out adaptive cases is increasing with the new methodologies, the ascertainment of the overall distribution of fitness in the genome and the mapping of genotype into phenotype and of phenotype into fitness has remained basically unknown since the era of the Modern Synthesis.

When talking about Darwinian gradualism I believe that we mean that 'adaptive evolution', driven by natural selection, proceeds by creeping, selecting mutations each of little phenotypic effect. This has been one of the most assumed premises in the Modern Synthesis, yet one of the greatest unknowns in evolutionary theory. The founding fathers of the synthesis assumed this Darwinian tenet based on some strong theoretical considerations and rather incomplete experimental evidence. Rooted in Darwin's idea that the organism's adjustment to the environment must be very precise, a micromutational view was widely, albeit not universally, adopted. Among those who sponsored this continuous view on underlying variation were the biometricians and also several theoreticians. Fisher was perhaps the most influential of them (see above) in convincing that each phenotypic trait could be treated as underlain by an infinite number of independent and additive genes, each with an infinitesimally small effect. Despite the mathematical power of the infinitesimal model to treat the response to selection, Orr (2005), a leading population geneticist, thinks that the 'micromutational' view hampered many empirical studies because 'there is little reason to ask non-trivial questions about the genes that underlie adaptation if one assumes that there are thousands of them, each with small and interchangeable effects on the phenotype'. It can also be argued that the genetic tools were not developed in the pre-molecular era and that when a large suite of mapped molecular markers were available the progress in deciphering the genetics of the adaptive traits started to accelerate. Be that as it may, in spite of claims in favour by early experimentalists of the Modern Synthesis, the rigorous genetic underpinnings of micromutationism had to wait several decades to be deciphered.

In the early 1980s, quantitative geneticists started to map chromosomal regions, named **Quantitative Trait Loci (QTL)**, that have a significant measured effect on continuous phenotypic traits. Since then this technique has provided a wealth of QTLs that are associated with each character, some of them with a major effect. The case of the monkeyflower (genus Mimulus) well illustrates the power of the QTL analysis. *M. cardinalis* and *M. lewisii* are two species that are pollinated by different species, the former primarily by hummingbirds, whereas the latter is a bumblebee-pollinated species. This natural prezygotic isolation (see Chapter 4) is determined by a large difference in flower shape and colour, each flower type attracting one specific pollinator (Plate 13). Since both species can cross under artificial conditions yielding fully fertile hybrids, which upon selfing can produce a suite of different flower phenotypes, the researchers (Bradshaw *et al.* 1998), using a set of about 200 molecular markers, could map the QTLs that underlie each of 12 flower morphological traits responsible for the species differences (and for their reproductive isolation). They found that the characters are generally associated to several QTLs (from 1 to 6), but in most characters (9 out of 12) at least a single major QTL explains more than 25% of the phenotypic variation.

Following this pioneer analysis of adaptive evolution in *Mimulus*, further analysis in plants and animals has revealed that evolution has often involved hereditary changes of relatively large effect in fitness and, in some cases, the number of these changes is rather modest. But QTLs of small effect in fitness also exist and contribute to adaptation. The problem is, how many of each class can be found? In other words, what is the distribution of fitness values in the genome? Some evolutionists think that small-effect mutations pave the way for ensuing explosive evolution, but a robust theory to catch up with the current molecular advances is badly needed. This theory would empirically bridge microevolution and macroevolution in a way unforeseeable by the founders of the synthesis.

The best worked-out case of large-effect substitution concerns the *Pitx*1gene (see Chapter 2, p. 77). David Kingsley at Stanford University, with his collaborators—among them Dolph Schluter, an ecologist and evolutionist at the University of British Columbia, and an expert in stickleback biology—set up crosses, by in vitro fertilization, between one saltwater female (armoured with a pair of pelvic spines) and one freshwater male (non-armoured) sticklebacks (Plate 14). This cross yielded a second generation suite of grandchildren ranging gradually from armoured to nude morphs that showed all sizes of pelvic spines. This situation is reminiscent of the phenotypic segregation found in the Mimulus plants and their QTL mapping analyses, and this is exactly what Kingsley and his collaborators did: map stickleback QTLs for the pelvic spine development and other armour plates. In 2004 they reported their *Pitx1* gene work (Shapiro *et al.* 2004). Although this gene is expressed in different parts of the embryo because it encodes a transcription factor—the pituitary homeobox transcription factor 1-further studies of the freshwater allele revealed that about 500 bp removed from the cis-regulatory DNA (the *Pel* enhancer region) were responsible for allele-specific down-regulation in the pelvic region (Chan *et al.* 2010) (Fig. 5.2). This **modular pleiotropy** is characteristic of the regulatory tool-kit genes of development as discussed in Chapter 2. Moreover, the researchers have been able to find molecular signatures of selection centred on the enhancer region that are paralleled by ecological studies in freshwater lakes (see Chapter 2). This case study convincingly shows a clear connection between an adaptive morphological trait and a DNA sequence alteration narrowed to a specific regulatory region that occurs repeatedly in similar ecological conditions in many different freshwater populations.

To many, the monkeyflower and the stickleback stories might cunningly epitomize a kind of 'hopeful monster' revenge (see Chouard 2010 for a longer discussion). Moreover, Darwinian discontents could argue that these stories bolster the leaping nature of adaptive evolution. Admittedly, these cases show that few genomic changes can impinge upon large adaptive steps. But how representative of the mutational fitness landscape they are is something still to be seen. Kingsley's group has identified other individual mutations that reduce the number of armour plates in sticklebacks. In this instance the freshwater and marine alleles differ at hundreds of their DNA positions, which have accumulated over millions of years. Since a single QTL is not necessarily caused by a single mutation, a QTL responsible for a large benefit in the short term could have evolved through previous accumulation of many mutations of small effect. Here, more whole-genome sequences from individuals in populations is needed if we are to understand the gradualness (or not) of adaptive evolution.

One caveat may exist in our present survey of adaptive case studies. Since detecting mutations with large effects, for instance loss of traits like pelvic fins, is far easier than seeing small phenotypic

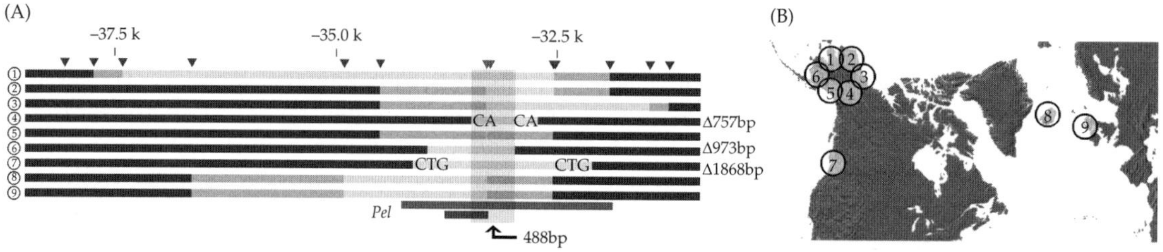

Fig. 5.2. (A) Mapping the regulatory region of *Pitx* gene by using SNPs (single nucleotide polymorphism markers) a series of deletions (light grey bars) were identified in nine (1–9) different pelvic-reduced surveyed populations; (B) shows the geographic location of surveyed populations. All deleted regions overlap in a common 488 bp stretch (shaded) that locates in the *Pel* enhancer region. From Chan *et al.* (2010), with permission from the American Association for the Advancement of Science.

changes, the abundance of large-effect mutations among the worked-out studies comes as no surprise. On the other hand, there are characters like enzyme function that require dozens of small-effect mutations to evolve in in vitro protein evolution experiments. While the ascertainment bias may be real, note that even in the most spectacular cases of large-effect mutations the contribution of the major QTL is a fraction of the total variation: about 25% in the Mimulus flower shape, about 60% in the stickleback pelvis spines, about 28% in the melanocortin receptor effect on coat colour, and so on. Some authors propose that the only way out of the uncertainties inherent in the difficulties of mapping genetic changes to phenotype effect and further to fitness value, is to approach the problem by combining evolutionary biology, molecular genetics, and structural biology, case by case. While this is a formidable effort, current technical advances make it feasible. In fact, some current results seem rather promising.

The functional synthesis in gradualism

Perhaps the best documented case of the pace of mutation effects through evolutionary time was provided by experimental studies in the bacterium *Escherichia coli*. Several decades ago, Richard Lenski started 12 parallel populations of this bacterium derived from a single cell. Due to *E. coli*'s fast reproductive rate, almost any possible mutation in its five million base pair genome can be produced in a few-day culture. Samples of each culture were daily transferred to a new culture. Since the culture was short of the sugar glucose there was competition among bacteria for sugar utilization and natural selection would favour reproduction of those more-efficient mutants, which will be represented by more progeny at each culture transfer. At first, the culture mean fitness, measured as the growth rate relative to the ancestor, increased in big leaps, but later these leaps levelled off. This could be interpreted as the rapid fixation of large advantageous mutations that respond to the initial drastic conditions (namely shortage of food) followed by fixation of small effect mutations.

However, there were significant exceptions to this rule. Although some of the earliest mutations provided small advantages in fitness, as the inactivation of the ribose breakdown gene complex, they were fixed in some few thousand generations. On the other hand, other large-effect mutational changes took tens of thousand generations to fixation. Among them the acquisition of the ability to use citrate as a new source of carbon appeared late in one of the 12 populations. Citrate was a component of the pH buffer permanently present in the growth medium since the inception of the experiment, but it took *E. coli* many generations to acquire the right gene changes to use it as a carbon source to replace the shortage of glucose to its fitness advantage. This was compared by Lenski and his collaborators (Blount *et al.* 2008) to a large phenotypic shift similar to that experienced by the earlier tetrapods that colonized the newly emerged land. This far-fetched comparison aside, it was important to analyse the genetic basis of this citrate-usage phenotype. The analysis was made by recovering hundreds of frozen samples that had been stored by the researchers in each generation since the start of the experiment. These revived samples were tested for their ability to use citrate in the absence of glucose. The results of thousands of millions of analysed cells showed that only some recent samples had the ability to use citrate, confirming that this capacity could not have evolved in one step from the original *E.coli* strain. Rather, it required several previous mutation events, a minimum of three, to pave the way for the large phenotypic effect.

While the comparison of these results with mutation pace in more complex organisms is contentious, this experiment remains intriguing. First, it suggests that large phenotypic effects could be the result of gradual small changes that interact with each other until a phenotypic shift becomes apparent. This phenomenon, known as **epistasis** with a threshold, may be operating in more complex regulatory networks (see below and Chapter 2). Second, it shows that the concept of gradualism (or saltationism) must be defined relative to its fitness context. Namely, although some mutations may affect phenotype, their effect on phenotypic fitness is minor and must be assessed before their evolutionary value is understood. By sequencing genomes sampled through 40,000 generations of an *E. coli* population, Lenski and his team observed discordances between the rates of genomic change and fitness advantages in that mutations accumulated at a near-constant rate, whereas fitness improvement decelerated. On the other hand genomic evolutionary rate accelerated when a new lineage established later. The mapping from genotype onto phenotype is important but not sufficient for understanding the gradualness of evolution; phenotypes must be mapped onto their fitness functions if we are to acquire a correct evolutionary synthesis. Unfortunately not much of either point is known for most organisms.

Some authors like Thornton and his associates (Dean and Thornton 2007) propose a functional synthesis combining evolutionary biology with molecular genetics and structural biology to fill the gap. They started to look at the evolutionary properties of proteins and found that mutational changes of apparently small effect on molecule structure may have a significant impact on evolutionary processes. For instance, they found that in steroid-hormone receptors, regulatory proteins that evolved for millions of years, small changes at distant sites in the molecule are responsible for large-effect fitness changes due to exchanging of amino acids that are directly in contact with the hormone. In particular, these researchers worked out the evolution of the corticosteroid-receptor specificity (Fig. 5.3), because these interactions are crucial for the cell's biological complexity. In many tetrapods, including humans, the hormone aldosterone activates the mineralocorticoid receptor (MR), which is related through an ancient gene duplication to the cortisol-activated glucocorticoid receptor (GR). Aldosterone and cortisol are two different hormones. Since aldosterone evolved in the tetrapod lineage much later than the MR, which is also activated in all vertebrates by other hormones like deoxycorticosterone (DOC), the question remains how did the recent MR–aldosterone specificity evolve? By inferring the sequence of the ancestral corticoid receptor (AncCR), reconstructing its structure, and testing in vitro for its hormone sensitivities, they inferred that AncCR was already preadapted for aldosterone activation and that the recently evolved tetrapod aldosterone co-opted the ancient MR for another function.

But, how did this co-option take place? Using a similar resurrection of the ancestral GR, the researchers showed that the evolution to cortisol specificity was mainly achieved by 5 changes in protein sequence of the 37 identified in the phylogenetic tree branch where this evolution took place. However, the introduction of two of those replacements in the AncCR molecule was responsible for most of the insensitivity, albeit not all, to aldosterone and DOC. Finally, to restore the receptor's full response to cortisol and full insensitivity to aldosterone and DOC, they had to introduce not only the remaining three changes but also two additional changes of the remaining 32, which by themselves do not have significant effects on function. It is beyond the scope of this text to detail the specific chemical background involved in these structural changes, but the researchers were able to work out all of them. Most importantly, their results underpin the fact that, contrary to Paley's watchmaker argument, a complex, highly integrated system can be constructed step by step by the integration of mutational changes which individually often have a negligible effect but when incorporated sequentially may interact epistatically to achieve a new function.

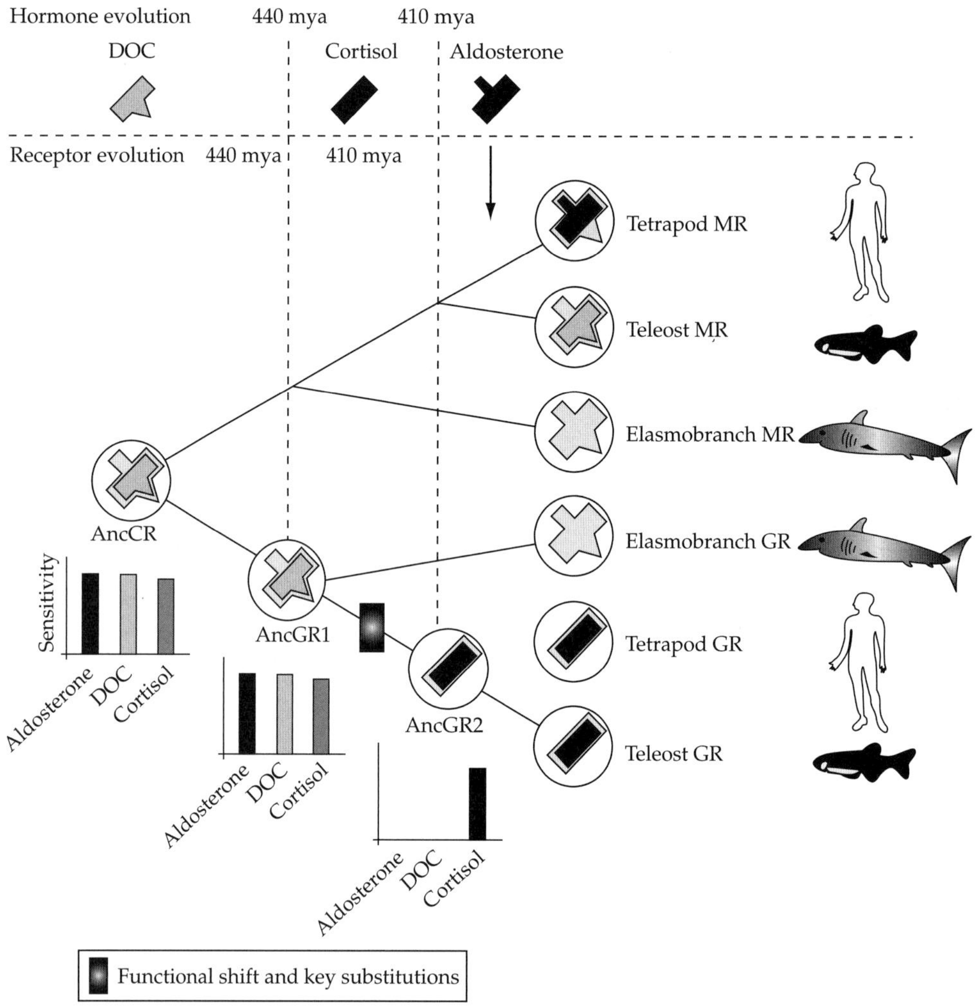

Fig. 5.3. Evolutionary tree of corticosteroid-receptor specificity. Deoxycorticosterone (DOC) is an ancient vertebrate hormone, whereas aldosterone evolved much later in the tetrapod lineage (as indicated by a black arrow). Modern mineralocorticoid receptors (MRs) can be activated by aldosterone, DOC, and to a lesser extent cortisol. The glucocorticoid receptor (GR) is activated only by cortisol in bony vertebrates. The resurrected ancestral corticoid receptor (AncCR) has MR-like sensitivity to all three hormones. Resurrection of GRs from the ancestral jawed vertebrate (AncGR1) and the ancestral bony vertebrate (AncGR2) show that GR's cortisol specificity evolved in the interval between AncGR1 and AncGR2 (represented as a vertical black box with a white dot in the centre). Dates from the fossil record are indicated in million of years ago (mya). See text for further details. From Dean and Thornton (2007) with permission from the Nature Publishing Group.

These two pieces of empirical research show how many subtleties encompass the concept of evolutionary gradualism. Namely, a mutation may have a large effect on phenotype but changes in fitness may be negligible, and the reverse may also be true. Most phenotypic characters whose fitness value has been assessed under an evolutionary context, have shown not only that they are genetically determined by several to many genome sequences, often interacting, but also that these are influenced by the environment. Since the environment changes through time, a small fitness effect can turn into a larger fitness effect in some future environment. The old epitome of 'one gene–one enzyme' to explain the mutation pace in evolution may not be true, but the 'hopeful monster' paradigm is also far from being corroborated as the panacea of evolutionary novelties through current detailed functional studies. In sum, though it should likely be finely tuned through present evidence, the gradualist view is a valid argument when applied to many complex biological systems.

Do we need to expand the Modern Synthesis?

Since the advent of the genome era some voices have periodically claimed that the time is ripe for a re-evaluation of the basic underpinnings of the Modern Synthesis. Based in high throughput technical advances in sequencing and bioinformatics, coupled with highly developed techniques on functional molecular analysis, studies in comparative and functional genomics have brought our understanding of evolution to unanticipated levels. At the same time they raise the question of whether a new paradigm in evolution is needed. Above I have presented the tenets of the Modern Synthesis and how it was achieved through the combination of findings from diverse scientific fields. Now I attempt to assess our genome-acquired knowledge within the frame of the Modern Synthesis.

Among the various tenets that support the epitome of descent (Mendelian inheritance) with modification (basically natural selection), three particulars in the Modern Synthesis emerge with greatest interest for the present proponents of an extended synthesis: gradualism, externalism, and gene centrism (see Pigliucci and Müller 2010 for a review). Gradualism was the topic of the last section. The present theory of Evo Devo, discussed in Chapter 2, and the molecular functional analysis sketched above provide a more lucid account of the gradualness of evolution than was available at the time the Modern Synthesis was formulated. Though gradualism in evolution remains a valid proposition, the new advances force a refinement of the current status of this concept. The externalism–internalism controversy has been largely discussed in Chapter 2, when dealing with the relevance of developmental studies at the genomic level for the unity of type. There, I proposed to relocate *sans souci* the 'chunk' of internal constraints from its almost invisible corner to a properly significant site, but I doubted that natural selection's leading role in adaptive evolution would be affected. In fact, the confrontation between development and selection seems to me rather artificial. Both are necessary to understand the evolutionary process; however, none of them can claim priority in the process: as in any human contest, what is it that makes the winner, the background or the competition superiority of the contender? The contender value is a necessary, but not a sufficient, condition, and his/her competitive superior ability is what will eventually decide the winner.

The gene centrism in the Modern Synthesis is another contentious point resurrected by some present challengers. Gene centrists have been accused of disregarding the role of **phenotypic plasticity** in evolutionary progress. In this section I want to clearly disentangle the complex issue of plasticity that developmental networks exhibit when confronted with external conditions and how this interaction influences adaptive evolution through gene fixation. Finally, once this contentious issue has been discussed, I will finish by attempting a comparison between our present understanding of the dynamic genome and the state of the theory of evolution in its twenty-first century version.

Soft inheritance and phenotypic plasticity

In the last edition of his book Darwin wrote:

'I have now recapitulated the facts and considerations which have thoroughly convinced me that species have been modified, during a long course of descent. This has been effected chiefly through the natural selection of numerous successive, slight, favourable variations; aided in an important manner by the inherited effects of the use and disuse of parts; and in an unimportant manner, that is in relation to adaptive structures, whether past or present, by the direct action of external conditions, and by variations which seem to us in our ignorance to arise spontaneously. It appears that I formerly underrated the frequency and value of these latter forms of variation, as leading to permanent modifications of structure independently of natural selection. But as my conclusions have lately been much misrepresented, and it has been stated that I attribute the modification of species exclusively to natural selection, I may be permitted to remark that in the first edition of this work, and subsequently, I placed in a most conspicuous position—namely, at the close of the Introduction—the following words: "I am convinced that natural selection has been the main but not the exclusive means of modification." This has been of no avail. Great is the power of steady misrepresentation; but the history of science shows that fortunately this power does not long endure.' (Darwin, *The Origin of Species*, 1872, p. 421).

I already discussed how this 'soft inheritance' concept, as dubbed by Mayr, was totally discredited in the Modern Synthesis (see the section: What was integrated and what left out in the Modern Synthesis, in this chapter), mainly due to the 'hard inheritance' validated by the Mendelian theory. Yet, I have here intentionally revived the concept of 'soft inheritance' because some new critical voices have recently been raised, invoking a kind of evolutionary vindication of the transmission of developmentally induced effects that have also been designated as soft inheritance (Jablonka and Lamb 2010, p. 137).

The developmental concept of soft inheritance is rooted in the concept of the **norm of reaction**: the set of phenotypes that a genotype can express in different environmental conditions (Box 5.2). Since the early systematic studies on biogeographic variability there has been a long dispute about how much of this observed variability can be assigned to genetic versus environmental causes. This is not a trivial question; it impinges upon the real nature of adaptation. **Clinal variation**, that is the phenotypic variation that correlates with gradation in an environmental factor, is the most publicized instance of adaptive (genetic) natural variation. Yet, this and other similar biogeographic patterns must be subjected to experiment before the adaptive explanation is fully accepted. Several processes other than direct adaptation, such as migratory patterns, can be responsible for similar patterns. Admittedly, the organisms must be adapted to their environments to survive, but how adaptation is achieved often remains contentious. Different genotypes can generate different phenotypes, each adapted to special environments, but also the same genotype can generate a suite of phenotypes each equally adapted to a variety of environmental conditions. This phenotypic plasticity has been well documented for a while, but its nature and evolutionary origin is only being recently unveiled.

Though phenotypic plasticity may be interpreted as purely somatic variation in response to environment, a more reasonable explanation was soon advanced: that the somatic response, a term used to distinguish non-heritable adaptive physiological changes from changes in the germ line that are directly passed to next generations, also has an underlying genetic basis that is therefore subject to natural selection. However, in this case the target of selection is a suite of physiological and developmental mechanisms that provide a range of phenotypes in response to several environments rather than a unique phenotype fitted for a specific condition; in other words, selection for a specific norm of reaction. But, still, is it not stretching it too far to think that adaptive evolution proceeds by selecting a 'genotype for all seasons'? Remember the old saying: 'Jack-of-all-trades, master of none'. In fact, when an organism is exposed to an unfamiliar environment a suite of reaction phenotypes ensues that, albeit not fully fitted, may allow it to survive. This initial phase of organismal survival was recognized by a few evolutionists as very important for facilitating the phase of fine-tuned population adaptation to the new environment that may follow. James Mark Baldwin, an experimental psychologist, perhaps influenced by the observation that behaviour shows all the facets of somatic adaptation, proposed early in the twentieth century that this stress-induced survival phase allows time for the appearance of the right adaptive mutations and/or new gene combinations, at least in a few members of the population. Then selection would fix these mutations and the organism would become stably adapted even in the absence of an environmental stimulus. This model has been designated since then as the 'Baldwin effect'.

Baldwin's model is a clever Darwinian explanation of a process that might otherwise be interpreted as Lamarckian. Moreover, it provides another example of the opportunistic nature of natural selection. Note that pre-existing adaptive variation is the key stone for the new genetic variation to stabilize the phenotype. Without this previous adaptability, the new stress condition would likely thrust the organism towards extinction. It is this already-present somatic adaptability that allows the organism to survive long enough for the new mutations to appear. Again, a novel complex adaptation does not emerge from scratch; rather, it is the result of modifying pre-existing processes in the genotype's reaction norm by new stabilizing genetic changes.

Though the Baldwin effect does conform to the basic Darwinian variation–selection mechanism, it

Box 5.2 Is phenotypic plasticity a driver of future evolution?

Empiricists have provided information about the evolutionary role of phenotypic plasticity. Early in the 1940s Jens Clausen's group at the Carnegie Institution approached this topic by performing transplantation experiments on genotypes across natural environments. These can be easily performed by transplanting cuttings of the same plant to environments alien from the original place. Figure A shows how the reaction norms of each plant (genotype) of yarrow (*Achillea millefolium*) grown at different altitudes are different, showing genetic variability for reaction norms. However, the adaptive nature of this variability is not easy to assess (see Schlichting and Pigliucci 1998 for a review).

Briefly, reaction norms can be adaptive or non-adaptive. In the first case the response to a new environment can be perfect or incomplete. In both cases, however, the survival in a new environment may be granted, allowing enough time for the population to become established by using standing genetic variation plus mutation and/or recombination (see text). When adaptive plasticity generates a phenotype close to the optimum in the new environment, due to a perfect response, selection stabilizes the population, no directional selection is needed and no further evolution ensues. Some introductions of plants, like the colonization of fountain grass (*Pennisetum setaceum*) in Hawaii, seem to follow this pattern as revealed by transplantation experiments. On the other hand, when the response is incomplete, the plastic phenotype approaches the optimum phenotype but remains below in fitness. Then the new population will be subjected to directional selection and evolution ensues. This is probably the most common form of adaptive plasticity that has been reported in many cases (see Ghalambor *et al.* 2007 for a survey).

But non-adaptive reaction norms are usually elicited when new environments fall outside the range historically experienced by the population, usually referred as 'stress environments'. In this case, most of the elicited phenotypes are largely unfit and their likelihood of persisting is very low. Contrary to the adaptive plasticity case where all individuals show a similar response, here the difference between stressed individuals may be high, generating an increase in phenotypic variance. Some case studies reveal that 'cryptic genetic variation' is released by environmental stress, of which the Hsp case is perhaps the most documented (see text). This case bolsters the idea that **canalization** by buffering the expression of new variation, performed in the Hsp case by chaperone proteins, is a mechanism of storing cryptic variation that can be released under stressful conditions. It is hard to image how in natural conditions these 'hopeful monsters' could persist, but if they can, adaptive evolution may act through **genetic assimilation**, as Waddington showed experimentally. Be that as it may, the value of phenotypic plasticity to drive future evolution depends on the adaptive degree of the reaction norm elicited. As Ghalambor *et al.* (2007) state, 'plasticity is not in itself an evolutionary mechanism on a par with natural selection . . . but rather provides the first step in the adaptive walk otherwise dependent on new mutations'.

All these plausible routes towards evolution facilitation notwithstanding, some evolutionists think that 'no study to date has actually provided empirical evidence for a major role of plasticity in facilitating adaptive evolution in natural populations', as Ghalambor asserts. There are studies suggesting that adaptive differences across populations have occurred rapidly from an initial non-adaptive plasticity stage, other works support the hypothesis that adaptive plasticity likely reduce the probability of extinction (see Ghalambor *et al.* 2007), but these are not definitive studies on the primary evolutionary role of phenotypic plasticity at a scale larger than ecological immediate adaptation. That phenotypic plasticity is a character under selection in nature has been suggested since Schmalhausen's studies on leaf morphologies in semi-aquatic plants (see text and Fig. 5.4) and was bolstered by showing that the degree of plasticity in leaf development in *Ranunculus flammula* depended on the previous environmental exposure. When plants are constantly exposed to a terrestrial or an aquatic environment, a change in their growing habitat in a future generation shows less phenotypic plasticity than when they have experienced an alternate environment (Cook and Johnson 1968). The question remains, however, how much weight in long-range evolution has this short-term evolutionary role of phenotypic plasticity apart from, perhaps, facilitating adaptive evolution in populations.

continues

Box 5.2 *(continued)*

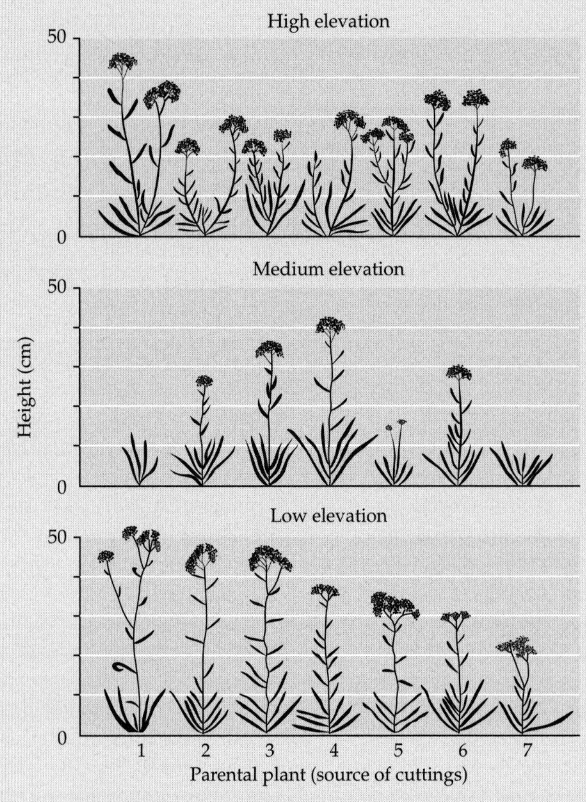

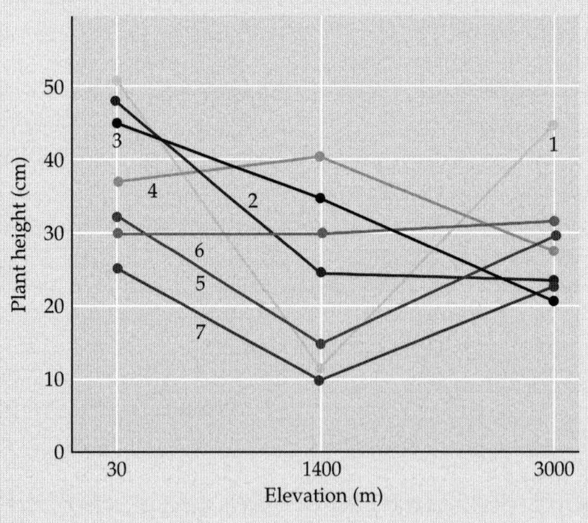

Fig. A. Height of plant cuttings of yarrow (*Achillea millefolium*) from 7 parental plants transplanted at different altitudes (above). The diagram below plots these heights across altitudes showing changes in slope among plants that depict the genotype-environment interactions. From Clausen, Keck and Hiesey (1948) with permission from Carnegie Institution for Science, Washington.

has been claimed recently that it provides a clear instance in which somatic adaptability precedes genetic adaptive variability in the course to evolutionary novelty (Kirschner and Gerhart 2005 p. 105; Pigliucci and Müller 2010). While this contention does not directly challenge Darwinism, it has been used as a strong argument to soften the 'genecentrism' tenet in the Modern Synthesis. As Pigliucci and Müller (2010, p. 14) state: 'Far from denying the importance of genes in organismal evolution, the extended theory gives less overall weight to genetic variation as a generative force.' These and other present-day evolutionists posit that time is ripe for an extension of the Modern Synthesis, which they call the Extended Synthesis. Among other propositions for revision, like the issues of gradualism and the externalism, which I have discussed above in this and the preceding chapters (Chapter 2 especially), they emphasize the view of 'genes as followers' in contrast to the 'genes-first' epitome. In other words, they propose that some mechanisms, like phenotypic plasticity and/or the epigenetic processes (see Chapter 3), may be the initial mobilizing agents of the phenotypic traits that are later fixed by progressive incorporation of the right genetic changes.

Baldwin's ideas, albeit later proved partially correct, had no empirical support at the time. It took other experimentalists to fill the empirical gap. But before explaining the empiricists' work, two evolutionists that were also interested in phenotypic evolution deserve some attention. The German geneticist Richard Goldschmidt, already mentioned in Chapter 2 for his strong position on 'macromutations', was one of them; the other was Schmalhausen. Both were very discontented with the Modern Synthesis, but while Schmalhausen's contention was that development was the missing chapter in the synthesis and that it should be incorporated in a prime position if the explanatory power of the synthesis was to be considered acceptable (see below), the radical revisionism proposed by Goldschmidt was taken as unacceptable defiance. Goldschmidt's book: *The Material Basis of Evolution*, published in 1940 as the major books about the Modern Synthesis were being released, epitomizes one of the most severe attacks on the Modern Synthesis. His main criticism is based on the concept of the gene as the substrate of evolutionary novelties, which, he believes, can only be elicited by the internal and external environments even within similar genetic backgrounds. He stressed the importance of environment and development for evolution against genetic composition. Goldschmidt sponsored the replacement of gene mutations by systemic chromosomal reorganizations with major phenotypic effects as the drivers of evolution. This concept of macromutation was influenced by his observations on the dramatic phenotypic differences between plants, like tobacco, that differed in the number and rearrangement of chromosomes. Both macromutations and environmentally induced developmental phenotypes led him to propose that macroevolution must be discontinuous (non-gradual) and interspersed with leaps highlighted by dramatic diverged phenotypes with high evolutionary value that he called 'hopeful monsters'. Goldschmidt's extreme ideas on macromutations and hopeful monsters as pillars of evolution are not supported by the present knowledge. They were scorned and derogated by the fathers of the Modern Synthesis. However, although Goldschmidt is superficially remembered by this evolutionary caricature, his view contains some points that closely relate to the concept of developmental reaction norm. Today our view of regulatory evolution is different from his, but we may agree that evolution of regulatory systems is more important than was envisaged by the founders.

By 1943, while Moscow was under Nazi siege, a Russian evolutionist, named Ivan Ivanovitch Schmalhausen, was completing his book *Factors of Evolution*. This book, that remained basically unknown until its English publication in 1949, was greatly respected by the founders of the Modern Synthesis, and was specially championed by the author's fellow countryman Dobzhansky, who significantly helped in the book's translation. Schmalhausen's argument is an extension of the Baldwin effect. He builds on the concept of 'reaction norm' to propose two kinds of phenotypic responses to stress: partially adaptive and non-adaptive morphoses (as coined by Schmalhausen). He presented several examples of both; for instance, the change of bristle numbers in the crustacean *Artemia salina* as a response to temperature is an example of the latter, whereas changes on leaf shape in semi-aquatic plants exemplifies the former. Leaf shape, which depends on the air–water exposure, is directly related to oxygen exchange under water or in the open air, and thus is likely adaptive (Fig. 5.4). This dual response to stress gives a wider view of the overall phenotype plasticity and broadens the Baldwin effect. Yet, it is not easy to explain the evolutionary significance of non-adaptive morphoses, except when they are adaptive for some condition other than the stress inducer. It is actually easy to explain the presence of non-adaptive phenotypic traits by resorting to trade-offs or developmental constraints. On the other hand, when the morphosis is physiologically adaptive, the process of stabilizing selection is very similar to the Baldwin effect; thus after the initial survival phase occurs following the stress response, heritable variations may accrue, but population adaptation and phenotype stabilization occur mainly through population reassortment of existing variation, rather than by incorporating new mutations. Note that these heritable variations may be fine-tuned changes in the regulatory networks that are likely not only responding to normal environmental changes, but also to unanticipated environmental challenges. This developmental perspective, based on the genetic fine tuning of 'small' regulatory processes that take advantage of the already present phenotypic plasticity, diverge from the Goldschmidtian macromutation tenet and bring Schmalhausen closer to our present concept of developmental reaction norm.

Schmalhausen's insight notwithstanding, his ideas were so far removed from the tenets of the Modern Synthesis that they were impossible to fit into this paradigm. A plausible justification might be that not enough empirically sound work was available to justify such incorporation. This might have been right in that pre-molecular era, but several highly endowed empiricists working as early as the 1950s were reporting critical results in favour of the evolutionary value of phenotypic stability. Conrad H. Waddington in England provided the experimental foundation for stabilizing selection, a term that he dubbed 'genetic assimilation' and that is widely used today. His approach was simple but clever: he exposed Drosophila flies to different environmental stresses (high salt concentration food, ether during embryogenesis, or high temperature in pupae or larvae), observed the elicited morphoses (tolerance to salt, an extra pair of wings, or absence of cross-veins on the wings, respectively), and selected for these morphs during 20–25 generations (Waddington 1953). Figure 5.5 depicts the progress of selection for cross-veinless wings. Note that in the initial heat-shocked generations the percentage of cross-veinless flies was low (about 20%), but increased significantly during

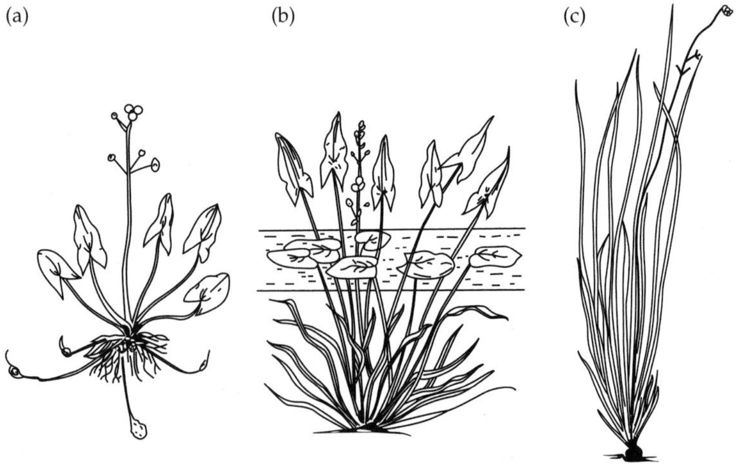

Fig. 5.4. *Sagittaria sagittifolia* leaves show several shapes (heterophylly). (a) Terrestrial plants only develop saggitate leaves; (b) however plants leaving in a mixed semi-aquatic habitat develop both kinds of leaves; (c) aquatic plants only produce elongated leaves. From Schmalhausen (1949) (reprinted in 1986 by University of Chicago Press).

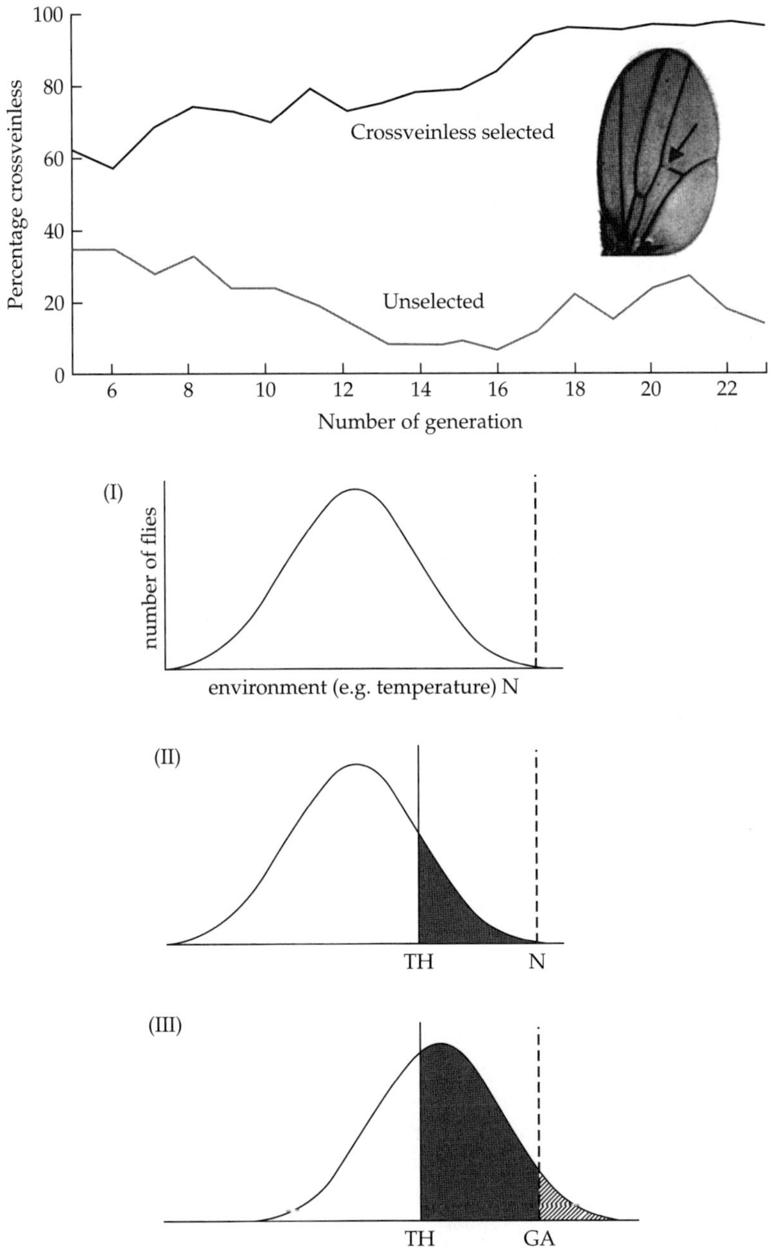

Fig. 5.5. (Top) Response to selection for crossveinless (CVL) flies. See the wing with a cross-vein gap (pointed by an arrow) produced by exposure of *Drosophila melanogaster* pupae to heat shock. In the fifth generation (start point in the graph) only about 60% of flies show the CVL phenotype, bur after 20 generations of selection almost 100% flies are CVL. (Bottom) Depiction of a putative threshold model. There is variation for underlying propensity to express gaps in venation. Only few flies whose propensity exceeds the N threshold are able to express these gaps under normal conditions (part I), but after heat shock (part II), flies with higher propensity than a lower threshold (TH) express the gap phenotype (shaded area). After several generations of selection for propensity (part III) the distribution of gap phenotypes shifts to the right, and the number of flies with high propensity (exceeding the GA threshold) that express wing gaps increases (striped area). This process was named genetic assimilation (GA) by Waddington and genetic stabilization by Schmalhausen. From Waddington (1952) (top graph) with permission from Nature Publishing Group, and from Gibson and Wagner (2000) (top gap wing) with permission from John Wiley and Sons.

later generations of selection, confirming that phenotypic plasticity is heritable. However, the interesting result was that after several rounds of selection the novel phenotypic morphosis needed no heat-shock treatment to be expressed. In fact, after about 20 generations over 90% of the flies lacked their cross vein. That this is largely genetic assimilation of previous genetic variation present in the heterogeneous initial population is buttressed by the lack of selective response when the original population was inbred (i.e. genetically homogeneous). Waddington interpreted his results in a similar way as Schmalhausen's ideas, but he was more radical, proposing that genetic assimilation, and therefore the missing developmental approach, should not just be incorporated into the synthesis, but obliged an in-depth revision of the whole Modern Synthesis edifice. This extreme view was strongly rejected by the founding fathers and other leading evolutionists. They stuck to the old tenet that since somatic adaptation is not heritable its evolutionary value is negligible, likely missing the point that reaction norm is as heritable as any other character trait.

Developmental plasticity in the twenty-first century

Reaction norms are the visible expression of phenotypic plasticity, but in development the most adaptive norm of reaction may be a constant phenotype, buffered against environmental and genetic perturbations. Then selection would favour developmental systems that are able to cope with these perturbations. Waddington called this phenomenon '**canalization**'. He used this concept to interpret his results of genetic assimilation. His reasoning goes like this: although phenotypes are usually canalized, their buffering is not all-powerful and can be disrupted when the genetic or the environmental impact reaches a certain strength. In other words, canalization works only when perturbations are slight. We are painfully aware of the dramatic effect that some drug administration to pregnant women, such as the thalidomide treatment in the 1960s, can inflict on the developing foetus. Extreme temperatures can also overpower the buffering capacity of developmental systems, allowing the pre-existing cryptic genetic variability to express a new phenotype.

Waddington produced circumstantial evidence that previous genetic variability was necessary for genetic assimilation to take place, but only in the past decade have direct molecular proofs been reported. The main actors of the cell heat buffering are the heat shock proteins (Hsps). Normally, Hsps are produced in the cell to stabilize the folding of important, inherently unstable, signal proteins involved in normal regulatory pathways. These proteins, that help other proteins in building their proper folded conformation, are called chaperones. However, overheating, like other stress environments, induces the unfolding of many cell proteins and the loss of their normal activity. If the heat stress is mild, Hsps can correctly refold damaged proteins to cure them, but when the organism is greatly heat-shocked not enough Hsps are available to stabilize the abundant newly damaged proteins as well as the normally unstable regulatory proteins. Then a suite of developmental aberrant phenotypes ensues that could be selected to become stabilized after several generations of heating, as Waddington showed with the cross-veinless phenotype about 50 years ago.

In the late 1990s, Susan Lindquist and her collaborators were able to show that Hsp90, a member of the Hsp family, has a major role as a developmental buffer not only in Drosophila but also in the model plant *Arabidopsis*. If Hsp90 is responsible for buffering development, reasoned the researchers, lowering the Hsp90 activity by any means should decrease the buffering ability. First they tested mutant alleles of *Hsp83* (the gene for Hsp90), which lowers the Hsp90 activity to the point that flies homozygous for *Hsp83* mutants are lethal in Drosophila. Although heterozygous flies are viable, they show an increase in abnormalities of the legs, eyes, and bristles (Plate 15) that could be stabilized after continuous selection. Interestingly, once the phenotype was stabilized not all abnormal phenotypes (80–90% of the progeny) had the *Hsp83* mutation, suggesting that selected cryptic variation unveiled by *Hsp83* could express the abnormal phenotype in the absence of impaired Hsp90. Since the abnormalities observed did not show simple Mendelian inheritance, the authors (Rutherford and Lindquist 1998) suggested that 'even though the founding populations were small, they contained a large amount of previously cryptic genetic variation that was capable of affecting these traits'. In sum, the results support the view that Hsp90 is likely a major component of a complex buffering system of canalization that encrypts many mutations, which are silenced to resist phenotypic fluctuations. Yet when the Hsp90 function is altered by mutation or environmental stress the expression threshold of these cryptic mutations is lowered, inducing a release of the hidden phenotypic variability. Most of this variability would probably be non-adaptive, but some 'may have been adaptive for particular lineages, perhaps allowing the rapid morphological radiations that are found in the fossil record', as the authors dare to advance.

Note that experiments on developmental plasticity acquire a high significance in evolutionary theory because they impinge upon the old contention confronting the sudden burst versus the gradual stepwise origin of novelties. The point is that this approach has revealed that many large-effect master genes in development hide large amounts of regulatory variability of minor effect. The presence of abundant cryptic variability in the Hsp system, well documented at the molecular level, is not unique—other analogous developmental systems show similar genomic structures. In Chapter 2, I discussed how gradual changes in *Ubx* function, mainly due to its CRE variability, have a role in the microevolution of Drosophila bristle patterns (Plate 16). The underlying genetic variability of *Ubx* has also become apparent in genetic assimilation studies on haltere morphologies. Waddington (1956) was able to produce flies with enlarged halteres by treating Drosophila embryos with ether (see above). This phenotype, which was stabilized after continuous selection in the same way as the cross-veinless phenotype, mimics that produced by some *Ubx* alleles (see Chapter 2). Thus the suspicion arose that the *Ubx* system was also hiding cryptic variability that was released by the ether treatment. Using currently available mapping techniques it was shown that this is the case: namely, a large proportion of phenotypic effects map to the *Ubx* site. This and other similar genetic assimilation experiments provided additional evidence for the presence of abundant variability that significantly contributes to phenotypic diversity. The interplay between this cryptic variability, essentially regulatory, and the role of master genes of large effect in development, is far from being understood, but some hints have been proposed.

Several paradoxes are still unresolved. One concerns the duality between developmental constraints and phenotypic novelties in evolution. We know that canalization prevents a mutation affecting development. But, on the other hand, canalization allows mutations to accrue in the genome without being expressed. This cryptic variability, momentarily escaping the natural selection sieve, may, in the long term, be elicited by changes in the environmental or in the genomic conditions and then become selectable. Thus, canalization has two contradictory sides: it both constrains and favours evolvability. But what is canalization? What are the mechanistic bases of canalization? It may be the result of redundancy in the genome. Through the past decade we have witnessed a major unanticipated proof of genome redundancy: namely that no phenotypic effect is detected after the deletion of developmentally important genes. Several explanations exist. In some instances, the loss of function in one gene is substituted by the activation of another gene; in other cases the genome already contains several redundant functions, namely functional proteins, that compensate one another's losses. Thus, if the muscle-forming *MyoD* gene is knocked out (silenced) in a mouse embryo, muscles still form. This is because *MyoD* usually down-regulates the synthesis of *Myf5*, a gene that can also direct muscle development. In the absence of *MyoD*, *Myf5* is allowed to be synthesized and can perform most of the functions of the lacking molecule. Finally, we know that the genome is redundant and probably hosts a wealth of buffering proteins (chaperones), like Hsps, which repair the disturbing loss of function provoked by environmental aggressions.

Another criticism against the importance of somatic adaptation argues that, after all, not many novelties can be achieved by its concourse. The idea that evolution is opportunistic, co-opting old structures for new functions, does not mean that novelties cannot arise. Somatic adaptation is just one of the many instances that use pre-existing core processes, ranging from simple transcriptional and structural mechanisms to large GRE networks. Some apparently counterintuitive examples of stability mechanisms that foster evolutionary change have been resolved when examined at the molecular level. For instance, haemoglobin is a highly plastic molecule whose structure is poised to load oxygen in the lungs, transport it, and unload it in the tissues, where it is loaded up with carbon dioxide, which haemoglobin transports and unloads in the lungs. This is not a simple function that depends on two molecular conformations: one for binding oxygen with high efficiency and one for binding oxygen at several-hundred-fold lower affinity. Both conformations interconvert rapidly (Fig. 5.6). When oxygen is at high levels (in the lungs) the loading state is induced, and at low levels of oxygen (in the tissues during exertion) the unloading state predominates. This two-state equilibrium must be modified when oxygen is scarce, such as at high altitudes. The loading of more oxygen is solved by increased breathing, but the unloading needs additional help from a molecule (2,3 diphosphoglycerate) manufactured by red cells under high oxygen demand (hypoxia) in the tissues. This molecule binds to haemoglobin and shifts its equilibrium towards the unloading state. Other non-human vertebrates use different corrector molecules under hypoxia, and this provides an excellent example of phenotypic plasticity in response to an internal hypoxic environment. Evolutionarily speaking haemoglobin also provides examples of how structural and molecular modifications have been stabilized to adapt to new physiological conditions using old adaptive mechanisms. One highly illustrative case is provided by foetal haemoglobins. Haemoglobins have evolved into a large family by gene duplication followed by nucleotide changes. In primates, foetal haemoglobin binds diphosphoglycerate poorly and shifts to the loading state, allowing the foetus to 'rob' oxygen from the maternal circulation to its advantage. In some birds that migrate over the Himalayan mountains, like some goose species, a single amino acid change shifts the haemoglobin toward the loading state under low oxygen tension.

This example shows that two strategies, physiological (somatic) and genetic, are available to modify the oxygen affinity. We might imagine that the physiological strategy preceded the evolutionary scenario. Thus, in the goose case, one can hypothesize that there was a strong selection for single amino acid changes that stabilized the inherent phenotypic plasticity of the early goose populations in search of a favourable high-altitude migration route. The lesson to take away would be that highly poised, conserved, complex adaptive systems responsive to simple signals (oxygen and phosphoglycerate) facilitate the replacement of these signals by simple mutations. This paradox about stability-induced change is apparently resolved if we redefine evolutionary robustness as not a rigid response to environmental stress. Rather, organisms are robust because they possess an adaptive physiology that allows them initially to cope with environmental stress, which is subsequently followed by genetic assimilation through genetic reassortment and/or new adaptively selected mutations. Although this simple Hb example conforms to a wishful general hypothesis to understand the evolutionary process of physiological adaptation, it is by no means proved for complex phenotypic traits. Obviously, more work is needed to establish a direct (or indirect) connection between DNA sequence and the phenotype, which stands as one of the greatest challenges in current evolutionary theory.

Coda: genome complexity as the playground for evolution

There are times when scientific hypothesis are starving for new empirical results to put them to test, and also times when there is such an avalanche of experimental data that it makes it difficult for them to be tested in a timely fashion against the current science paradigm. In biological science we are at present witnessing an era of overwhelming information that stems mainly from genomic studies. During the last decade the number of sequenced nucleotides in the databases of finished sequences has grown from 8 billion, at the time of the first human genome announcement in 2000, to almost 280 billion in 2010. This exponential increase means doubling the size of finished sequences every 2 years. In terms of number of deposited completed genomes to date we have access to the genome sequence of more than

3,800 organisms (including about 2,500 non-cellular viruses), while no more than 600 organisms had been sequenced (mostly viruses) by 2000 (Fig. 5.7 and Plate 17). Moreover, if we add to these figures the total number of shorter sequences deposited by researchers all over the world, the numbers rise spectacularly up to trillions of nucleotides. The interesting point for evolutionary science is that this sky-rocketing suite of genome sequences is generating a wealth of comparative studies of genomes that provide unprecedented opportunities to test the principles of current evolutionary theory. In particular and most importantly, these analyses allow us to shed light on putative new evolutionary mechanisms and to propose changes, amendments, or extensions on those mechanisms that are (or are not) currently accepted.

How a scientific synthesis is born: the pioneer synthesizers

That is the good news about this genome explosion. The less good news is that this massive generation of data reveals so many unanticipated insights that we are still unable to synthesize. This scenario has been declared by many as unprecedented in the history of biology. But I believe this opinion is unjustified. Historians of science know that cases of mass data production, without a sound theory of synthesis, are not rare. For instance, throughout most of the nineteenth and twentieth centuries geologists amassed a large body of knowledge that comprised data on volcanoes, earthquakes, sediment strata, deep earth structure, mountain formations, and earth history. Yet they were unable to relate all this information by a unified theory. Alfred Wegener proposed the unifying hypothesis of continental drift in 1915, but it was not until long past the middle of the twentieth century, after obtaining much empirical evidence from disparate fields, that his ideas were accepted by the scientific community. Memories of my undergraduate geology classes, pictured as a series of unconnected concepts on crystallography, stratigraphy, and earth catastrophes, make me recall how eager (and yet distrusting) I was to find a thread that would unite all these

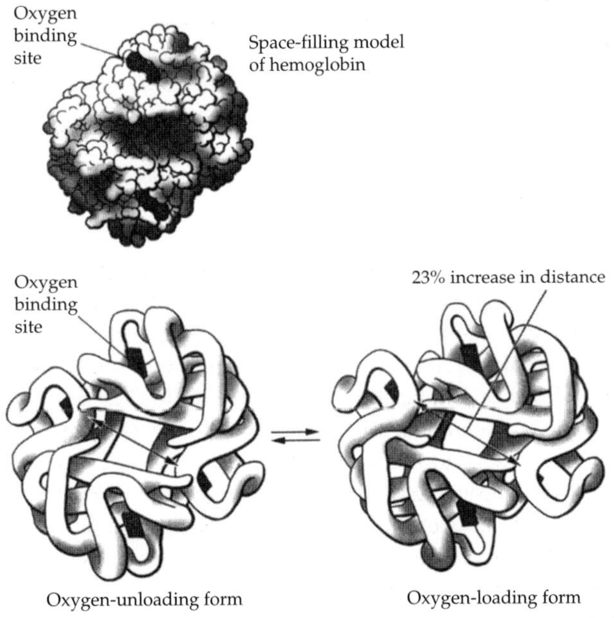

Fig. 5.6. Scheme of the haemoglobin molecule showing its oxygen binding site (above). The two equilibrium states (below) depict that shifting from each other, induced either by oxygen concentration or any other molecular stimulus, is produced by changing the internal distances in the molecule polypeptide spatial conformation. From Kirschner and Gerhart (2005), with permission from the illustrator, John Norton.

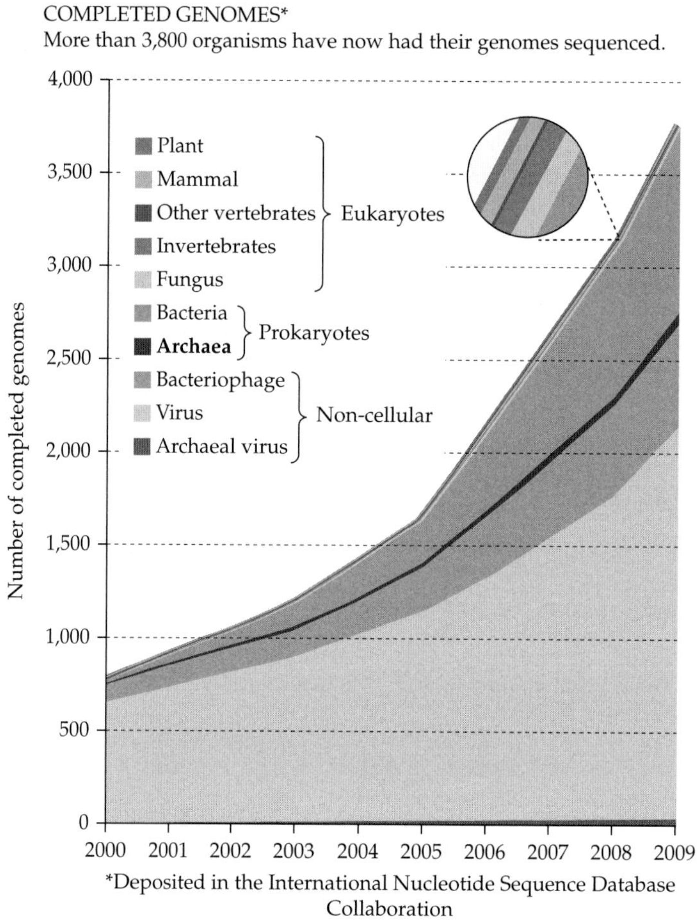

Fig. 5.7. The figure depicts the ever-increasing number of deposited completed genomes in the past decade. From Venter (2010) with permission from the Nature Publishing Group. See also Plate 17.

pieces of the puzzle. Fortunately, the plate tectonics theory converted geology into a unified science, vindicating Wegener's ideas on continental drift due to the continuous movement of plates in the lithosphere, the solid outer layer of the Earth. At certain zones, for instance the mid-Atlantic ridge and the rift valley of the great African lakes, magma from the inner Earth rises to the surface, cools, and the solidified new crust pushes the existing plates to either side. When two drifting plates eventually collide one plate plunges under the other, generating tremendous pressures that may produce earthquakes, volcanoes, and mountain-building. Thus, all the formerly disconnected geological events combined in a unified theory.

This episode is an excellent example of how a massive chunk of information needs to be integrated into a plausible theory to acquire significance. Interestingly, plate tectonics contributed decisively to understanding why animal and plant distributions show sharp discontinuities in close geographic areas and similarities in widely separated continents. For instance, kangaroos and other marsupials are abundant in Australia, a continent devoid of native placental mammals. Yet, marsupials are absent in Borneo, an island not far from Australia, where apes and monkeys and other placental mammals, predominate in the absence of marsupials. On the other hand, marsupials are abundant in South America, a far-away continent.

This paradox remained a conundrum until historical plate tectonics revealed that about 200 mya Australia, Antarctica, and South America were united in a unique land mass that then split in these three continents, which were moved towards their present geographical location by continental drift. Obviously, the evolutionary explanation is that marsupials evolved in isolation in this continent, which later disintegrated. This continental drift story explains why we find marsupials in Australia and South America, now two remote continents but united in ancient times, and not in Borneo and South-east Asia, only recently located close to Australia.

Although plate tectonics helped to disentangle this marsupial evolution story and also many others of similar difficulty, it is worth noting that the marsupial conundrum was elicited by another massive inflow of information on the planetary distribution of living beings. Even now that Earth faunas and floras have been studied for centuries, we are still discovering many new species in remote (and not so remote) areas. But the situation was much more dramatic just about 200 years ago. The idea of evolution owes much to eighteenth- and nineteenth-century naturalists who led tremendous expeditions to remote places to collect thousands of new exotic species, as living or fossil specimens. They were humans of endurance with great eagerness and vision who often left behind the pleasant home environment for years of extreme loneliness. Their effort was not in vain. People such as Humboldt, Wallace, Bates, and Darwin, among many others, flooded the museums and universities of Europe with massive containers stuffed with thousands of specimens. As an example, Wallace reports in his magnum opus *The Malay Archipelago*, that he could ship to England 125,600 specimens, comprising mammals, reptiles, birds, shells (molluscs), and insects, after about six years of collecting in that archipelago. The problem was that this material was not only to be classified, but its distribution interpreted by the naturalists too. This took time and, most importantly, scientific skill. This was a time of massive data production that hardly needed a unifying synthesis of the natural world, which did not exist. The more carefully nature was studied, the more difficult it became to fit the species into the creationist paradigm that was in force. Contrary to the picture of a stable, unchanging universe cherished by their creationist mentors, these 'explorers' were faced with a dynamic world in which God's creatures were competing, adapting, and, in sum, evolving, in modern parlance. Moreover, those creatures that failed were thrust to extinction, as evidenced by the fossils. How could extinction match the perfection of created organisms?

Although by the end of the eighteenth century many naturalists had worked out the adaptation of species to their environment, an idea contrary to the belief that all species were independently originated in the same place (the garden of Eden) and then dispersed worldwide, most eminent naturalists like Linaeus and Buffon still interpreted the data under the creationist paradigm. It took men of vision like Wallace and Darwin, and also Lamarck, albeit in a different way, to create a new evolutionary paradigm to understand and order the massive data provided by the naturalists. Since then, these pioneering men, turned from naturalists to scientists, have been joined by those like Dozhansky, Mayr, Simpson, Stebbins, and many others, who have followed and refined their ideas, and also their lust for nature, incorporating in a synergic way the Mendelian paradigm in the Modern Synthesis. Following the consolidation of the Modern Synthesis spectacular advances in molecular biology, coached by theoretical insights (see Box 1.1) have shaped a new view of evolution in which non-adaptive processes have been formally described. In the previous chapters of this book I have summarized some of these advances and their significance. After a period where the production of new data seemed to levelling off, this ongoing evolutionary saga has reached a point of a new explosive era of biological information.

In this book, I have given reasons that support the idea that the genome is not a static, stable repository of genetic information—rather, that dynamism is its chief characteristic. Yet, the dynamic genome analysis is inaugurating a new era of massive empirical information that recalls those described above. The question remains whether we have the right evolutionary theory to understand the unanticipated structures and mechanisms revealed by the genomic science. In particular, how much of the

Modern Synthesis has to be changed to cope with the genome insights? Are these genome changes footnotes, additions, extensions, or substitutions of the Modern Synthesis principles? In other words, do they require a new paradigm? The situation, I believe, is not new in evolutionary science. But, do we really need a new Darwin?

The survival of the fittest, the arrival of the fittest, and . . . of the luckiest

Answering these questions is not an easy task. Let me begin with the easiest assumptions, listed in Table 5.1 as the material for evolution and the fixation mechanisms. Ever since Darwin everybody has agreed that heritable variation is a must for evolution to proceed. In spite of the absence in his time of a sound hereditary mechanism, Darwin was sharp in this assertion. His proven intuition notwithstanding, Darwin's adherence to 'pangenesis' and 'blending inheritance' was misguided and gave him many problems in understanding the origin and the maintenance of inherited variation. In fact, the prevailing belief in blending inheritance at the time posited that hereditary factors were body fluids that mixed, and diluted, in the progenies; ironically we still talk about 'pure blood' and 'blue blood' to designate gene inbreeding in animals and gene purity in aristocracy. Dilution of hereditary factors across generations makes it extremely difficult to maintain the needed hereditary variation for natural selection to act upon. Perhaps, this difficulty shifted Darwin towards embracing the hypothesis of pangenesis, positing that all parts of an organism contribute to heredity in the sex cells by means of hypothetical particles he called gemmules. Darwin was wrong in that, as Weissman later showed in his theory of the separation of soma and germ lines (see above: From Darwinism to the Modern Synthesis). These gaps were filled in the Modern Synthesis by the incorporation of the Mendelian theory of particulate genetics. Although Mendelian genes were the 'prima donna' factors in genetic variation for decades, their material nature only started to be understood after the DNA structure and dynamics were unveiled in the mid-1950s. Since then, molecular and genome studies have unveiled an intricate gene structure of regulatory, coding, and non-coding sequences that makes the gene a difficult concept to define. These ongoing studies have introduced in evolutionary theory a wealth of genetic variation elements ranging from simple and larger nucleotide changes to unanticipated small mobile DNA stretches (Chapter 3) and regulatory elements (Chapters 1 and 2) whose role in evolution is important but still surrounded by controversy. Moreover, much of the non-coding DNA, originally qualified as junk, seems to transcribe into RNA sequences that likely play an important regulatory role and represent a crucial material for evolution (see Chapter 1).

That Darwin's intuition could have led him to propose such a wrong inheritance hypothesis, yet, at the same time, to retain the essentially right hypothesis on how heritable variation could be fixed by selection, shows that the continuous presence of genetic variation, regardless of its mechanism of transmission, is the primary requisite for evolution by natural selection. However, under the Darwinian paradigm genetic variation is undirected with regard to adaptation and it is natural selection that decides which variants will stay and which ones will be eliminated. While nobody objects to natural selection as an evolutionary keystone, its prime role is still challenged by some internalists from the field of development that view developmental constraints as the leading factors in channelling evolution. The ancient origin of this controversy and its present state were discussed in Chapter 2. There I argue that natural selection, Darwin's great revolutionary insight, is the leading force in adaptive evolution, a view sponsored by the great majority of current evolutionists and an ever-increasing number of developmental biologists. However, natural selection is not the only mechanism that may drive the genetic material to fixation. Genetic drift was recognized, albeit to a different degree, as a fixation mechanism by the fathers of the Modern Synthesis. For instance, while Fisher was sceptical over the importance of genetic drift, Wright contended that genetic drift was a first-rate mechanism in the evolution of populations. The power of genetic drift has become apparent in many small populations, particularly when the selective value of genes is small (see Box 1.1 for a further insight), as it is the rule in many molecular changes. Recently, these ideas have been applied to

Table 5.1. Some comparisons between Darwinism, the Modern Synthesis, and several putative current additions, for several important topics in the evolutionary theory.

Topic	Darwinism	Modern Synthesis	Current status with additions
1. Material for evolution	Unknown heritable factors	Heritable Mendelian gene variation	Yes. New DNA variation including single nucleotide insertions and deletions, small duplications (CNVs), regulatory elements, like CREs, miRNAs, TEs, etc. must be added.
2. Fixation mechanisms	Natural selection as the driver of evolution (soft inheritance secundary). Some hints about fixation of neutral traits	Natural selection and genetic drift (no soft inheritance)	Natural selection and genetic drift. Neutral fixation is important at the molecular level. Epigenetic evolution is an expanding field still under scrutiny.
3. Developmental evolution and phenotypic plasticity	Important but largely unknown	Not considered	Developmental biology explains the evolution of form. Genetic assimilation is an important factor in phenotypic plasticity because facilitates adaptation to stress environments.
4. Gradualism: *natura non facit saltum*	Evolution proceeds by small steps	Yes, but at different rates	Gradualism is the rule even in most regulatory changes of development; yet some leaps are possible as in HGT and symbiosis.
5. Externalism: environment as the inducer of adaptation	Yes	Yes	Yes. Constraints do not direct evolution, but positive constraints are deep homologies that channel some lineages to converge at the molecular level.
6. Origin of species	Natural selection as the main cause, but may be helped by isolation	Isolation is the primary cause. Allopatric speciation is ubiquitous	Speciation is a pluralistic mechanism: sympatric speciation is frequent in some groups. Hybrid speciation is important.
7. The tree of life (TOL)	Yes: Darwin's innovation idea supporting the unity of type by lineage diversification	Yes	The web of life: in prokaryotes and in some eukaryotes lineages the HGT is frequent. Yet, the TOL remains a central working scheme in the origin of diversity.
8. The target of selection	The individual in populations	Yes	The individual in most instances, but group selection, especially in kin groups, is real.
9. Complexity	Evolves by accumulation of new adaptive mutations	Yes	Selection of adaptive mutations is important, but complexity is most likely the result of interactions between genes. Evolution is facilitated by the robustness ensuing from gene networks that buffer responses under stress due to compensating mechanisms (i.e. redundancy and complex interactions).

understanding the increasing size in genomes of complex organisms. Lynch argues that most of the disproportionate 'excess' of non-coding genome size (the C-value paradox) could be the outcome of non-adaptive drift processes rather than of adaptive fixation mechanisms (see Chapter 1), but some evolutionists object that more empirical genome comparisons between species and more functional genomic studies are badly needed before this contention can be settled.

Other fixation mechanisms (Lamarckism, orthogenesis, use and disuse, etc) framed inside the general term of soft inheritance by Mayr, were also adopted, albeit in a rank secondary to natural selection, by Darwin in the last editions of *The Origin* (see above). However, 'soft inheritance' sensu Mayr was wholly dismissed by the Modern Synthesis. Although it may seem unnecessary to raise this concept anew, I purposely bring it up here because recently the term 'soft inheritance' has been used in a different, non-Lamarckian way that can cause some confusion. Jablonka and Lamb (2010), for instance, defined soft inheritance as 'the transmission of variations acquired during development' adding immediately that it 'not only exists, it is found in every type of organism and seems to be common'. To these authors, soft inheritance includes both non-DNA variations and developmentally induced variations in DNA sequences. All these variations are the consequence of epigenetics in its broad sense, including not only those changes that modify DNA by adding or removing chemical groups to nucleotide bases, like methylation or demethylation of cytosines (see Chapter 3), but also all other cell variant states that are transmitted through cell division, including self-sustaining regulatory loops and RNA-mediated inheritance. As the defenders of these cellular epigenetic variations admit, it is not known how frequently these epigenetic variants are transmitted between generations, but at least some cases have been documented (see the examples of genetic methylation in mice of Chapter 3). Since these epigenetic systems are closely involved in the regulation of gene expression and phenotype production, they may be subjected to genetic assimilation. For instance, I argued in Chapter 3 that some epigenetic cell mechanisms like methylation of DNA and acetylation of histones likely evolved from a defence mechanism against genome parasites, including transposable elements, that was later recruited as a genome-wide regulatory process. Whether this genomic view of epigenetic inheritance mirrors the ideas of Darwin and his contemporaries on soft inheritance is still debatable, but this is an expanding field that should probably be incorporated into the genomic synthesis.

Developmental biology was absent in the Modern Synthesis. Some reasons for this void were alluded to in Chapter 2 and in the present chapter. Perhaps, it has been emphasized too much that 'synthesizers', albeit not Darwin, were unaware of the importance of embryology in evolution, and that they were just defining evolution as the result of 'changes in gene frequencies' by default. This view has become a rather derogative and, I believe, unfounded caricature. Rather, I adhere to the more conciliatory view that both neo-Darwinists and embryologists were leading characters in a play of ignoring each other. Darwinism was highly dependent on comparative embryology to establish the unity of type, one of the Darwinian pillars, as explained in Chapter 2. Is it not ironic for anti-Darwinists that Cuvier, the founder of comparative anatomy, was considered by Darwin, together with Linnaeus as 'one of his two Gods'? But initial success in allocating difficult animal groups, like ascidians and barnacles, to their correct systematic place using comparative embryology tools was immediately proved to be inadequate for distinguishing true homologies from convergences in the majority of embryos. That complete ignorance of hereditary mechanisms at the time was the reason of this failure comes as no surprise. At the time embryologists abandoned comparative studies for experimental embryology, and turned their interest to how embryo differences, more than similarities (homologies), between embryos come about. In sum, when the Modern Synthesis was born embryologists were not interested in evolution because embryologic genetic studies were practically absent. This scenario discouraged population geneticists, the drivers of the synthesis, who likely decided that, given the then poor knowledge of embryological genetics coupled with the molecular ignorance of the gene, developmental mechanisms were irrelevant to an evolutionary theory based on natural selection. In other words, they agreed, maybe using a thinly reasoned argument, that the genetic processes underlying development should not be so different from those observed for other genes in populations of organisms, and could be ignored.

This disconnection is fortunately coming to an end. The concept of evolution as the result of hereditary changes in development was largely lost in the Modern Synthesis. This traditional view was superseded in the Modern Synthesis by the population concept, based on genetically determined fitness differences in adult organisms in populations. This population approach emphasizes the survival of the fittest, while the developmental approach concentrates more on the arrival of the fittest. The former is basically interested in differential adaptive mechanisms as the conveyors of environmentally tuned traits across generations, the latter aims at deciphering how morphological traits are built to cope with the adaptive challenges posed by the environment. Neither view is an irreconcilable alternative to evolution; rather, they complement each other in a unified way. The population genetic account focuses on variation within populations, the material stuff for natural selection, whereas the developmental genetic account concentrates on variation between evolutionary groups above the population level. These operational disparate approaches notwithstanding, both views are united by the same fundamental Darwinian epiphany, namely that evolution is descent with modification.

Once the nature and the intricate function of genes at the molecular level were deciphered, developmental geneticists have been able to disentangle the complexities of regulatory pathways in which genes are immersed. This produced an appreciation for population studies by some developmental geneticists, who started to study how regulatory networks may initiate their evolution by comparing closely related species and/or relating regulatory responses to environmental challenges (see above for wing eye-spots, pelvis spine reductions, eye-induction by eyeless gene, etc). Similarly, population geneticists are becoming increasingly aware that genes in morphology often do not perform fixed functions, contradicting the traditional perception of the classic Mendelian gene; rather, they can be equated to informational bits that are turned on and off under a suite of cell and tissue signals, as discussed in Chapter 2. For instance, the well studied gene *Pax-6* encodes a transcription factor that not only intervenes in inducing eye formation (see p. 40; Chapter 2), but is also needed for developing other systems and organs such as the nervous system and the pancreas. This pleiotropic gene action is not new to population geneticists, but what is new is perhaps the concept of modular pleiotropy, which obliges them to rethink new population models. This is an open field for exploration that should contribute to the convergence of population and developmental studies.

As population geneticists could benefit from developmental thinking, so evolutionists with a developmental background could gain some crucial ideas from population geneticists to interpret evolutionary processes. Population geneticists are not only interested in the survival of the fittest; they have long discovered that some population structures may lead to the survival of genetic neutral variants that do not increase the organismal fitness (see Chapter 1 for neutral evolution). Moreover, even some mutants that slightly increase fitness may be eliminated in relatively small populations, behaving as effectively neutral. The mechanism, known as genetic drift (see above), and the neutral (or quasi-neutral) theory of molecular evolution, has been discussed in Chapter 1 and relates to the C-value paradox of the genome. So, although it is very appealing to attribute to natural selection the apparently well-fitted regulatory mechanisms with functional significance that we find in complex organisms, other alternatives must be contemplated (and tested) in which natural selection might not be the main actor. In light of this view at least two of these scenarios, the genome size evolution and the origin of new gene functions by duplication, have been discussed (Chapter 1) in which neutral evolution could play an important role. Those new genetic effectively neutral structures that are lucky enough to escape selection, regardless of their future co-opted value, deserve a close scrutiny if we are to gain a better understanding of evolution. In that sense I daresay that all kinds of evolutionists should not only be interested in the arrival of the fittest but also in the arrival of the luckiest.

Finale: Allegro maestoso *cum tinkering,* ma non troppo

Here, I will try to compile in a few closing paragraphs some of the ideas developed throughout this abridged treatise about the impact of some genome-centred insights on evolutionary theory. Since evolution is a multi-faceted theory, I was often forced to diverge from this genome centrism to other evolutionary-related fields, for which I must beg your patience. My defence, and also my justification, is the impossibility of giving a solid evolutionary interpretation of many genome facts without an understanding of the historical and present-day ideas underlaying these facts. Moreover, I have attempted to compare three evolutionary views: Darwin's ideas, post-Darwinian evolutionary principles, and propositions for a future synthesis (see Table 5.1). Again, I must assume that your benevolence will forgive my presumptuous attempt. This approach is guided more by my lust for communicating to the widest readership a general, albeit concise, overview of current evolutionary principles than by assuming the air of an expert in all fields. Take, then, my account as an attempt void of any suspicious boasting.

In an attempt to summarize, three main approaches to the understanding of genome complexity as the playground for evolution can presently be envisaged. One, dubbed systems biology, is derived in part from the overwhelming input of massive data on genome sequences that induces some evolutionists to reduce all this information to a few general rules. Complexity is just one of the emergent properties that systems biology tries to understand; among the others, modularity, robustness, and evolvability may be the most studied. I have discussed some of them above, but the focus of systems biology is beyond the scope of the present treatise. For those interested in the field and its connection to the general evolutionary theory there are some interesting reviews (see for example Koonin 2009). Systems biology tries to connect evolution with the genome function; in particular it aims at relating genome dynamics with the phenotype, a long sought, but still largely incomplete, endeavour of evolutionists. An illustrated accomplishment of this genome–phenome link, in the '-omics' parlance, which may be attributed to the systems biology approach, refers to protein evolution. While applying statistical methods to large data sets, a significant negative correlation has been found between sequence evolution rate and expression level. Systems biologists claim that this finding shifts the traditionally assigned prevalent roles of structural constraints and the biological function of the protein in sequence evolution towards the pre-eminence of protein expression, a phenomic variable. Then, based on computer simulations, they continue arguing that selection for robustness against protein misfolding (see above the haemoglobin and the Hsp case studies) could explain why highly expressed proteins show low sequence evolution rate. This and other revealed emergent regularities in the form of correlations and distributions notwithstanding, there is a certain scepticism about the real value of systems biology. For instance, Lynch (2007, p. 386), after pointing out that a similar attempt in the 1960s in the field of systems ecology ended with a total failure, asks himself whether systems biology will 'suffer the same fate, or is there really something special about the properties of organisms?'. This sceptical view, I believe, would be subscribed today by many evolutionists.

The second attempt to understanding genome complexity is deeply rooted in the neutral (or quasi-neutral) theory of molecular evolution. Pioneered by the work of Sewall Wright (see above), genetic drift has since become one of the fundamental forces of evolution. Throughout the book I have discussed, in different contexts, the role of population size in shaping many gene and chromosomal distributions, and genome evolution cannot be alien to this mechanism. The question is, however, not whether drift is present, something that no one would dare deny, but how important it is relative to other major forces of evolution, mainly natural selection. The non-adaptive directionality of evolution has been a refrain that has accompanied the long symphony of the theory of evolution since its inception. Even Darwin recognized that 'neutral' variation can exist as a companion of selective variation when he states:

'This preservation of favourable individual differences and variations, and the destruction of those which are injurious, I have called Natural Selection, or the Survival of the Fittest. Variations neither useful nor injurious would not be affected by natural selection, and would be left either a fluctuating element, as perhaps we see in certain polymorphic species, or would ultimately become fixed, owing to the nature of the organism and the nature of the conditions'. (Darwin, *The Origin of Species*, 1872, p. 63)

Under a genome perspective, Lynch (2007) is promoting the latest non-adaptive argument that aims at explaining most, if not all, of the origins of the present genome architecture. In Chapter 1, I briefly discussed many of his arguments. Here I want to emphasize, for brevity, two points. I think that Lynch is right, I daresay extremely right, in his defence of population genetics as the key pillar to understand evolution. He may seem hyperbolic when he asserts that 'nearly all of the mathematical foundations [of population genetics] developed prior to the molecular era not only still stand today, but are making major contributions to our understanding [of molecular evolution, animal and plant breeding, and human genetic disorders] (terms in brackets summarized by me)'. I fully agree with him, except perhaps when he qualifies as just 'embellishments' all additions to the theory after the molecular era. The second point is his storytelling about the pre-eminence of non-adaptive mechanisms in the origins of the genome architecture, followed by an acrimonious critique on selection-centred evolution and its defenders. That many adaptations we now see did not originate as such is not news in evolutionary literature. That non-adaptive mechanisms, like genetic drift, 'set the stage for future paths of adaptive evolution', in Lynch's words, is not new either. In a different way, Lynch's defence of non-adaptive forces as the builders of genome architecture reminds me of the acid critiques against panselectionism sponsored in the 1970s by evolutionists like Gould and Lewontin (1979). An abridged exposé of Lynch's argument on population size can be found in Chapter 1. There I give credit to the importance of Lynch's well-documented arguments, but I also urge restraint until more functional experiments can validate their conclusions based mainly on correlations; and, remember, correlation is not necessarily causation.

Finally, my third point on genome complexity, and its ensued dynamism, is centred on what I consider the keystone, which all other pieces of the evolving structure depend on. Let me refer the reader to the opening of the present chapter. There I insisted on the difficulties in understanding the way natural selection acts. Among the critics that Darwin complained about, the public misunderstanding of the concept of natural selection and its opportunistic nature stands first. The keystone of adaptive evolution is a 'tinkerer', as Darwin himself tried to communicate, albeit not always successfully. Since then it seems to me that this principle, epitomized by many illustrious evolutionists like Jacob (see the introductory chapter quote), when deeply comprehended, explains most, if not all, difficulties in evolution. I believe that the recurrent controversy over what comes first, adaptation or noise, is secondary to the meaning of adaptive evolution. Admittedly, non-adaptive original inputs in the genome, albeit not always, should be accepted in light of current evidence. But, is this sufficient to relegate natural selection as a subsidiary leading force in evolution? Then, which force is first?

The origin of novelties, a greatly debated issue, encompasses a wide range of processes like small changes in the expression of single-gene products, regulatory changes and interactions in developmental networks, horizontally acquiring new functions as in gene transfer and symbiosis, among others. These processes may escape selection, albeit not always, and may even become fixed in the genome, but many of them will be likely detected in future internal and external organism conditions by natural selection, that will 'tinker' with them towards producing new rearranged novelties. In each chapter of this book I have produced, hopefully in an understandable language, a set of 'tinkering' examples. In Chapter 1, I discussed the non-adaptive origin of genome structure; in Chapter 2, I explained how development is now understood as a set of complex networks composed by interacting, usually ancient, regulatory genes; in Chapter 3, I gave a genome-wide view on the 'tinkering' adaptive value of transposable elements versus their traditional view of genome parasites; and in Chapter 4, I exposed a radical new concept of the horizontal species. The unanticipated role of horizontally transferred

DNA stretches in the origin of species has been taken by some as an anti-Darwinian epiphany against the tree of life concept. I gave reasons for the shift from a tree to a web of life, but this new concept does not invalidate the idea of descent with modification. Descent is the rule that assures transmission of characters and the unity of type.

We have come a long way through all the chapters of this book. The key message I would like to communicate is two-fold: first, the genome is a dynamic entity, still largely unexplored, that provides a wealth of evolutionary insights, some of them unanticipated; second, while these new insights must be incorporated into the evolutionary theory, I do not think that they are changing the Darwinian paradigm, although they may extend the Modern Synthesis. In fact, some of the new genome knowledge comes closer to Darwin's original ideas than to the Modern Synthesis, whose tenets at times seem too inflexible. For instance, the almost universally accepted origin of species by geographical isolation (the allopatric model), a universal mechanism never accepted by Darwin, is now weakened by the presence of many other mechanisms, such as sympatric and hybrid speciation. These mechanisms give a more important role to natural selection in species origin than allopatric speciation does, a view that is closer to Darwin's. In sum, we should think of reconstructing Darwin, but never of deconstructing him.

Paraphrasing Monod's words 'from a noise source (chance mutations) selection is able, by itself, to extract all the music of the biosphere', I may wax lyrical in the closing of this symphony about evolutionary genomics. Yet, as any symphony must have a *finale*, I believe that I must put an end here with a grand *finale* ('*maestoso*' in opera parlance, but not so *maestoso* for some non-mystic people) with an emphasis on the 'tinkering' nature of adaptation. We, humans, should start to appreciate the grandeur of the universe within us, the genome. Yet this grandeur is the result of the evolution of other genomes, which may make us less unique in the inner universe but more significant in the whole cosmos. This meaning may come from realizing that the whole universe has not only attained a stage in which one of its evolutionary products can think about the universe, but also that, looking through us, the universe can appreciate its beauty. Impressive as it may seem, this *allegro* finale should be taken as a sign of happiness '*ma non troppo*'.

Glossary

Agamospermy A form of reproduction without fertilization in which embryos and seeds are produced asexually. It does not include vegetative reproduction.

Allele Each of the different forms of a gene. A diploid organism has two alleles of each gene; when both forms are identical the genotype is said homozygote, when they are different it is designated a heterozygote.

Allopatric speciation A model that proposes the necessity of a geographical barrier between two populations to allow sufficient population differentiation to generate two reproductively isolated species.

Allopatry Occupancy by two species or populations of two geographical regions separated by a geographic barrier.

Allopolyploidy A polyploid condition in which the multiplication of the genome is produced from a hybrid genome between two species. See **polyploidy**.

Allozyme Each form of an enzyme encoded by each one of the alleles of a gene. Usually allozymes can be detected by their different mobility in gel electrophoresis.

Alternative splicing Differential splicing of the same mRNA precursor by which different sets of exons form a suite of mature mRNAs that translate into various proteins, usually specific of tissue. Thus a single gene can encode different proteins, contributing to organismic complexity.

Amplified fragment length polymorphism (AFLP) A technique that combines methods of PCR and RFLP to detect length polymorphisms due to modifications in DNA sequence. It is perceived as a specific pattern of DNA electrophorelic bands. See **PCR** and **RFLP**.

Apicomplexan A member of the eukaryotic phylum Apicomplexa, which includes many parasites such as *Plasmodium falciparum*, the agent of malaria.

Archaea With bacteria and eukarya, one of the three domains of life. These unicellular organisms without nucleus are phylogenetically closer to eukaryotes than bacteria and usually inhabit extreme habitats like highly hot and saline localities.

Assortative mating Non-random mating induced by phenotype similarity (positive) or dissimilarity (negative) between mates. Positive assortative mating is often known as homogamic selection.

Bacteria (also **eubacteria)** One of the three domains of life. Together with archaea comprise all organisms without nuclei in their cells, designated as prokaryotes.

Balancing selection Any form of selection that maintains polymorphism in a locus.

Bauplan Often used as a synonym of body plan, but bauplans are reminiscent of the idealized archetypal body plans that were thought to be independently created.

Body plan The general morphological structure of an organism or taxon, usually referred to the basic, common structures shared by all members of a phylum in animals (or a division in plants).

Bottleneck A temporary reduction in population size that normally causes an important episode of genetic drift. It may be produced by a founder event.

Breeding allochrony The non-overlapping breeding season that experiences two species or populations.

Canalization The buffering of development against environmental and genetic perturbations that produces a constant standard phenotype.

cDNA (complementary DNA) DNA sequence that is complementary to an RNA sequence (for instance a messenger RNA), generated by a process named reverse transcription.

Centromere The chromosome region that unites to the spindle fibres at mitosis and meiosis and assures the symmetrical chromosome segregation.

Character identity networks (ChINs) Those GRNs that specify the character identity. See **Gene regulatory network** or **GRN**.

Chloroplast A photosynthetic DNA-containing organelle present in plants, algae, and some single-celled eukaryotes that evolved from cyanobacteria.

Chromosome In eukaryotes each of the cell organelles included in the nucleus that host the whole or part of the DNA organism content. It contains molecules, like histone proteins, other than DNA that help to pack the DNA and regulate its expression and transmission. In prokaryotes the DNA forms a long circular molecule, termed by extension chromosome, but lacks the eukaryotic protein components and it is not included in a nucleus.

Cis-regulatory element (CRE) Each of the DNA stretches that surround the gene, frequently upstream, where the homeodomain of Hox proteins bind to activate or repress gene expression.

Clinal variation Phenotypic or genotypic variation that correlates with gradation in an environmental transect. Often genetic variation underlies the phenotypic clinal variation.

CNV See **copy-number variant.**

Coalescence The process that merges the current gene copies into a single common ancestor copy.

Compartment A group of cells that descend from a single cell and form a discrete developmental unit in which specific developmental patterns exist. Each compartment is independent from the others in that a different set of genes are expressed or combined in different GRNs.

Conjugation The passage by means of a pilus of DNA from a prokaryote to another.

Co-option When an ancestral structure (including a gene) is modified to perform a new adaptive function it is said that this structure has been 'co-opted' and the co-opted new adaptive structure is named 'exaptation'.

Copy-number variant (CNV) Sub-microscopic DNA structural variants that appear at variable copy number among individuals of a population. See **structural variants**.

CRE See **cis-regulatory element.**

C-value paradox Lack of positive correlation between phenotype complexity and DNA content of eukaryotic genomes.

Directional selection Selection for a character trait whose value was higher or lower than its current value. Sometimes taken as synonym of positive selection. See **positive selection**.

Diversifying selection (syn. Disruptive selection) Selection favouring the extreme values of a trait. See **underdominant selection.**

Division A high rank category within the hierarchy of plant classification just below the kingdom that groups all organisms with the same body plan.

DNA base sequence The ordering of the four bases (A, T, C, G) along the DNA chain molecule.

DNA transcription The process by which DNA is copied into the mRNA precursor. In eukaryotes the primary copied precursor mRNA is spliced to produce mature mRNA by removing intron-copied sequences.

Domain A discrete portion of a DNA or a protein molecule that can usually adopt a specific function and/or structure independently of other domains.

Duplication The generation of another copy of any DNA stretch. It comprises short non-coding DNA sequences to whole genes and even to large and whole genome sequences. It is an important mechanism of genome expansion and diversification. See **gene duplication**.

Ectopic recombination Recombination between repetitive DNA sequences that occupy different positions in the genome.

Effective population size The size of an idealized population (in which all individuals reproduce equally) that experiences the same drift effect as the real population.

Endosymbiosis A kind of symbiosis in which an organism lives within the body (usually the cells) of another. See **symbiosis.**

Epigenetics A term loosely defined as the study of all interactions between genes and the non-genic developmental factors which bring the phenotype into existence through genetic expression. More precisely to-date it means the conformational changes in DNA and chromatin that regulate gene expression without changes in nucleotide sequence.

Epistasis The phenotypic effect of the interaction between several gene loci so that this effect is different from the addition of the individual effects of each locus.

Eukaryotes Members of the domain Eukarya, including single-celled (like protists), and multicellular organisms (like many plants, fungi, and animals). Their cells deploy internal membrane-bound organelles including a nucleus, secretory vesicles, mitochondria, and chloroplasts (in plants), plus DNA-containing chromosomes. They also they have a dynamic, internal cytoskeleton that supports cell shape and organization, and also movement of materials inside the cell.

Evolvability The ability to use genetic variation to evolve into new adaptive forms.

Exon A stretch of a eukaryotic gene that is first transcribed into RNA and later spliced out and incorporated into the mature messenger RNA. Exons encode protein sequences.

Exon shuffling The production of new genes by combining exons from different genes. The processes underlying new formation of genes by exon shuffling include recombination of intervening introns from two genes and retrotransposition of alien exons into genes.

Founder effect The change in gene frequencies and loss of genetic variation that accompanies the foundation of a new population by a small number of individuals due to the ensuing genetic drift.

Gene duplication The production of an additional copy of a gene. Its repetition through evolutionary time may generate a gene family. See **gene family**.

Gene family A set of highly paralogous genes that evolved by serial duplications from an ancestral gene. See **paralogous.**

Gene flow The transfer of genes into the gene pool of a population or species from another gene pool. Usually refers to movement in space, but can also include introgression.

Gene pool The whole set of genes of a given sexual population or species.

Gene regulatory networks (GRNs) Networks of genes that play a crucial role in development due to their mutual intricate relationships through regulatory pathways.

Genetic assimilation The stabilization of a phenotype, expressed only under a particular environment, through selection of the underlying genetic variability that eventually allows this phenotype also to be expressed under normal environmental conditions.

Genetic drift Random change in gene frequencies caused by the finite sample of gametes that passes to the next generation.

Genome The entire DNA nucleotide sequence in an organism cell. In eukaryotes genome usually refers to the cell nucleus DNA, but other DNA-containing cell organelles, like mitochondria and plastids, also have their genomes. Genome is a general term for the genetic contents of a species, while genotype applies to genetic contents of whole or parts (like genes) of individual organisms. See also **genotype.**

Genotype The set of genes (alleles) in an individual.

Gradualism An evolutionary process that posits that large differences among living beings are the result of the accretion through time of many slightly different intermediate forms.

GRN See **gene regulatory networks.**

Group selection Selection of characters that increase the survival of groups of individuals (populations and, by extension, species) rather than that of particular individuals.

Habitat preference The ability of a phenotype to choose the habitat to which it is fitted.

Habitat selection A form of diversifying selection in which the extreme phenotypes, usually homozygotes, occupy different habitats to which they are fitted, and the intermediates (usually heterozygotes) show less fitness because no habitat suits their needs.

Haltere Round-shaped sense organ found on the second thoracic segment of Diptera.

Haplotype A set of linked alleles in a specific genome region. Also refers to each kind of DNA sequence that differs from its homologous sequences by one or more nucleotides.

Heterozygote A diploid genotype in which both alleles of a gene (or both haplotypes) are different.

HGT See **horizontal gene transfer**.

Homeobox A DNA sequence that encodes a DNA-binding domain (homeodomain) of transcriptional regulatory proteins (i.e. transcription factors) that play a role in development. It is about 180 nucleotides long and is contained in the homeobox genes.

Homeotic mutation A class of mutation that transforms a structure of an organism into another organism's structure (e.g. antenappedia homeotic mutation transforms the antenna into legs in Drosophila).

Homogamy Positive **assortative mating**.

Homologous Used for structures, including genes and other DNA stretches, that share a common ancestry but are currently not necessarily similar in structure or function. See **ortologous, paralogous**.

Homoploid species Species that originated by hybridization between species with no posterior genome duplication.

Homozygote A diploid genotype in which both alleles of a gene (or haplotype) are identical.

Hopkins effect The propensity of insects whose larvae feed on a certain type of plant to lay their eggs in the same plant species.

Horizontal gene transfer (HGT) The transfer of genetic material from an organism to another that is not its offspring. Also known as **lateral gene transfer (LGT)**.

***Hox* genes** Genes with homeobox.

Imprinting The differential expression of alleles according to their maternal or paternal origin.

Instant speciation Refers to the rapid recovery from sterility and/or inviability of hybrids when their genome is doubled by allopolyploidy. This process generates a chromosome homologue for each chromosome that allows a perfect pairing in meiosis that results in a normal gametogenesis. See **allopolyploidy**.

Introgression The incorporation of genes into a gene pool of a population or species by hybridization between individuals of two distinct populations or species.

Intron A non-coding DNA stretch of a eukaryotic gene that is first transcribed into RNA but later is spliced out and not incorporated into the messenger RNA.

Island model The structure of a local population (the island) that receives a fraction m of the genes from the mainland population and the remaining 1-m genes come from the local population.

Junk DNA Derogative designation of all non-coding DNA. An outmoded term because much DNA that does not encode proteins has important regulatory functions.

Kernel GRN component that because of its developmental role and particular internal structure, is most impervious to evolutionary change.

Kin selection Increase in population frequency of an allele present in an individual because of the favourable effect from other individuals of the population (kin individuals) that share the same allele. This often applies to the cooperation of relatives on the reproductive success of an individual.

Lateral gene transfer (LGT) See **horizontal gene transfer (HGT)**.

Ligand A molecule that binds to a receptor. See **receptor**.

Lineage sorting Process by which, in the process of a population split to become two species, alleles from different genes are sorted out in the splitting populations and thereafter distantly (closely) related genes may appear in the same (distant) species. Lineage sorting may be responsible for the fact that species trees do not correspond with gene trees, mimicking the introgression effect by hybridization or lateral transfer.

Lobopodians Fossil group of animals with unjointed limbs (lobopods) that are thought to be the ancestors of arthropods. They became extinct in the Cambrian but members of the phylum Onychophora (velvet worms) are currently thought to be their descendants.

Locus (plural loci) In classical genetics, the place occupied by a gene in the chromosome. It may also refer to the location in the genome of any stretch of DNA or even of a single nucleotide. Locus is often loosely referred as the gene itself.

Long ncRNA Transcripts longer than 200 nucleotides that have little or no protein-coding capacity, but can regulate gene expression.

LTR (long terminal repeat) Each of the two long terminal repeats that possess some retrotransposons at both of their ends.

Macroevolution Evolutionary processes that occur at or above the species level, to be distinguished from those that occur at the population level (designated as **microevolution**).

Messenger RNA (mRNA) The mature gene transcript, produced in eukaryotes after removing the introns, that, like a messenger, carries the protein-coding information to the cytoplasmic ribosomes to be translated into protein.

Methylation The binding of methyl groups to DNA, usually to megabase stretches of CpG dinucleotides. Histone can also be methylated. Methylation serves as an inhibitor of downstream gene expression.

MicroRNA (miRNA) A class of small RNAs (usually 20–30 bp long) that act as specific regulators of mRNA translation and transcription. They are processed from longer RNA transcribed precursors which form hairpin loops.

Microevolution Evolutionary processes that occur in populations, to be distinguished from those that occur at higher levels (designated as macroevolution).

Microsatellite A short, untranslated DNA stretch of many repeated few-base sequences whose number is highly variable among individuals in populations. This high polymorphism makes microsatellites excellent genetic markers.

Mitochondria Cell organelles derived from proteobacteria that are responsible for aerobic respiration and contain a non-recombining autonomous DNA genome (mtDNA) which is amply used as a genetic marker.

Mobile element See **transposable element**.

Modular pleiotropy See **mosaic pleiotropy**.

Modularity The property of different parts (i.e. compartments) of an organism to develop, function, and evolve independently from one another.

Monophyly The condition of a taxon or a gene tree whose members comprise all those that derive from a common ancestor. See **paraphyly** and **polyphyly**.

Mosaic pleiotropy A kind of pleiotropy that allows pleiotropic genes to only express in certain compartments, facilitating the evolution of variation in those compartments without impairing the viability in others.

Neutral mutation A mutation that does not change the organism's fitness that carries it, which means that is not selected in favour or against.

Norm of reaction The set of phenotypes that a genotype can express in different environments.

Orthologous Used of homologous structures, including genes and other DNA stretches, that originated by diversification in two different lineages (i.e. species) from a structure present in the most recent common ancestor. Cf. **paralogous**.

Paedomorphosis The developmental process that generates sexually mature adults which retain juvenile characters of the ancestor organism.

Paralogous Used of homologous structures, including genes and other DNA stretches, that arose by duplication and later diversified from each other within a single lineage. See **orthologous.**

Parapatry Occupancy of contiguous, but not overlapping, geographic regions by each of two species or populations.

Paraphyly The condition of a taxon or a gene tree whose members have a common ancestor, but does not include all descendants of this ancestor. See **monophyly** and **polyphyly**.

PCR (polymerase chain reaction) A technique for amplifying a target DNA sequence.

Phagotrophic A way of feeding by ingesting organic particles.

Phenotypic plasticity The capacity of a genotype to generate different phenotypes as a result of the environment encountered during development. Usually this capacity is considered to be adaptive to the environment.

Phylum (plural **phyla**) A high rank category within the hierarchy of animal classification, just below the kingdom, that groups all organisms with the same body plan.

Pleiotropy A condition of a gene that affects more than one phenotypic character.

Pleistocene The geological epoch between 1.8 mya and about 11,000 years ago that includes several glaciations.

Polyphyly Refers to a taxon or a gene tree whose members do not have a common ancestor. See **monophyly** and **paraphyly**.

Polyploidy A cellular condition in which the genome is duplicated one or several times from an ancestral diploid condition.

Position effect Change in the expression of a gene due to a change in its spatial position relative to other genes or DNA regions. A classic example of position effect is the influence of heterochromatin in the regulation of nearby genes like eye-colour genes generating eyes that in Drosophila appear mottled red and white, a condition named 'position effect variegation'.

Positive selection Selection for alleles that increase fitness in populations.

Promoter The DNA stretch usually located upstream of a gene that is recognized and bound by the RNA polymerase whose activity in gene transcription it regulates.

Pseudogene A non-functional gene derived from a previous functional gene. Normally pseudogenes are inactivated copy-members of gene families originated by gene duplications. See **processed pseudogenes**.

Processed pseudogenes Pseudogenes that originate by retrotransposition of mRNAs.

Purifying selection A form of natural selection that favours the removal of deleterious alleles from populations. Cf. **positive selection**.

Quantitative trait locus (QTL) A chromosome region containing genetic factors that explain some variation of a quantitative trait. These loci are used to map the chromosome positions of regions that influence the variability in a quantitative trait, a procedure named QTL mapping.

Random amplification of polymorphic DNA (RAPD) A type of PCR reaction, but the segments of DNA that are amplified are random. See **PCR.**

Receptor The binding site situated on the cell membrane or in some of the molecular components of the cell where a hormone or other ligand attaches to initiate its activity.

Recombination Any mechanism that produces a new assortment of hereditary material (i.e. DNA; chromosomes), such as in meiosis through crossing-over. Recombination can be homologous when two identical (or nearly) stretches of DNA align and exchange parts as in the chromosomal alignment in meiosis. See **ectopic recombination**.

Recombinational speciation Model of hybrid speciation that proposes that fertility in hybrids is restored only by those hybrid recombining gametes carrying a whole, balanced genome, which are able to produce homozygous fertile progeny.

Refuge An isolated area of a large region that escapes major environmental changes and serves as an appropriate site for species to persist in the ancient habitat.

Reinforcement The increase of prezygotic isolation through selection in favour of matings between similar genotypes (homogamic) by discarding low-fitted crossbred genotypes.

Repetitive DNA A fraction of DNA whose sequence is repeated, ranging from a small to a large number of copies. Copies can be located in tandem (e.g. microsatellites) or be found dispersed (e.g. transposable elements).

Reproductive isolation The keeping apart of two distinct gene pools due to incompatibilities that prevent interbreeding.

Restriction fragment length polymorphism (RFLP) Variation in the length of fragments generated by cutting DNA with restriction enzymes that is employed to detect variability in DNA sequences among species or populations.

Retrotransposon A member of a large group (class 1) of transposable elements, which transpose through an intermediate RNA copy that is retrotranscribed to DNA before its insertion.

RFLP See **restriction fragment length polymorphism**.

RNA-coding sequences DNA sequences that encode RNA molecules which have their own cell functions and do not translate into proteins (i.e. ribosomal RNA).

RNAi (RNA-mediated interference) A suite of mechanisms of gene regulation based on the inhibition of gene expression through the generation of double-stranded RNA molecules.

SDs See segmental duplication.

Segmental duplications (SDs) These duplications, formerly named low copy repeats (LCR), are >1 Kb long DNA segments that occur in two or more copies that share >90% sequence identity. They are fixed but can often vary in copy number, in which case they are CNVs.

Selfish DNA Alien, invasive DNA that can replicate by itself, or via other self-replicating DNA, usually faster than the DNA of the host in which resides. Normally, selfish DNA has no positive function or is detrimental to the host organism. Viruses are considered selfish elements.

Sex chromosome Any chromosome that determines the sex of an organism. The other chromosomes are designated autosomes.

Sexual selection Selection for differential reproduction due to the ability to obtaining a mate.

Single-nucleotide polymorphism (SNP: pronounced 'snip') A polymorphism produced by variation among population individuals in the nucleotide that occupies a specific position in a DNA sequence.

siRNAs See **small-interfering RNAs**.

Small RNAs (sRNAs) A generic term for non-translating small RNA molecules, like siRNAs and miRNAs, that participate directly in cell regulatory pathways.

Small-interfering RNAs (siRNAs) A class of small RNAs (usually 20–30 bp long), similar to miRNAs, processed from longer double-stranded RNAs. They act in post-transcriptional gene silencing. See **RNAi**.

SNP See **single-nucleotide polymorphism**.

Solo LTR Dispersed LTR that appears isolated from any retrotransposon in the genome.

sRNAs See **small RNAs**.

Structural variants Genome regions larger than 1 Kb that are found in variable number, localization, and orientation in different individuals of a species. They include microscopic (i.e. karyotypic and numerical chromosomal changes) and submicroscopic (ranging from 1 Kb to 3Mb in size) variants. See **copy-number variants**.

Symbiosis An intimate association between two organisms that usually describes all relationships between two different species. Often restricted to associations in which both partners are mutually beneficial.

Sympatry Occupancy of the same geographic region by two species or populations.

Syngameon A set of distinct and related taxa, ranging from subspecies to species rank, that are able to interbreed and exchange genes. This is not uncommon in plants.

Telomere The chromosome region situated at its end. In many organisms telomeres are composed of highly repetitive DNA.

Transcription factor A molecule (often a protein) that binds to a regulatory DNA sequence and plays a crucial regulatory role in gene transcription.

Transduction Formerly exclusively referred as the transfer of genes from one donor cell to a recipient cell via a virus. Currently, this term includes the TE-mediated transposition of stretches of DNA inside the genome.

Transformation The incorporation of a stretch of foreign DNA into a genome.

Transposable (or mobile) element (TE) A stretch of DNA that can itself, or copies of which can, move (transpose) and become inserted into different sites in the genome. See **transposon** and **retrotransposon**.

Transposon A member of a large group (class 2) of transposable elements, which transpose by a 'cut-and-paste'mechanism (i.e. P elements).

Underdominant selection Selection against the heterozygote and in favour of the two homozygotes in a segregating locus.

Unequal recombination Usually refers to the incorrect pairing of two homologous sequences that ensues by recombination to duplications and deletions. When two tandemly repeated sequences pair unequally one product of recombination increases and the other diminishes the number of copies.

References

Agrawal A., Eastman Q.M., and Schatz D.G. (1998) Implications of transposition mediated by V(D) J-recombination proteins RAG1 and RAG2 for origins of antigen-specific immunity. *Nature* **394**, 51.

Allis, C.D., Jenuwein, T., Reinberg, D. (eds.); and Marie-Laure Caparros, (ass.ed.) (2007) *Epigenetics*. Cold Spring Harbor Laboratory Press, New York.

Almeida, R. and Allshire, R.C. (2005) RNA silencing and genome regulation. *Trends in Cell Biology* **15**, 251–8.

Aminetzach Y.T., Macpherson J.M., and Petrov D.A. (2005) Pesticide Resistance via Transposition-Mediated Adaptive Gene Truncation in Drosophila. *Science* **309**, 764–7.

Anderson, E. (1949) *Introgressive Hybridization*. John Wiley and Sons, New York.

Andersson, J.O. (2005) Lateral gene transfer in eukaryotes. *Cell Molecular and Life Sciences* **62**, 1182.

Anxolabéhère, D., Kidwell, M.G., and Périquet, G. (1988) Molecular characteristics of diverse populations are consistent with the hypothesis of a recent invasion of Drosophila melanogaster by P elements (1988). *Mol. Biol. Evol.* **5**, 252–269.

Armour, J.A.L. (2009) Sharp focus on the variable genome. *Nature* **461**, 735–6.

Arnold, M.L. (1993) *Iris nelsonii*: Origin and genetic composition of a homoploid hybrid species. *American Journal of Botany*, **80**, 577–83.

Arnold, M.L. (1997) *Natural Hybridization and Evolution*. Oxford University Press, Oxford.

Arnold, M.L. (2006) *Evolution Through Genetic Exchange*. Oxford University Press, Oxford.

Arnold, M.L. and Bennett, B.D. (1993). Natural hybridization in Louisiana irises: Genetic variation and ecological determinants. In R.G. Harrison (ed.), *Hybrid Zones and the Evolutionary Process*, pp. 115–39. Oxford University Press, Oxford.

Averof, M. and Cohen, S.M. (1997). Evolutionary origin of insect wings from ancestral gills. *Nature* **385**, 627–30.

Avise, J.C. (2000). *Phylogeography: The History and Formation of Species*, p. 318. Harvard University Press, Cambridge, Mass.

Bailey J.A., Carrel L., Chakravarti A., and Eichler E. (2000) Molecular evidence for a relationship between LINE-1 elements and X chromosome inactivation: The Lyon repeat hypothesis. *Proc. Natl Acad. Sci. USA* **97**, 6634–9.

Bailey, J.A., Liu, G., and Eichler, E.E. (2003) An *Alu* transposition model for the origin and expansion of human segmental duplications. *Am. J. Hum. Genet.* **73**, 823–34.

Barlow, G.W. (2002) *The Glchlid Fishes*. Perseus Publishing, Cambridge, MA.

Barluenga, M., Stölting, K.N., Salzburger, W., Muschick, M. and Meyer, A. (2006) Sympatric speciation in Nicaraguan crater lake cichlid fish. *Nature* **439**, 719–23.

Bartolomé C., Bello X., and Maside X. (2009) Widespread evidence for horizontal transfer of transposable elements across *Drosophila* genomes. *Genome Biology* **10**, R22 (doi:10.1186/gb-2009-10-2-r22).

Barton, N.H., Briggs, D.E.G., Eisen, J.A., Goldstein, D.B., and Patel, N.H. (2007) *Evolution*. Cold Spring Harbor Laboratory Press, New York.

Bateson, W (1922) Evolutionary faith and modern doubts. *Science* **55**, 55–61.

Beldade, P. and Brakefield, P.M. (2002). The genetics and Evo-Devo of butterfly wing patterns. *Nature Reviews Genetics* **3**, 442–52.

Bergthorsson, U., Adams, K.L., Thomason, B., and Palmer, J.D. (2003) Widespread horizontal transfer of mitochondrial genes in flowering plants. *Nature* **424**, 197–201.

Berlocher, S.H. and Feder, J.L. (2002) Sympatric speciation in phytofagous insects: Moving beyond controversy? *Annu. Rev. Entomol.* **47**, 773–815.

Biemont, C. and Vieira, C. (2005) What transposable elements tell us about genome organisation and evolution: The case of *Drosophila*. In Volf, J-N. (ed.), *Retrotransposable Elements and Genome Evolution*, pp. 25–34. Reprint of *Cytogenetic and Genome Research* **110**, No. 1–4, Karger, Basel.

Birks, H.H. and Amman, B. (2000) Two terrestrial records of rapid climatic change during the glacial-Holocene transition (14,000–9,000 calendar years B.P.) from Europe. *Proc. Natl Acad. Sci. USA* **97**, 1390–94.

Blair J.E. and Hedges S.B. (2005) Molecular clocks do not support the Cambrian explosion. *Mol. Biol. Evol.* **22**, 387–90.

Blount, Z.D., Borland, C.Z., and Lenski, R.E. (2008) *Proc. Natl Acad. Sci. USA* **105**, 7899–906.

Bowen, N.J. and McDonald, J.F. (1999) Genomic analysis of *Caernorhabditis elegans* reveals ancient families of retroviral-like elements. *Genome Research* **9**, 924–35.

Bowen, N.J. and McDonald, J.F. (2001) *Drosophila* euchromatic LTR retrotransposons are much younger than the host species in which they reside. *Genome Research* **11**, 1527–40.

Bradshaw, Jr, H.D., Otto, K.G., Frewen, B.E., McKay, J.K., and Schemske, D.W. (1998) Quantitative trait loci affecting differences in floral morphology between two species of monkeyflower (Mimulus). *Genetics* **149**, 367–82.

Brennecke, J., Aravin, A.A., Stark, A., *et al.* (2007) Discrete small RNA-generating loci as master regulators of transposon activity in Drosophila. *Cell* **28**: 1089–103.

Brüggemann, H., Cazalet, C., and Buchrieser, C. (2006) Adaptation of *Legionella pneumophila* to the host environment: Role of protein secretion, effectors and eukaryotic-like proteins. *Current Opinion in Microbiology* **9**, 86–94.

Burt, A. and Trivers, R. (2006) *Genes in Conflict.* Belknap Press, Cambridge, Mass.

Busack, S.D. (1986). Biogeographic analysis of the herpetofauna separated by the formation of the Strait of Gibraltar. *Natl Geogr. Res.* **2**, 17–36.

Bush, G.L. (1993). Host race formation and sympatric speciation in *Rhagoletis* fruit flies (Diptera: Tephritidae). *Psyche* **99**, 335–57.

Bush, G.L. (1994). Sympatric speciation in animals: New wine in old bottles. *Trends Ecol. Evol.* **9**, 285–8.

Bush, G.L. (1998). The conceptual radicalization of an evolutionary biologist. In D.J. Howard and S. Berlocher (eds.) *Endless Forms: Species and Speciation*, pp. 425–38. Oxford University Press, Oxford.

Carninci, P. (2009) The long and short of RNAs. *Nature* **457**, 974–5.

Carroll, S.B. (1995) Homeotic genes and the evolution of arthropods and chordates. *Nature* **376**, 479–85.

Carroll, S.B. (2005) *Endless Forms Most Beautiful.* W.W. Norton & Co., New York.

Carroll, S.B. (2006) *The Making of the Fittest.* W.W. Norton & Co., New York.

Carroll, S.B. (2008) Evo-Devo and an expanding evolutionary synthesis: A genetic theory of morphological evolution. *Cell* **134**, 25–36.

Carroll, S.B., Gates, J., Keys, D.N. *et al.* (1994) Pattern formation and eyespot determination in butterfly wings. *Science* **265**, 109–14.

Carroll, S.B., Weatherbee, S.D., and Langeland, J.A. (1995) Homeotic genes and the regulation and evolution of insect wing number. *Nature* **375**, 58–61.

Carroll, S.B., Grenier, J.K., and Weatherbee, S.D. (2001) *From DNA to Diversity. Molecular Genetics and the Evolution of Animal Design* (2nd edn). Blackwell Publishing, Malden.

Carson, H. (2000). Sexual selection in populations: The facts require a change in the definition of the species. In R.S. Singh and C.B. Krimbas (eds.), *Evolutionary Genetics: From Molecules to Morphology*. Cambridge University Press, Cambridge.

Cartwright, P. and Collins, A. (2007) Fossils and phylogenies: integrating multiple lines of evidence to investigate the origin of early major metazoan lineages. *Integrative and Comparative Biology* **47**, 744–51.

Cavalier-Smith, T. (1985) *The Evolution of Genome Size.* John Wiley and Sons, Chichester, USA.

Cavalier-Smith, T. (1997) The blind biochemist. *Trends Ecol. Evol.* **12**, 162–3.

Chan, Y.F., Marks, M.E., Jones, F.C. *et al.* (2010) Adaptive evolution of pelvic reduction in sticklebacks by recurrent deletion of a Pitx1 enhancer. *Science* **327**, 302–5.

Charlesworth, B. (2008) The origin of genomes—not by natural selection? *Current Biology* **18**, R140–1.

Charlesworth, B., and Langley C.H. (1989) The population dynamics of Drosophila transposable elements. *Annu. Rev. Genet.* **23**, 251–87.

Chouard, T. (2010) Revenge of the hopeful monster. *Nature* **463**, 864–7.

Chung, H., Bogwitz, M.R., McCart, C. *et al.* (2007) Cis-regulatory elements in the accord retrotransposon result in tissue-specific expression of the Drosophila melanogaster insecticide resistance gene Cyp6g1. *Genetics* **175**, 1071–7.

Clausen, J., Keck, D., and Hiesey, W.M. (1948) *Experimental Studies of the Nature of Plant Species. III. Environmental Responses of Climatic Races of Achillea.* Carnegie Institution of Washington, Pub. No. 581, Washington DC.

Conant, G.C. and Wolfe, K.H. (2008) Turning a hobby into a job: How duplicated genes find new functions. *Nature Reviews.Genetics* **9**, 938–50.

Conrad, D.F., Pinto, D., Redon, R. *et al.* (2009) Origins and functional impact of copy number variation in the human genome. *Nature* doi:10.1038/nature08516.

Cook, S.A., and Johnson, M.P. (1968) Adaptation to heterogeneous environments. I. Variation in heterophylly in Ranunculus flammula L. *Evolution* **22**, 496–516.

Coope, G.R. (1979) Late Cenozoic fossil Coleoptera: Evolution, biogeography, and ecology. *Annual Review of Ecology and Systematics* **10**, 247–67.

Coope, G.R. (2004) Several million years of stability among insect species because of, or in spite of, Ice Age climatic instability? *Phil. Trans. R. Soc. Lond.* B **359**, 209–14.

Cordaux, R. and Batzer, M.A. (2009) The impact of retrotransposons on human genome evolution. *Nature Reviews. Genetics* **10**, 691–703.

Coyne, J.A. and Orr, H.A. (2004) *Speciation*. Sinauer Associates: Sunderland, Mass.

Cracraft, J. (1989) Speciation and its ontology: The empirical consequences of alternative species concepts for understanding patterns and processes of differentiation. In D. Otte, and J. Endler, (eds.), *Speciation and its Consequences*, pp. 28–59. Sinauer Associates, Sunderland, Mass.

Cropley, J.E., Suter, C.M., Kenneth, B., Beckman, K.B., David, I.K., and Martin, D.I.K. (2006) Germ-line epigenetic modification of the murine *Avy* allele by nutritional supplementation. *Proc. Natl Acad. Sci. USA* **103**, 17308–12.

Daborn, P.J., Yen, J.L., Bogwitz, M.R. *et al.* (2002) A single P450 allele associated with insecticide resistance in Drosophila. *Science* **297**, 2253–6.

Daniels, S.B., Peterson, K.R., Strausbaugh, L.D., Kidwell, M.G., and Chovnik, A. (1990) Evidence for horizontal transmission of the *P* transposable element between Drosophila species. *Genetics* **124**, 339–55.

Darwin, C.R. (1859, 1st edn; 1861, 3rd edn; 1866, 4th edn; 1869, 5th edn 1872, 6th edn) *On the Origin of Species by means of Natural Selection, or the Preservation of Favoured Races in the Struggle for Life*. John Murray, London.

Darwin, C.R. (1871) *The Descent of Man and Selection in Relation to Sex*, 1st edn. John Murray, London.

David, J.P. and Capy, P. (1988) Genetic variation of Drosophila melanogaster natural populations. *Trends in Genetics* **4**, 106–11.

Davidson, E.H. (2006). *The Regulatory Genome: Gene Regulatory Networks in Development and Evolution*. Academic Press, Oxford.

Davidson, E.H. and Erwin, D.H. (2006) Gene regulatory networks and the evolution of animal body plans. *Science* **311**, 796–800.

Davis, Ch. C. and Wurdack, K.J. (2004) Host-to-parasite gene transfer in flowering plants: Phylogenetic evidence from Malpighiales. *Science* **305**, 676–8.

Dawe, R.K. and Henikoff, S. (2006) Centromeres put epigenetics in the driver´s seat. *Trends in Biochemical Sciences* **31**, 662–9.

Dean, A.M., Thornton, J.W. (2007) Mechanistic approaches to the study of evolution: The functional synthesis. *Nature Rev. Genet.* **8**, 675–88.

Delprat, A., Negre, B., Puig, M., and Ruiz, A. (2009) The transposon Galileo generates natural chromosomal inversions in Drosophila by ectopic recombination. *PLoS One*. **4** (11),e7883 19936241.

Dobzhansky, Th. (1935). A critique of the species concept in biology. *Philosophy of Science* **2**, 344–55.

Dobzhansky, Th. (1937) *Genetics and the Origin of Species*. Columbia University Press, New York.

Dobzhansky, Th. (1980) The birth of the genetic theory of evolution in the Soviet Union in the 1920s. In: E. Mayr and W.B. Provine (eds.), pp. 229–42, *The Evolutionary Synthesis*. Harvard University Press, Cambridge, Mass.

Dolgin, E. (2009) The genome finishers. *Nature* **462**, 843–5.

Doolittle W.F. (1998) You are what you eat: A gene transfer ratchet could account for bacterial genes in eukaryotic nuclear genomes. *Trends Genet.* **14**, 307–11.

Doolittle, W.F. and Sapienza, C. (1980) Selfish genes, the phenotype paradigm and genome evolution. *Nature* **284**, 601–3.

Dowling, T.E., and DeMarais, B.D. (1993) Evolutionary significance of introgressive hybridization in cyprinid fishes. *Nature* **362**, 444–6.

Dunning Hotopp, J.C., Clark, M.E., Oliveira, D.C.S.G., *et al.* (2007) Widespread lateral gene transfer from intracellular bacteria to multicellular eukaryotes. *Science* **317**, 1753–6.

Eckenwalder, J.E. (1984) Natural intersectional hybridization between North American species of Populus (Salicaceae) in sections Aigeiros and Tacamahaca. III. Paleobotany and evolution. *Canadian Journal of Botany* **62**, 336–42.

Edwards, K.A., Doescher, L.T., Kaneshiro, K.Y., *et al.* (2007) A database of wing diversity in the Hawaiian Drosophila. *PLoS ONE* **2**(5), e487. doi:10.1371/journal.pone.0000487.

Eickbush, T.H. and Malik, H.S. (2002) Origins and evolution of retrotransposons. In: Craig, N.L., Craigie, R., Gellert, M., and Lambowitz, A.M. (eds.), *Mobile DNA II*, pp 1111–44. ASM Press, Washington DC.

Elinson, R.P. (1987) Changes in developmental patterns: Embryos of amphibians with large eggs. In R.A. Raff and E.C. Raff (eds): *Development as an Evolutionary Process*, pp. 1–21. Alan R. Liss, New York.

Feder, J.L., Chilcote, C.A., and Bush G.L. (1988) Genetic differentiation between sympatric host races of Rhagoletis pomonella. *Nature* **336**, 61–4.

Feder, J.L., Stolz, U., Lewis, K.M., Perry, W.M., Roethele, J.B., and Rogers, A. (1997) The effects of winter length on the genetics of apple and hawthorn races of Rhagoletis pomonella (Diptera: Tephritidae). *Evolution* **51**, 1862–76.

Feder, J.L., Berlocher, S.H., Roethele, J.B., *et al.* (2003) Allopatric genetic origins for sympatric host-plant shifts and race formation in Rhagoletis. *Proc. Natl Acad. Sci. USA.* **100**, 10314–19.

Feder, J.L., Xie, X., Rull, J., *et al.* (2005) Mayr, Dobzhansky, and Bush and the complexities of sympatric speciation in Rhagoletis. *Proc. Natl Acad. Sci., USA.* **102**, 6573–80.

Feschotte, C. (2008) Transposable elements and the evolution of regulatory networks. *Nature Reviews Genetics* **9**, 397-405 *doi: 10.1038/nrg2337.*

Feschotte, C., Jiang, N., and Wessler, S.R. (2002) Plant transposable elements: Where genetics meets genomics. *Nature Reviews Genetics* **3**, 329–41.

Feuk, L., Carson, A.R., and Scherer, S.W. (2006) Structural variation in the human genome. *Nature Review Genetics* **7**, 85–97.

Finston, T.L. and Peck, S.B. (1995) Population structure and gene flow in Stomion: A species swarm of flightless beetles of the Galapagos Islands. *Heredity* **75**, 390–7.

Fisher, R.A. (1918) The correlation between relatives on the supposition of Mendelian inheritance. *Transactions of the Royal Society of Edinburgh* **52**, 399–433.

Fisher, R.A. (1930) *The Genetical Theory of Natural Selection.* Oxford University Press, Oxford.

Fontdevila, A. (1988) The evolutionary potential of the unstable genome. In G. de Jong (ed.), *Populations Genetics and Evolution*, pp. 251–63. Springer-Verlag, Berlin.

Fontdevila, A. (1992) Genetic instability and rapid speciation: are they coupled? *Genetica* **86**, 247–58.

Fontdevila, A. (1995) Genetics and ecology of natural populations. In R. Levine (ed.), *Genetics of Natural Populations: The Continuing Importance of Theodosius Dobzhansky*, pp. 198–221. Columbia University Press, New York.

Fontdevila, A. (2004) Introgression and hybrid speciation via transposition. In A. Moya and E. Font (eds.), *Evolution: From Molecules to Ecosystems*, pp. 182–94. Oxford University Press, Oxford.

Fontdevila, A. (2005) Hybrid genome evolution by transposition. *Cytogenetics and Genome Research* **110**, 49–55.

Force, A., Lynch, B., Pickett, A., Amores, A., Yan, Y.-L., and Postlethwait, J. (1999) Preservation of duplicate genes by complementary, degenerative mutations. *Genetics* **151**, 1531–45.

Forrest, B. and Gross, P.R. (2005) *Creationism's Trojan Horse: The Wedge of Intelligent Design.* Oxford University Press, Oxford.

Fukagawa, T., Nogami, M., Yoshikawa, M. *et al.* (2004) Dicer is essential for formation of heterochromatin structure in vertebrate cells. *Nature Cell Biology* **6**, 784–91.

Futuyma, D.J. (1998): *Evolutionary Biology.* Sinauer Associates, Sunderland, Mass.

Garcia-Bellido, A. (1998) The engrailed story. *Genetics* **148**, 539–44.

García-Fernández, J. and Holland, P. (1994) Archetypal organization of the amphioxus Hox gene cluster. *Nature* **421**, 526–9.

Ghalambor, C.K., McKay, J.K., Carroll, S.P., and Reznick, D.N. (2007) Adaptive versus non-adaptive phenotypic plasticity and the potencial for contemporary adaptation in new environments. *Functional Ecology* **21**, 394–407.

Gibson, G. and Wagner, G. (2000) Canalization in evolutionary genetics: a stabilizing theory? *Bioessays* **22**, 372–80.

Gilbert, S.F. (2006) *Developmental Biology* (8th edn). Sinauer Associates, Sunderland, Mass.

Gilbert. W. (1987) The exon theory of genes. *Cold Spring Harbor Symp. Quant. Biol.* **52**, 901–5.

Goldenfeld, N. and Woese, C. (2007) Biology's next revolution. *Nature* **445**, 369.

Goldschmidt, R.B. (1940) *The Material Basis of Evolution.* Yale University Press, New Haven.

Gompel, N., Prud'homme, B., Wittkopp, P.J., Kassner, V.A., and Carroll, S.B. (2005) Chance caught on the wing: cis-regulatory evolution and the origin of pigment patterns in Drosophila. *Nature* **433,** 481–7.

González, J., Lenkov, K., Lipatov, M., Macpherson, J.M., and Petrov, D.A. (2008) High rate of recent transposable element–induced adaptation in Drosophila melanogaster. *PLoS Biol* **6(10):** e251. doi:10.1371/journal.pbio.0060251.

Gould, S. J. (1989) *Wonderful Life: The Burguess Shale and the Nature of History.* W.W. Norton, New York.

Gould, S.J. (2002) *The Structure of Evolutionary Theory.* The Belknap Press of Harvard University Press, Cambridge, Mass.

Gould, S.J. and Lewontin, R.C. (1979) The spandrels of San Marco and the Panglossian paradigm: acritique of the adaptationist programme. *Proc. R. Soc. Lond. B* **205**, 581–98.

Grandbastien, M-A. (1998) Activation of plant retrotransposons under stress conditions. *Trends Plant Sci.* **3**, 181–7.

Grant, V. (1957) The plant species in theory and practice. In E. Mayr (ed.), *TheSspecies Problem*, pp. 39–80. American Association for the Advancement of Science, Publication No. 50, Washington, D.C.

Grant, V. (1958) The regulation of recombination in plants. *Cold Spring Harb. Symp. Quant. Biol.* **23**, 337–63.

Grant, V. (1981) *Plant Speciation.* Columbia University Press, New York.

Gregory, T.R. (2005a) Synergy between sequence and size in large-scale genomics. *Nature Reviews Genetic* **6**, 699–708.

Gregory, T.R. (2005b) Macroevolution and the genome. In T.R. Gregory (ed.), *The Evolution of the Genome,* pp. 679–729. Elsevier, Burlington, Mass.

Gregory, T.R. (2005c) Genome size evolution in animals. In T.R. Gregory (ed.), *The Evolution of the Genome*, pp. 3–87. Elsevier, Burlington, Mass.

Gregory, T.R. and Witt J.D.S. (2008) Population size and genome size in fishes: a closer look. *Genome* **51**, 309–13.

Grenier, J.K. and Carroll, S.B. (2000) Functional evolution of the Ultrabithorax protein. *Proc. Natl Acad. Sci. USA* **97**, 704–9.

Grenier, J.K., Garber, T.L., Warren, R., Whitington, P.M., and Carroll, S. (1997) Evolution of the entire arthropod *Hox* gene set predated the origin and radiation of the onychophoran/arthropod clade. *Current Biology* **7**, 547–53.

Gross, B.L, and Rieseberg, L.H. (2005) The ecological genetics of homoploid hybrid speciation. *Journal of Heredity* **96**, 241–52.

Grosshans, H. and Filipovicz, W. (2008) The expanding world of small RNAs. *Nature* **451**, 414–16.

Hales, C.N. and Barker, D.J. (2001) The thrifty phenotype hypothesis. *Br. Med. Bull.* **60**, 5–20.

Herr, A.J., Jensen, M.B., Dalmay, T., and Baulcombe, D.C. (2005) RNA polymerase directs silencing of endogenous DNA. *Science* **308**, 118–20.

Hewitt, G.M. (2000) The genetic legacy of the Quaternary ice ages. *Nature* **405**, 907–13.

Hewitt, G.M. (2004). Using molecules to understand the distribution of animal and plant diversity. In A. Moya and E. Font (eds.), *Evolution: From Molecules to Ecosystems*, pp.157–70. Oxford University Press, Oxford.

Hittinger, C.T. and Carroll, S.B. (2007) Gene duplication and the adaptive evolution of a classic genetic switch. *Nature* **449**, 677–81.

Hoeckstra, H.E., and Coyne, J.A. (2007) The locus of evolution: Evo Devo and the genetics of adaptation. *Evolution* **61**, 995–1016.

Hopkins, A.D. (1917) Entomologists' discussions. *Journal of Economic Entomology*, **10**, 92–3.

Huang, J., Mullapudi, N., Lancto, C.A., *et al.* (2004) Phylogenomic evidence supports past endosymbiosis, intracellular and horizontal gene transfer in Cryptosporidium parvum. *Genome Biology* **5**, R88.

Huxley, J.S. (1942) *Evolution, The Modern Synthesis.* Allen and Unwin, London.

Jablonka, E. and Lamb, M.J. (2010) Transgenerational epigenetic inheritance. In: M. Pigliucci and G. Muller (eds.), *Evolution: The Extended Synthesis*, pp. 138–74. MIT Press, Cambridge, Mass.

Jacob, F. (1977) Evolution and tinkering, *Science* **196**, 1161–6.

Jacob, F. and Monod, J. (1961) Genetic regulatory mechanisms in the synthesis of proteins. *J. Mol. Biol.* **3**, 318.

Jain, R., Rivera, M.C., and Lake, J.A. (1999). Horizontal gene transfer among genomes: The complexity hypothesis. *Proc. Natl Acad. Sci. USA* **96**, 3801–6.

Jiang, Y. and Doolittle, R.F. (2003) The evolution of vertebrate blood coagulation as viewed from a comparison of puffer fish and sea squirt genomes. *Proc. Natl Acad. Sci. USA* **100**, 7527–32.

Jordan, D.S. (1908) The law of geminate species. *American Naturalist* **42**, 73–80.

Jordan, I.K. and McDonald, J.F. (1998) Evidence for the role of recombination in the regulatory evolution of Sacharomyces cerevisiae Ty elements. *Journal of Molecular Evolution* **47**, 14–20.

Jordan, I.K., Rogozin, I.B., Glazko, G.V., and Koonin, E.V. (2003) Origin of a substancial fraction of human regulatory sequences from transposable elements. *Trends Genet.* **19**, 68–72.

Josefsson, C., Dilkes, B., and Comai, L. (2006) Parent-dependent loss of gene silencing during interspecies hybridization. *Curr. Biol.* **16**, 1322–8.

Josse, T., Teysset, L., Todeschini, A-L., Sidor, C.M., Anxolabéhère, D., and Ronsseray, S. (2007) Telomeric trans-silencing: An epigenetic repression combining RNA silencing and heterochromatin formation. *PLoS Genet.* **3**(9), e158. doi:10.1371/journal.pgen.0030158.

Kaneshiro K.Y. (1988) Speciation in the Hawaiian Drosophila: Sexual selection appears to play an important role. *BioScience* **38**: 258–63.

Kaneshiro, K. and Val, F.C. (1977) Natural hybridization between a sympatric pair of Hawaiian Drosophila. *American Naturalist* **111**, 897–902.

Kashkush, K., Feldman, M., and Levy, A.A. (2002) Gene loss, silencing and activation in a newly synthesized wheat allotetraploid. *Genetics* **160**,1651–9.

Kawakami, T., Strakosh, S.C., Zhen Y., and Ungerer, M.C. (2010) Different scales of Ty1/copia-like retrotransposon proliferation in the genomes of three diploid hybrid sunflower species. *Heredity* **104**, 341–50.

Kidwell, M.G. (2005) Transposable elements. In T.R. Gregory, (ed.), *The Evolution of the Genome*, p. 165. Elsevier, Burlington, Mass.

Kidwell, M.G. and Lisch, D.R. (2000) Transposable elements and host genome evolution. *Trends Ecol. Evol.* **15**, 95–9.

Kidwell, M.G., Kidwell, J. F., and Sved, J.A. (1977) Hybrid dysgenesis in Drosophila melanogaster: A syndrome of aberrant traits including mutation sterility and male recombination. *Genetics* **86**, 813–33

Kimura, M. (1968) Evolutionary rate at the molecular level. *Nature* **217**, 624–6.

Kirschner, M.W. and Gerhart, J.C. (2005) *The Plausibility of Life.* Yale University Press, New Haven.

Klicka, J. and Zink, R.M. (1997) The importance of recent ice ages in speciation: A failed paradigm. *Science* **277**, 1666–9.

Klicka, J. and Zink, R.M. (1999) Pleistocene effects on North American songbird evolution. *Proc. R. Soc. Lond. B* **266**, 695–700.

Knowlton, N. and Weigt, L.A. (1998). New dates and new rates for divergence across the Isthmus of Panama. *Proc. R. Soc. Lond. B* **265**, 2257–63.

Koonin, E.V. (2009) Darwinian evolution in the light of genomics. *Nucleic Acids Research*, 1–24, doi:10.1093/nar/gkp089.

Krol, A.R. van der, Mur, L.A., Beld, M., Mol, J.N.M., and Stuitje, A.R. (1990) Flavonoid genes in Petunia: Addition of a limited number of gene copies may lead to a suppression of gene expression. *The Plant Cell* **2**, 291–9.

Kwan, T., Liu, J., DuBow, M., Gros, P., and Pelletier, J. (2005) The complete genomes and proteomes of 27 Staphylococcus aureus bacteriophages. *Proc. Natl Acad. Sci. USA* **102**, 5174–9.

Labrador, M. and Fontdevila, A. (1994) High transposition rates of Osvaldo, a new Drosophila buzzatii retrotransposon. *Mol. Gen. Genet.* **245**, 661–74.

Labrador, M., Farré, M., Utzet, F., and Fontdevila, A. (1999) Interspecific hybridization increases transposition rates of *Osvaldo*. *Mol. Biol. Evol.* **16**, 931–7.

Lagercrantz, U. and Lydiate, D.J. (1996) Comparative genome mapping in Brassica. *Genetics* **144**, 1903–10.

Lander, E.S. *et al.* (+ 242 others). International Human Genome Sequencing Consortium (2001) Initial sequencing and analysis of the human genome. *Nature* **309**, 860–921.

Lawrence J.G. and Ochman H. (1998) Molecular archaeology of the Escherichia coli genome. *Proc. Natl Acad. Sci. USA* **95**, 9413–17.

Lee, J.T. (2003) Molecular links between X-inactivation and autosomal imprinting: X inactivation as a driving force for the evolution of imprinting? *Current Biology* **13**, R242–R254.

Leitch, A.R. (2007) The dynamic interplay between genomic DNA and the outside world. *Heredity* **98**, 61–2.

Lessios, H.A. (1998). The first stage of speciation as seen in organisms separated by the Isthmus of Panama. In D.J. Howard and S. Berlocher (eds.). *Endless Forms: Species and Speciation*, pp. 186–201. Oxford University Press, Oxford.

Lewis, E.B. (1978) A gene complex controlling segmentation in Drosophila. *Nature* **276**, 565–70.

Lheureux, F., Carreel, F., Jenny C., Lockhart B.E.L., and Iskra-Caruana, M.L. (2003) Identification of genetic markers linked to banana streak disease expression in inter-specific Musa hybrids. *Theor. Appl. Genet.* **106**, 594–8.

Li, A., Beard, C., and Jaenisch, D. (1993) Role for DNA methylation in genomic imprinting. *Nature* **366**, 362–5.

Li, W. and Shaw, J.E. (1993) A variant Tc4 transposable element in the nematode C. elegans could encode a novel protein. *Nucleic Acids Res.* **21**, 59–67.

Lichtneckert, R. and Reichert, H. (2005) Insights into the urbilaterian brain: Conserved genetic patterning mechanisms in insect and vertebrate brain development. *Heredity* **94**, 465–77.

Lippman, Z. and Martienssen, R. (2004) The role of RNA interference in heterochromatic silencing. *Nature* **431**, 364–70.

Lippman, Z., Gendrel, A.V., Black, M., *et al.* (2004) Role of transposable elements in heterochromatin and epigenetic control. *Nature* **430**, 471–6.

Liu, Y. (2007) Like father like son. *EMBO reports* **8(9)**, 798–803.

Liu, B. and Wendel, J.F. (2000) Retrotransposon activation followed by rapid repression in introgressed rice plants. *Genome* **43**, 874–80.

Liu, B. and Wendel, J.F. (2003) Epigenetic phenomena and the evolution of plant allopolyploids. *Molecular Phylogenetics and Evolution* **29**, 365–79.

Loftus, B., Anderson, I., Davies, R., *et al.* (2005) The genome of the protist parasite *Entamoeba histolytica*. *Nature* **433**, 865–8.

Luikenhuis, S., Wutz, A., and Jaenisch, R. (2001) Antisense transcription through the *Xist* locus mediates *Tsix* function in embryonic cells. *Molecular and Cell Biology* **21**, 8512–20.

Lyko, F. (2001) DNA methylation learns to fly. *Trends in Genetics* **17**, 169–72.

Lynch, M. (2007) *The Origins of Genome Architecture*. Sinauer Associates, Sunderland, Mass.

Lynch, M. and Conery, J.S. (2003) The origins of genome complexity. *Science* **302**, 1401–4.

Lyon, M.F. (1998) X-Chromosome inactivation: A repeat hypothesis. *Cytogenet. Cell Genet.* **80**, 133–7.

Machado, C.A. and Hey, J. (2003) The causes of phylogenetic conflict in a classic Drosophila species group. *Proc. Roy. Soc. Lond. B* **270**, 1193–202.

Macpherson, J.M., Sella, G., Davis, J.C., and Petrov, D.A. (2007) Genomewide spatial correspondence between nonsynonymous divergence and neutral polymorphism reveals extensive adaptation in Drosophila. *Genetics* **177**, 2083–99.

Madlung, A., Tyagi, A.P., Watson, B., *et al.* (2005) Genomic changes in synthetic Arabidopsis polyploids. *The Plant Journal* **41**, 221–30.

Mallet, J. (2005a) Speciation in the 21st century. *Heredity* **95**, 105–9.

Mallet, J. (2005b) Hybridization as an invasión of the genome. *Trend Ecol. Evol.* **20**, 229–37.

Manak, J.R., Dike, S., Sementchenko,V. *et al.* (2006) Biological function of unannotated transcription during the early development of Drosophila melanogaster. *Nature Genetics* **38**, 1151–8.

Margalef, R. (1968) *Perspectives in Ecological Theory*. University of Chicago Press, Chicago.

Marhold, J., Rothe, N., and Pauli, A. (2004) Conservation of methylation in dipteran insects. *Insect Mol. Biol.* **13**: 117–23.

Maside, X., Bartolome, C., Assimacopoulos, S., and Charlesworth, B. (2001) Rates of movement and distribution of transposable elements in Drosophila melanogaster: In situ hybridization vs Southern blotting data. *Genet. Res.* **78**, 121–36.

Mattick, J.S. (2009) Deconstructing the dogma: A new view of the evolution and genetic programming of complex organisms. *Ann. N. Y. Acad. Sci.* **1178** 29–46.

Matzke, M.A., Mette, M.F., Aufsatz, W., Janowitsch, J., and Matske, A.J.M. (2000) Host defenses to parasitic sequences and the evolution of epigenetic control mechanisms. In J. F. McDonald, (ed.), *Transposable Elements and Evolution*, pp. 271–8. Kluwer, Dordrecht.

Mavárez, J. and Linares, M. (2008) Homoploid hybrid speciation in animals. *Mol. Ecol.* **17**, 4181–5.

Maynard Smith, J. (1966) Sympatric speciation. *The American Naturalist* **104**, 487–90.

Mayr, E. (1942) *Systematics and the Origin of Species*. Columbia University Press, New York. (Reprinted in 1999 by Harvard University Press).

Mayr. E. (1957) Species concepts and Definitions. In E. Mayr (ed.), *The Species Problem Am. Assoc. Adv. Sci. Publ.* **50**, pp. 1–22.

Mayr, E. (1963) *Animal Species and Evolution*. Belknap Press, Cambridge, Mass.

Mayr, E. (1970) *Populations, Species, and Evolution*. Belknap Press, Cambridge, Mass.

Mayr, E. (1980) Prologue: Some thoughts on the history of the Evolutionary Synthesis In E. Mayr and W.B. Provine (eds.), *The Evolutionary Synthesis*. Harvard University Press, Cambridge, Mass.

McCarthy, E.M., Asmussen, M.A., and Anderson, W.W. (1995) A theoretical assessment of recombinational speciation. *Heredity* **74**, 502–9.

McClintock, B. (1950) The origin and behaviour of mutable loci in maize. *Proc. Natl Acad. Sci. USA* **36**, 344–55.

McClintock, B. (1953) Induction of instability of selected loci in maize. *Genetics* **38**, 579–99.

McClintock, B. (1980) Modified gene expressions induced by transposable elements. In W.A. Scott, R. Werrner, D.R. Joseph, and J. Schultz (eds.), *Miami Winter Symp.* **17**, 11–19. Academic Press, New York.

McClintock, B. (1984) The significance of response of the genome to challenge. *Science* **226**, 792–801.

McDonald, J.F. (1990) Macroevolution and retroviral elements. *BioScience* **40**, 183–91.

McDonald, J.F., Matyunina, L.V., Wilson, S., Jordan, I.K., Bowen, N.J. and Miller, W.J. (1997) LTR retrotransposons and the evolution of eukaryotic enhancers. *Genetica* **100**, 3–13.

McDonald, J.F., Matzke, M.A., and Matzke, A.J. (2005) Host defenses to transposable elements and the evolution of genomic imprinting. In J.N. Volff (ed.), *Retrotransposable Elements and Genome Evolution*, pp. 242–9. Karger, Basel.

Menken, S.B.J. and Roessingh, P. (1998) Evolution of insect-plant associations: Sensory perceptions and receptor modifications, direct food specializations, and host shifts in phytophagous insects. In D.J. Howard and S.H. Berlocher (eds.), *Endless Forms: Species and Speciation*, pp. 145–56. Oxford University Press, Oxford.

Mercer, T.R., Dinger, M.E., and Mattick, J.S. (2009) Long non-coding RNAs: insights into functions. *Nature Reviews Genetics* **10**, 155–9.

Metcalfe, C.J., Bulazel, K.V., Ferreri, G.C. *et al.* (2007) Genomic instability within centromeres of interspecific marsupial hybrids. *Genetics* **177**, 2507–17.

Michalak, P. (2009) Epigenetic, transposon and small RNA determinants of hybrid dysfunctions. *Heredity* **102**, 45–50.

Michalak, P. (2010) An eruption of mobile elements in genomes of hybrid sunflowers. *Heredity* **104**, 329–30.

Monod, J. (1970) *Le hasard et la nécessité*. Editions du Seuil, Paris.

Moore, J.A. (1984) Science as a way of knowing. Evolutionary Biology. *American Zoologist* **24**: 467–534.

Moore, J.A. (ed.) (1987) *The Discovery and Characterization of Transposable Elements. The Collected Papers of Barbara McClintock.* Garland Publishing, New York & London.

Moran, J.V., DeBernardinis, R.J., and Kazazian H.H. (1999) Exon shuffling by L1 retrotransposition. *Science* **283**, 1530–4.

Moran, N.A., and Jarvik, T. (2010) Lateral transfer of genes from fungi underlies carotenoid production in aphids. *Science* **328**, 624–7.

Morgan, H.D., Sutherland, H.G.E., Martin, D.I.K., and Whitelaw, E. (1999) Epigenetic inheritance at the agouti locus in the mouse. *Nature Genetics* **23**, 314–18.

Mower, J.P., Stefanovic, S., Young, G.J., and Palmer, J.D. (2004) Gene transfer from parasitic to host plants *Nature* **432**, 165–6.

Murad, L., Bielawski, J.P., Matyasek, *et al.* (2004) The origin and evolution of geminivirus-related DNA sequences in Nicotiana. *Heredity* **92**, 352–8.

Napoli, C., Lemieux, C., and Jorgensen, R. (1990) Introduction of a chimeric chalcone synthase gene into Petunia results in reversible co-suppression of homologous genes *Ín* trans. *The Plant Cell* **2**, 279–89.

Navarro, A. and Barton, N.H. (2003) Chromosomal speciation and molecular divergence—accelerated evolution in rearranged chromosomes. *Science* **300**, 321–4.

Naveira, H. and Fontdevila, A. (1985) The evolutionary history of *Drosophila buzzatii*. IX. High frequencies of new chromosome rearrangements induced by introgressive hybridization. *Chromosome* **91**, 87–94.

Nekrutenko, A. and Li, W.H. (2001) Transposable elements are found in a large number of human protein-coding genes. *Trends. Genet.* **17**, 619–21.

Noor, M.A.F., and Chang, A.S. (2006) Evolutionary genetics: Jumping into a new species. *Current Biology* **16**, R890-1.

Noor, M.A.F. and Feder, J.L. (2006) Speciation genetics: Evolving approaches. *Nature Reviews Genetics* **7**, 851–61.

Noor, M.A.F., Grams, K.L., Bertucci, L.A., and Reiland, J. (2001) Chromosomal inversions and the reproductive isolation of species. *Proc. Natl Acad. Sci. USA* **98**, 12084–8.

O'Neill, R.J.W., O'Neill M.J., and Graves J.A.M. (1998) Undermethylation associated with retroelement activation and chromosome remodelling in an interspecific mammalian hybrid. *Nature* **393**, 68–72.

Ohno, S. (1970) *Evolution by Gene Duplication*. Springer-Verlag, Berlin.

Ohta, T. (1992) The nearly neutral theory of molecular evolution. *Annual Review of Ecology and Systematics* **23**, 263–86.

Orgel, L.E. and Crick, F.H.C. (1980) Selfish DNA: The ultimate parasite. *Nature* **284**, 604–7.

Orr, H.A. (2005) The genetic theory of adaptation: A brief history. *Nature Reviews Genetics* **6**, 119–27.

Pallen, M.J. and Matzke, N.J. (2006) From the origin of species to the origin of bacterial flagella. *Nature Reviews Microbiology* **4**, 784–90.

Panganiban, G., Irvine, S.M., Lowe, C., *et al.* (1997) The origin and evolution of animal appendages. *Proc. Natl Acad. Sci. USA* **94**, 5162–6.

Pantazidis, A., Labrador, M., and Fontdevila, A. (1999) The retrotransposon Osvaldo from Drosophila buzzatii displays all structural features of a functional retrovirus. *Mol. Biol. Evol.* **16**, 909–21.

Patton, J.L. and da Silva, M.N.F. (1998) Rivers, refuges, and ridges: The geography of speciation of Amazonian mammals. In D. Howard and S.Berlocher (eds.), *Endless Forms: Species and Speciation*, pp. 202–13. Oxford University Press, Oxford.

Petrov, D.A. (2002) DNA loss and evolution of genome size in Drosophila. *Genetica* **115**, 81–91.

Petrov, D.A. and Wendel, J.F. (2006) Evolution of eukaryotic genome structure. In Ch. Fox and J.B. Wolf (eds.), *Evolutionary Genetics*, pp.144–56. Oxford University Press, Oxford.

Picard, G. (1976) Non-Mendelian female sterility in *Drosophila melanogaster*: Hereditary transmission of *I* factor. *Genetics* **83**, 107–123.

Pickeral, O.K., Makalowski, W., Boguski, M.S., and Boeke, J.D. (2000) Retrotransposition requent human genomic DNA transduction driven by LINE-1. *Genome Res.* **10**, 411–15.

Pigliucci, M., and Muller, G. (eds.) (2010) *Evolution: The Extended Synthesis*. MIT Press, Cambridge, Mass.

Pires, J.C., Zhao, J., Schranz, M.E., *et al.* (2004) Flowering time divergence and genomic rearrangements in resynthesized Brassica polyploids (Brassicaceae). *Biological Journal of the Linnean Society* **82**, 675–88.

Plasterk, R.H. (2002) RNA silencing: The genomes's immune system. *Science* **296**,1263–5.

Powell, J.R. (1983) Interspecific cytoplasmic gene flow in the absence of nuclear gene flow: Evidence from Drosophila. *Proc. Natl Acad. Sci. USA* **80**, 492–5.

Powell, J.R. (1991) Monophyly/paraphyly/polyphyly and gene/species trees: An example from Drosophila. *Mol. Biol. Evol.* **8**, 892–6.

Prud'homme, B., Gompel, N., Rokas, A. *et al.* (2006) Repeated morphological evolution through cis-regulatory changes in a pleiotropic gene. *Nature* **440**, 1050–3.

Quesada, H., Posada, D., Caballero, A., Morán, P., and Rolán-Alvarez, E. (2007) Phylogenetic evidence for multiple sympatric ecological diversification in a marine snail. *Evolution* **61**, 1600-12. doi:10.1111/j.1558-5646.2007.00135.x.

Quiring, R., Walldorf, U., Kloter, U., and Gehring, W.J. (1994) Homology of the eyeless gene of Drosophila to the Small eye gene in mice and Aniridia in humans. *Science* **265**, 785–9.

Raff, R.A. (1996) *The Shape of Life*. The University of Chicago Press, Chicago.

Randolph, L.F. (1966) *Iris nelsonii*, a new species of Louisiana iris of hybrid origin. *Baileya* **14**, 143–69.

Rebollo, R., Horard, B., Hubert, B., and Vieira, C. (2010) Jumping genes and epigenetics: Towards new species. *Gene* doi:10.1016/j.gene.2010.01.003.

Rensch, B. (1959) *Evolution Above the Species Level*. Columbia University Press, New York.

Richardson, A.O. and Palmer, J.D. (2007) Horizontal transfer in plants. *J.Exp. Botany* **58**: 1–9.

Rieseberg, L.H. (1991) Homoploid reticulate evolution in Helianthus (Asteraceae): Evidence from ribosomal genes. *Am. J. Bot.* **78**, 1218–37.

Rieseberg, L.H. (2001) Chromosomal rearrangements and speciation. *Trends Ecol. Evol.* **16**, 351–8.

Rieseberg, L.H. and Noyes, R.D. (1998) Genetic map-based studies of reticulate evolution in plants. *Trends Plant Sci.* **3**, 254–9.

Rieseberg, L.H., Raymond, O., Rosenthal, D.M., *et al.* (2003). Major ecological transitions in wild sunflowers facilitated by hybridization. *Science* **301**, 1211–16.

Rieseberg, L.H., Sinervo, B., Randal Linder, C., Ungerer, M., and Arias, D.M. (1996) Role of gene interactions in hybrid speciation: Evidence from ancient and experimental hybrids. *Science* **272**, 741–5.

Rio, D.C. (2002) P transposable elements in Drosophila melanogaster. In N.L. Craig, R. Craigie, M. Gellert, and A.M. Lambowitz (eds.), *Mobile DNA II*, pp 484–518. ASM Press, Washington DC.

Robertson, H.M. and Lampe, D. J. (1995) Recent horizontal transfer of a *mariner* transposable element among and between Diptera and Neuroptera. *Mol.Biol. Evol.* **12**, 850–62.

Rolán-Alvarez, E. (2007) Sympatric speciation as a by-product of ecological adaptation in the Galician Littorina saxatilis hybrid zone. *Journal of Molluscan Studies* **73**, 1–10.

Roth, G., Blanke, J., and Wake, D.B. (1994) Cell size predicts morphological complexity in the brains of frogs and salamander. *Proc. Natl Acad. Sci. USA* **91**, 4796–800.

Roth, G., Nishikawa, K.C., and Wake, D. (1997) Genome size, secondary simplification, and the evolution of the brain in salamanders. *Brain Behavior and Ecology* **50**, 50–9.

Rundle, H.D. and Schluter, D. (2004) Natural selection and ecological speciation in sticklebacks. In U. Dieckmann, M. Doebeli, J.A.J Metz, and D. Tautz, (eds.), *Adaptive Speciation*, pp. 192–209. Cambridge University Press, Cambridge.

Rushforth, A.M. and Anderson, P. (1996) Splicing removes the *Caenorhabditis elegans* transposon Tc1 from most mutant pre-mRNAs. *Mol. Cell. Biol.* **16**, 422–9.

Rutherford, S.L., and Lindquist, S. (1998) Hsp90 as a capacitor for morphological evolution. *Nature* **396**, 336–42.

Salazar, C., Baxter, S.W., Pardo-Diaz, C. *et al.* (2010) Genetic evidence for hybrid trait speciation in Heliconius butterflies. *PLoS Genet.* **6**:e1000930.

Salmon, A., Ainouche, M.L., and Wendel, J.F. (2005) Genetic and epigenetic consequences of recent hybridization and polyploidy in Spartina (Poaceae). *Molecular Ecology* **14**, 1163–75.

Salvini-Plawen, L.V. and Mayr, E. (1977) On the evolution of photoreceptors and eyes. *Evo. Biol.***10**, 207–63.

SanMiguel, Ph., Gaut, B.S., Tikhonov, A., Nakajima, Y., and Bennetzen, J.L. (1998) The paleontology of intergene retrotransposons of maize. *Nat. Genet.* **20**, 43–5.

Schemske, D.W. and Bradshaw, Je., H.D. (1999) Pollinator preference and the evolution of floral traits in monkeyflowers (Mimulus). *PNAS* **96**:11910–15.

Schilthuizen, M. (2001) *Frogs, Flies and Dandelions*. Oxford University Press, Oxford.

Schilthuizen, M. and Scott, B. (2004). Allopatric speciation: Not so simple after all. In A. Moya and E. Font (eds.), *Evolution: From Molecules to Ecosystems*, pp.173–81. Oxford University Press, Oxford.

Schlenke, T.A. and Begun, D.J. (2004) Strong selective sweep associated with a transposon insertion in Drosophila simulans. *Proc. Natl Acad. Sci. USA* **101**, 1626–31.

Schlichting, C.D, and Pigliucci, M. (1998) *Phenotypic Evolution: A Reaction Norm Perspective.* Sinauer Associates, Sunderland, Mass.

Schmalhausen, I. I. (1949) *Factors of Evolution: The Theory of Stabilizing Selection* (trans. from the Russian by I. Dordich, ed. Th. Dobzhansky, 1949). Nauka Science, reprinted by University of Chicago Press in 1986.

Schmidt, E.R. (1984) Clustered and interspersed repetitive DNA sequence family in *Chironomus*. *Journal of Molecular Biology* **178**, 1–15.

Shan, X., Liu, Z., Dong, Z. *et al.* (2005) Mobilization of the active MITE transposons mPing and Pong in rice by introgression from wild rice (*Zizania latifolia* Griseb.) *Mol. Biol. Evol.* **22**, 976–90.

Shapiro, M.D., Marks, M.E., Peichel, C.L. *et al.* (2004) Genetic and developmental basis of evolutionary pelvic reduction in threespine stickleback. *Nature* **428**, 717–23.

Shubin, N., Tabin, C., and Carroll, S. (1997) Fossils, genes, and the evolution of animal limbs. *Nature* **388**, 639–48.

Shubin, N., Tabin, C., and Carroll, S. (2009) Deep homology and the origins of evolutionary novelty. *Nature* **457**, 818–23.

Simpson, G.G. (1944) *Tempo and Mode in Evolution*. Columbia University Press, New York.

Simpson, G.G. (1961). *Principles of Animal Taxonomy*. Columbia University Press, New York.

Slotkin, R.K. and Martienssen, R. (2007) Transposable elements and the epigenetic regulation of the genome. *Nature Reviews Genetics* **8**, 272–85, | doi:10.1038/nrg2072

Sober, E. (1980). Evolution, population thinking and essentialism. *Philos. Sci.* **47**, 350–83.

Stebbins, G.L. (1950) *Variation and Evolution in Plants*. Columbia University Press, New York.

Stebbins, G.L. (1957) The hybrid origin of microspecies in the Elymus glaucus complex. *Cytologia Suppl* **36**, 336–40.

Strachan, T. and Read, A. (2004) *Human Molecular Genetics*, Garland Publishing, New York & London.

Strand, D.J. and McDonald, J.F. (1985) Copia is transcriptionally responsive to environmental stress. *Nucl. Acids Res.* **13**, 4401–10.

Sturtevant, A.H. (1939) High mutation frequency induced by hybridization. *Proc. Natl Acad. Sci. USA* **25**, 308–10.

Syvanen, M. (2007) Some computational problems associated with horizontal gene transfer. *Systems Biology, Volume 1: Genomics*, pp. 248–68. Oxford University Press, Oxford.

Tarlinton, R.E., Meers, J., and Young, P.R. (2006) Retroviral invasion of the koala genome. *Nature* **442**, 79–81.

Taylor, J.S. and Raes, J. (2004) Duplication and divergence: The evolution of new genes and old ideas. *Annual Review of Genetics* **38**, 615–43.

Templeton, A.R. (1989) The meaning of species and speciation: A genetic perspective. In D. Otte and J. Endler (eds.), *Speciation and its Consequences*, pp. 3–27. Sinauer Associates, Sunderland, Mass.

Ungerer, M.C., Strakosh, S.C., and Zhen, Y. (2006) Genome expansion in three hybrid sunflower species is associated with retrotransposon proliferation. *Curr. Biol.* **16**, R872–R873.

Venter, C. (2010) Multiple personal genomes await. *Nature* **464**, 676–7.

Venter, J.C., Adams, M.D., Myers, E.W, *et al.* (2001) The sequence of the human genome. *Science* **291**, 1304–51.

Vieira, C., Aubry, P., Lepetit, D., and Biémont, C. (1998) A temperature cline in copy number for 412 but not *roo/B104* retrotransposons in populations of Drosophila simulans. *Proc. Biol. Sci.* **265**,1161–5.

Waddington, C.H. (1942) The epigenotype. *Endeavour* **1**, 18–20.

Waddington, C.H. (1952) Selection of the genetic basis for an acquired character. *Nature* **169**, 278.

Waddington, C.H. (1953) Genetic assimilation of an acquired character. *Evolution* **7**, 118.

Waddington, C.H. (1956) Genetic assimilation of the bithorax phenotype. *Evolution* **10**, 1–13.

Wagner, G.P. (2007). The developmental genetics of homology. *Nature Reviews Genetics* **8**, 473–9.

Wallace, A.R. (1852) On the monkeys of the Amazon. *Proc. Zool. Soc. London* **20**, 107–10.

Wallace, A.R. (1869) *The Malay Archepelago.* Macmillan and Co., London; reprinted in 1989 by Oxford University Press, New York.

Wang, R.L., Wakeley, J., and Hey, J. (1997) Gene flow and natural selection in the origin of Drosophila pseudoobscura and close relatives. *Genetics* **147**, 1091–106.

Waterland, R.A. and Jirtle, R.L. (2003) Transposable elements: targets for early nutritional effects on epigenetic gene regulation. *Molecular and Cellular Biology* **23**, 5293–300.

Waterston, R.H., Lander, E.S., and Sulston, J.E. (2002) On the sequencing of the human genome. *Proc. Natl Acad. Sci. USA* **99**, 3712–16.

Watson, J.D. and Crick, F.H. (1953) Molecular structure of nucleic acids; A structure for deoxyribose nucleic acid. *Nature* **171**, 737–8.

Weaver, I.C.G., Cervoni, N., Champagne, F.A. *et al.* (2004) Epigenetic programming by maternal behaviour. *Nature Neuroscience* **7**, 847–54.

Wessler, S.R. (1989) The splicing of maize transposable elements from pre-mRNA—a minireview. *Gene* **82**, 127–33.

Wessler S.R. (1996) Plant retrotransposons: Turned on by stress. *Curr. Biol.* **6**, 959–61.

Wessler S.R. (2006) Eukaryotic transposable elements: Teaching old genomes new tricks. In L.H. Caporale (ed.), *The Implicit Genome*, pp. 138–62. Oxford University Press, New York.

Wicker, T., Sabot, F., Hua-Van, A. *et al.* (2007) A unified classification system for eukaryotic transposable elements. *Nature Reviews Genetics* **8**, 973–82.

Wiley, E.O. (1978) The evolutionary species concept reconsidered. *Syst. Zool.* **27**, 17–26.

Woese, C.R. (2002) On the evolution of cells. *Proc. Natl Acad. Sci. USA* **99**, 8742–7.

Wolfe, K.H. and Shields, D.C. (1997) Molecular evidence for an ancient duplication of the entire yeast genome. *Nature* **387**, 708–13.

Wood, T.K. (1993) Speciation in the Enchenopa binotata complex (Insects: Homoptera: Membracidae). In D.R. Lees and D. Edwards (eds.), *Evolutionary Patterns and Processes* (Linnean Society Symposium N. 14), pp. 299–318. Linnean Society, London.

Wray, G.A., Levinton, J.S., and Shapiro, L.H. (1996) Molecular evidence for deep precambrian divergences among metazoan phyla. *Science* **274**, 568–73.

Wright, S.I. and Andolfatto, P. (2008) The impact of natural selection on the genome: Emerging patterns in Drosophila and Arabidopsis. *Annu. Rev. Ecol. Evol. Syst.* **39**, 193–213.

Xie, X., Lu, J., Kulbokas, E.J., *et al.* (2005) Systematic discovery of regulatory motifs in human promoters and 3′ UTRs by comparison of several mammals. *Nature* **434**, 338–45.

Xing, J., Wang, H., Belancio, V.P. *et al.* (2006) Emergence of primate genes by retrotransposon-mediated sequence transduction. *Proc. Natl Acad. Sci. USA* **103**, 17608–13.

Yi, S. (2006) Non-adaptive evolution of genome complexity. *BioEssays* **28**, 979–82.

Yi, S. and Streelman, J.T. (2005) Genome size is negatively correlated with effective population size in ray-finned fish. *Trends Genet.* **21**, 643–6.

Yoder, J.A., Walsh, C.P., and Bestor, T.H. (1997) Cytosine methylation and the ecology of intragenomic parasites. *Trends Genet.* **13**, 335–40.

Zamore, P.D. (2002) Ancient pathways programmed by small RNAs. *Science* **296**, 1265–9.

Zhao, X.P., Si, Y., Hanson, R.E. *et al.* (1998) Dispersed repetitive DNA has spread to new genomes since polyploid formation in cotton. *Genome Res.* **8**, 479–92.

Index

Note: numbers in italics indicate the position of a keyword in the glossary, and numbers in bold indicate the main page for the keyword.